Second Edition

Preparation *for* College MATHEMATICS

Guided Notebook

Editors:
S. Rebecca Johnson,
Barbara Miller

Copy Editors:
Mary Janelle Cady,
Rebecca Lebeaux,
Anna Tavormina

Creative Director:
Tee Jay Zajac

Designers:
Natalie Ezabele,
Trudy Gove,
Patrick Thompson

Cover Design:
Patrick Thompson,
Tee Jay Zajac

Cover Illustration:
Jameson Deichman
www.behance.net/jdeichman

Composition and Answer Key:
QSI (Pvt.) Ltd

VP Research & Development: Marcel Prevuznak

Director of Content: Kara Roché

A division of Quant Systems, Inc.

546 Long Point Road, Mount Pleasant, SC 29464

Copyright © 2020, 2019 by Hawkes Learning / Quant Systems, Inc. All rights reserved.

No part of this publication may be reproduced, stored in a retrieval system, or transmitted in any form or by any means, electronic, mechanical, photocopying, recording, or otherwise, without the prior written consent of the publisher.

Printed in the United States of America

10 9 8 7 6 5 4 3 2

ISBN: 978-1-64277-006-3

Table of Contents

Preface

Strategies for Academic Success

CHAPTER 1

Whole Numbers

CHAPTER 2

Integers

CHAPTER 3

Fractions, Mixed Numbers, and Proportions

CHAPTER 4

Decimal Numbers

CHAPTER 5

Percents

CHAPTER 6

Measurement and Geometry

CHAPTER 7

Solving Linear Equations and Inequalities

CHAPTER 8

Graphing Linear Equations and Inequalities

CHAPTER 9

Systems of Linear Equations

CHAPTER 10

Exponents and Polynomials

CHAPTER 11

Factoring Polynomials

CHAPTER 12

Rational Expressions

CHAPTER 13

Roots, Radicals, and Complex Numbers

CHAPTER 14

Quadratic Equations

CHAPTER 15

Exponential and Logarithmic Functions

CHAPTER 16

Conic Sections

Math@Work

Math Knowledge Required for Math@Work Career Explorations

The following table summarizes the math knowledge required for each Math@Work career exploration. Use this table to determine when you are ready to explore each career.

Math@Work Career	Whole Numbers	Fractions	Integers	Decimal Numbers	Averages	Percents	Simple Interest	Ratios	Proportions	Geometry	Statistics	Graphing	Linear Equations	Systems of Equations	Mixture Problems	Scientific Notation	Greatest Common Factor	Rational Expressions	Radicals
Basic Inventory Management	✓																		
Hospitality Management	✓	✓			✓														
Bookkeeper				✓															
Pediatric Nurse				✓				✓	✓										
Architecture				✓						✓									
Statistician: Quality Control				✓							✓	✓							
Dental Assistant				✓		✓													
Financial Advisor				✓		✓	✓						✓						
Market Research Analyst				✓									✓						
Chemistry				✓										✓	✓				
Astronomy				✓												✓			
Math Education			✓														✓		
Physics		✓																✓	
Forensic Scientist				✓		✓													✓
Other Careers in Mathematics																			

Support

If you have questions or comments we can be contacted as follows:

24/7 Chat: chat.hawkeslearning.com

Phone: (843) 571-2825

E-mail: support@hawkeslearning.com

Web: hawkeslearning.com

Our support hours are 8:00 a.m. to 10:00 p.m. (ET), Monday through Friday.

How to Use the Guided Notebook

There are a variety of elements in this Guided Notebook that will help you on your way to mastering each topic. Here is a rundown of how to use the elements as you work through this notebook.

Fill-in-the-Blanks

1. When there is an incomplete sentence, you will need to write in the ___*missing*___ word(s).

The ___*missing*___ words can be found by reading through the Learn screens.

Boxed Content

Definitions and **procedures** are highlighted within a box, like the ones shown here. The missing content will vary from box to box. Sometimes an entire definition is missing and sometimes only part of a sentence is missing. Here are two examples of the box variations.

Definition

First term to define: ___*Write the definition here.*___

Second term to define: ___*If there is another term, define it the same way as above.*___

DEFINITION

Terms Related to Probability

Outcome	An individual result of an experiment.
Sample Space	The set of all possible outcomes of an experiment.
Event	Some (or all) of the outcomes from the sample space.

DEFINITION

Properties and **Procedure** boxes are completed in a similar way:

Commutative Property of Multiplication

The order of the numbers in multiplication can be reversed without changing the product.

For example, $3 \cdot 4 = 12$ and $4 \cdot 3 = 12$.

PROPERTIES

Subtracting Whole Numbers

1. Write the numbers vertically so that the place values are lined up in columns.
2. Subtract only the digits with the same place value.
3. Check by adding the difference to the subtrahend. The sum must be the minuend.

PROCEDURE

Watch and Work

For each Watch and Work, you will need to watch the corresponding video in Learn mode and follow along while completing the example in the space provided.

Example 5 Multiplying Whole Numbers

Multiply: $12 \cdot 35$

Solution

The standard form of multiplication is used here to find the product $12 \cdot 35$.

$$\begin{array}{rl} \overset{1}{1}2 & \\ \times\ 35 & \\ \hline 60 & 12 \cdot 5 = 60 \\ 360 & 12 \cdot 30 = 360 \\ \hline 420 & \text{Product} \end{array}$$

Now You Try It!

After working along with the example video, work through a similar exercise on your own in the space provided.

Example A Multiplying Whole Numbers

Multiply:
$$\begin{array}{r} 25 \\ \times 42 \\ \hline 1050 \end{array}$$

1.1 Exercises

Each section has exercises to offer additional practice problems to help reinforce topics that have been covered. The exercises include Concept Check, Practice, Application, and Writing & Thinking questions. The odd answers can be found in the Answer Key at the back of the book.

Concept Check

True/False. Determine whether each statement is true or false. If a statement is false, explain how it can be changed so the statement will be true. (**Note:** There may be more than one acceptable change.)

1. When the given statement is true, you write "True" for the answer.

True

Practice

For each set of data, find **a.** the mean, **b.** the median, **c.** the mode (if any), and **d.** the range.

2. ***Presidents:*** The ages of the first five US presidents of the 20th century on the date of their inaugurations were as follows. (The presidents were Roosevelt, Taft, Wilson, Harding, and Coolidge.)

42, 51, 56, 55, 51

a. 51 b. 51 c. 51 d. 14

Applications

Solve.

3. ***Grades:*** Suppose that you have taken four exams and have one more chemistry exam to take. Each exam has a maximum of 100 points and you must average between 75 and 82 points to receive a passing grade of C. If you have scores of 85, 60, 73, and 76 on the first four exams, what is the minimum score you can make on the fifth exam and receive a grade of C?

81

Writing & Thinking

4. State how to determine the median of a set of data.

The first step to finding the median is always to arrange the data in order. Once the data is in order, the median is the number in the middle. If there is an even number of items, average the two middle numbers to find the median.

Strategies for Academic Success

Strategies for Academic Success

How to Read a Math Textbook

Reading a textbook is very different than reading a book for fun. You have to concentrate more on what you are reading because you will likely be tested on the content. Reading a math textbook requires a different approach than reading literature or history textbooks because the math textbook contains a lot of symbols and formulas in addition to words. Here are some tips to help you successfully read a math textbook.

Don't Skim

When reading math textbooks, look at everything: titles, learning objectives, definitions, formulas, text in the margins, and any text that is highlighted, outlined, or in bold. Also pay close attention to any tables, figures, charts, and graphs.

Minimize Distractions

Reading a math textbook requires much more concentration than a novel by your favorite author, so pick a study environment with few distractions and a time when you are most attentive.

Start at the Beginning

Don't start in the middle of an assigned section. Math tends to build on previously learned concepts and you may miss an important concept or formula that is crucial to understanding the rest of the material in the section.

Highlight and Annotate

Put your book to good use and don't be afraid to add comments and highlighting. If you don't understand something in the text, reread it a couple of times. If it is still not clear, note the text with a question mark or some other notation so you can ask your instructor about it.

Go through Each Step of Each Example

Make sure you understand each step of an example. If you don't understand something, mark it so you can ask about it in class. Sometimes math textbooks leave out intermediate steps to save space. Try working through the examples on your own, filling in any missing steps.

Take Notes

< This is important!

Write down important definitions, symbols or notation, properties, formulas, theorems, and procedures. Review these daily as you do your homework and before taking quizzes and tests. Practice rewriting definitions in your own words so you understand them better.

Notes 9-25-17:

- *The opposite of a negative integer is a positive integer.*
- *To add two integers with the same signs add their absolute values and use their common sign*

Use Available Resources

Many textbooks have companion websites to help you understand the content. These resources may contain videos that help explain more complex steps or concepts. Try searching the internet for additional explanations of topics you don't understand.

Read the Material Before Class

Try to read the material from your book before the instructor lectures on it. After the lecture, reread the section again to help you retain the information as you look over your class notes.

Understand the Mathematical Definitions + × =

Many terms used in everyday English have a different meaning when used in mathematics. Some examples include equivalent, similar, average, median, and product. Two equations can be equivalent to one another without being equal. An average can be computed mathematically in several ways. It is important to note these differences in meaning in your notebook along with important definitions and formulas.

Try Reading the Material Aloud

Reading aloud makes you focus on every word in the sentence. Leaving out a word in a sentence or math problem could give it a totally different meaning, so be sure to read the text carefully and reread, if necessary.

Questions

1. Explain how taking notes can help you understand new concepts and skills while reading a math textbook.
2. Think of two more tips for reading a math textbook.

Strategies for Academic Success

Tips for Success in a Math Course

Read Your Textbook/Workbook

One of the most important skills when taking a math class is knowing how to read a math textbook. Reading a section before class and then reading it again afterwards is an important strategy for success in a math course. If you don't have time to read the entire assigned section, you can get an overview by reading the introduction or summary and looking at section objectives, headings, and vocabulary terms.

Take Notes

Take notes in class using a method that works for you. There are many different note-taking strategies, such as the Cornell Method and Concept Mapping. You can try researching these and other methods to see if they might work better than your current note-taking system.

Review

While the information is fresh in your mind, read through your notes as soon as possible after class to make sure they are readable, write down any questions you have, and fill in any gaps. Mark any information that is incomplete so that you can get it from the textbook or your instructor later.

Stay Organized

As you review your notes each day, be sure to label them using categories such as definition, theorem, formula, example, and procedure. Try highlighting each category with a different colored highlighter.

Use Study Aids

Use note cards to help you remember definitions, theorems, formulas, or procedures. Use the front of the card for the vocabulary term, theorem name, formula name, or procedure description. Write the definition, the theorem, the formula, or the procedure on the back of the card, along with a description in your own words.

Practice, Practice, Practice!

Math is like playing a sport. You can't improve your basketball skills if you don't practice—the same is true of math. Math can't be learned by only watching your instructor work through problems; you have to be actively involved in doing the math yourself. Work through the examples in the book, do some practice exercises at the end of the section or chapter, and keep up with homework assignments on a daily basis.

Do Your Homework

When doing homework, always allow plenty of time to finish it before it is due. Check your answers when possible to make sure they are correct. With word or application problems, always review your answer to see if it appears reasonable. Use the estimation techniques that you have learned to determine if your answer makes sense.

Understand, Don't Memorize

Don't try to memorize formulas or theorems without understanding them. Try describing or explaining them in your own words or look for patterns in formulas so you don't have to memorize them. For example, you don't need to memorize every perimeter formula if you understand that perimeter is equal to the sum of the lengths of the sides of the figure.

Study

Plan to study two to three hours outside of class for every hour spent in class. If math is your most difficult subject, then study while you are alert and fresh. Pick a study time when you will have the least interruptions or distractions so that you can concentrate.

Manage Your Time

Don't spend more than 10 to 15 minutes working on a single problem. If you can't figure out the answer, put it aside and work on another one. You may learn something from the next problem that will help you with the one you couldn't do. Mark the problems that you skip so that you can ask your instructor about it during the next class. It may also help to work a similar, but perhaps easier, problem.

Questions

1. Based on your schedule, what are the best times and places for you to study for this class?
2. Describe your method of taking notes. List two ways to improve your method.

Strategies for Academic Success

Tips for Improving Math Test Scores

Preparing for a Math Test

- Avoid cramming right before the test and don't wait until the night before to study. Review your notes and note cards every day in preparation for quizzes and tests.
- If the textbook has a chapter review or practice test after each chapter, work through the problems as practice for the test.
- If the textbook has accompanying software with review problems or practice tests, use it for review.
- Review and rework homework problems, especially the ones that you found difficult.
- If you are having trouble understanding certain concepts or solving any types of problems, schedule a meeting with your instructor or arrange for a tutoring session (if your college offers a tutoring service) well in advance of the next test.

Test-Taking Strategies

- Scan the test as soon as you get it to determine the number of questions, their levels of difficulty, and their point values so you can adequately gauge how much time you will have to spend on each question.
- Start with the questions that seem easiest or that you know how to work immediately. If there are problems with large point values, work them next since they count for a larger portion of your grade.
- Show all steps in your math work. This will make it quicker to check your answers later once you are finished since you will not have to work through all the steps again.
- If you are having difficulty remembering how to work a problem, skip it and come back to it later so that you don't spend all of your time on one problem.

After the Test

- The material learned in most math courses is cumulative, which means any concepts you miss on each test may be needed to understand concepts in future chapters. That's why it is extremely important to review your returned tests and correct any misunderstandings that may hinder your performance on future tests.
- Be sure to correct any work you did wrong on the test so that you know the correct way to do the problem in the future. If you are not sure what you did wrong, get help from a peer who scored well on the test or schedule time with your instructor to go over the test.
- Analyze the test questions to determine if the majority came from your class notes, homework problems, or the textbook. This will give you a better idea of how to spend your time studying for the next test.
- Analyze the errors you made on the test. Were they careless mistakes? Did you run out of time? Did you not understand the material well enough? Were you unsure of which method to use?
- Based on your analysis, determine what you should do differently before the next test and where you should focus your time.

Questions

1. Determine the resources that are available to you to help you prepare for tests, such as instructor office hours, tutoring center hours, and study groups.
2. Discuss two additional test taking strategies.

Strategies for Academic Success

Practice, Patience, and Persistence!

Have you ever heard the phrase "practice makes perfect"? This saying applies to many things in life. You won't become a concert pianist without many hours of practice. You won't become an NBA basketball star by sitting around and watching basketball on TV. The saying even applies to riding a bike. You can watch all of the videos and read all of the books on riding a bike, but you won't learn how to ride a bike without actually getting on the bike and trying to do it yourself. The same idea applies to math. Math is not a spectator sport.

Math is not learned by sleeping with your math book under your pillow at night and hoping for osmosis (a scientific term implying that math knowledge would move from a place of higher concentration—the math book—to a place of lower concentration—your brain). You also don't learn math by watching your professor do hundreds of math problems while you sit and watch. Math is learned by doing. Not just by doing one or two problems, but by doing many problems. Math is just like a sport in this sense. You become good at it by doing it, not by watching others do it. You can also think of learning math like learning to dance. A famous ballerina doesn't take a dance class or two and then end up dancing the lead in The Nutcracker. It takes years of practice, patience, and persistence to get that part.

Now, we aren't suggesting that you dedicate your life to doing math, but at this point in your education, you've already spent quite a few years studying the subject. You will continue to do math throughout college—and your life. To be able to financially support yourself and your family, you will have to find a job, earn a salary, and invest your money—all of which require some ability to do math. You may not think so right now, but math is one of the more useful subjects you will study.

It's important not only to practice math when taking a math course, but also to be patient and not expect immediate success. Just like a ballerina or NBA basketball star, who didn't become exceptional athletes overnight, it will take some time and patience to develop your math skills. Sure, you will make some mistakes along the way, but learn from those mistakes and move on.

Practice, patience, and persistence are especially important when working through applications or word problems. Most students don't like word problems and, therefore, avoid them. You won't become good at working word problems unless you practice them over and over again. You'll need to be patient when working through word problems in math since they will require more time to work than typical math skills exercises. The process of solving word problems is not a quick one and will take patience and persistence on your part to be successful.

Just as you work your body through physical exercise, you have to work your brain through mental exercise. Math is an excellent subject to provide the mental exercise needed to stimulate your brain. Your brain is flexible and it continues to grow throughout your life span—but only if provided the right stimuli. Studying mathematics and persistently working through tough math problems is one way to promote increased brain function. So, when doing mathematics, remember the 3 P's—Practice, Patience, and Persistence—and the positive effects they will have on your brain!

Questions

1. What is another area (not mentioned here) that requires practice, patience, and persistence to master? Can you think of anything you could master without practice?
2. Can you think of an example in your study of math where practice, patience, and persistence have helped you improve?

Strategies for Academic Success

Note Taking

Taking notes in class is an important step in understanding new material. While there are several methods for taking notes, every note-taking method can benefit from these general tips.

General Tips

- Write the date and the course name at the top of each page.
- Write the notes in your own words and paraphrase.
- Use abbreviations, such as ft for foot, # for number, def for definition, and RHS for right-hand side.
- Copy all figures or examples that are presented during the lecture.
- Review and rewrite your notes after class. Do this on the same day, if possible.

There are many different methods of note taking and it's always good to explore new methods. A good time to try out new note-taking methods is when you rewrite your class notes. Be sure to try each new method a few times before deciding which works best for you. Presented here are three note-taking methods you can try out. You may even find that a blend of several methods works best for you.

Note-Taking Methods

Outline

An outline consists of several topic headings, each followed by a series of indented bullet points that include subtopics, definitions, examples, and other details.

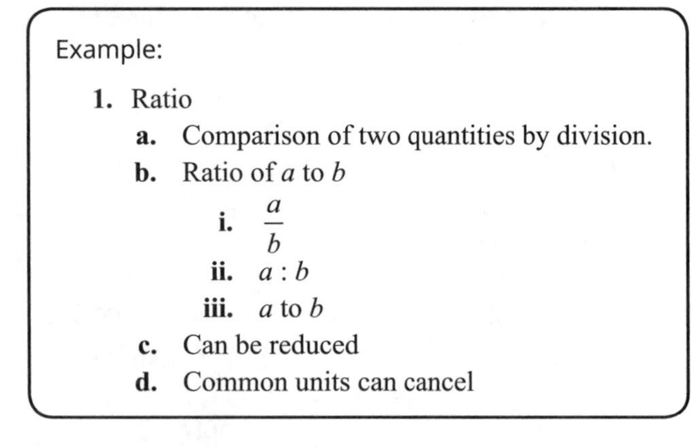

Example:

1. Ratio
   a. Comparison of two quantities by division.
   b. Ratio of a to b
      i. $\frac{a}{b}$
      ii. $a : b$
      iii. a to b
   c. Can be reduced
   d. Common units can cancel

Split Page

The split page method divides the page vertically into two columns with the left column narrower than the right column. Main topics go in the left column and detailed comments go in the right column. The bottom of the page is reserved for a short summary of the material covered.

Example:

Keywords:	**Notes:**
Ratios	1. Comparison of two quantities by division 2. $\frac{a}{b}$, $a : b$, a to b 3. Can reduce 4. Common units can cancel
Summary: Ratios are used to compare quantities and units can cancel.	

Mapping

The mapping method is the most visual of the three methods. One common way to create a mapping is to write the main idea or topic in the center and draw lines, from the main idea to smaller ideas or subtopics. Additional branches can be created from the subtopics until all of the key ideas and definitions are included. Using a different color for subtopic can help visually organize the topics.

Example:

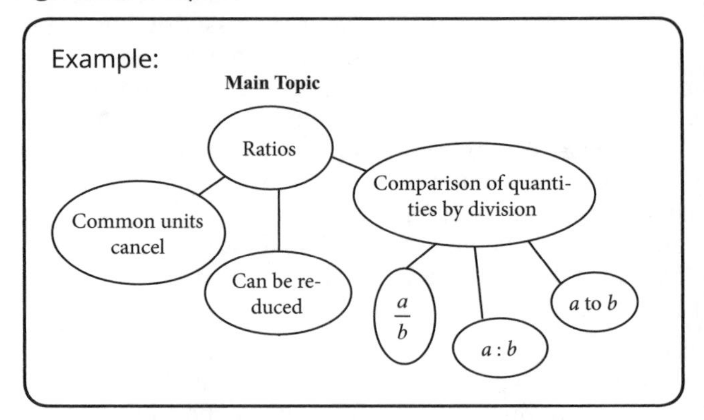

Questions

1. Find two other note taking methods and describe them.
2. Write five additional abbreviations that you could use while taking notes.

Strategies for Academic Success

Do I Need a Math Tutor?

If you do not understand the material being presented in class, if you are struggling with completing homework assignments, or if you are doing poorly on tests, then you may need to consider getting a tutor. In college, everyone needs help at some point in time. What's important is to recognize that you need help before it's too late and you end up having to retake the class.

Alternatives to Tutoring

Before getting a tutor, you might consider setting up a meeting with your instructor during their office hours to get help. Unfortunately, you may find that your instructor's office hours don't coincide with your schedule or don't provide enough time for one-on-one help.

Another alternative is to put together a study group of classmates from your math class. Working in groups and explaining your work to others can be very beneficial to your understanding of mathematics. Study groups work best if there are three to six members. Having too many people in a study group may make it difficult to schedule a time for all group members to meet. A large study group may also increase distractions. If you have too few people and those that attend are just as lost as you, then you aren't going to be helpful to each other.

Where to Find a Tutor

Many schools have both group and individual tutoring available. In most cases, the cost of this tutoring is included in tuition costs. If your college offers tutoring through a learning lab or tutoring center, then you should take advantage of it. You may need to complete an application to be considered for tutoring, so be sure to get the necessary paperwork at the start of each semester to increase your chances of getting a tutoring time that works well with your schedule. This is especially important if you know that you struggle with math or haven't taken any math classes in a while.

If you find that you need more help than the tutoring center can provide, or your school doesn't offer tutoring, you can hire a private tutor. The hourly cost to hire a private tutor varies significantly depending on the area you live in along with the education and experience level of the tutor. You might be able to find a tutor by asking your instructor for references or by asking friends who have taken higher-level math classes than you have. You can also try researching the internet for local reputable tutoring organizations in your area.

What to Look for in a Tutor

Whether you obtain a tutor through your college or hire a personal tutor, look for someone who has experience, educational qualifications, and who is friendly and easy to work with. If you find that the tutor's personality or learning style isn't similar to yours, then you should look for a different tutor that matches your style. It may take some effort to find a tutor who works well with you.

How to Prepare for a Tutoring Session

To get the most out of your tutoring session, come prepared by bringing your text, class notes, and any homework or questions you need help with. If you know ahead of time what you will be working on, communicate this to the tutor so they can also come prepared. You should attempt the homework prior to the session and write notes or questions for the tutor. Do not use the tutor to do your homework for you. The tutor will explain to you how to do the work and let you work some problems on your own while he or she observes. Ask the tutor to explain the steps aloud while working through a problem. Be sure to do the same so that the tutor can correct any mistakes in your reasoning. Take notes during your tutoring session and ask the tutor if he or she has any additional resources such as websites, videos, or handouts that may help you.

Questions

1. It's important to find a tutor whose learning style is similar to yours. What are some ways that learning styles can be different?
2. What sort of tutoring services does your school offer?

Strategies for Academic Success

Tips for Improving Your Memory

Experts believe that there are three ways that we store memories: first in the sensory stage, then in short term memory, and finally in long term memory.[1] Because we can't retain all the information that bombards us daily, the different stages of memory act as a filter. Your sensory memory lasts only a fraction of a second and holds your perception of a visual image, a sound, or a touch. The sensation then moves to your short term memory, which has the limited capacity to hold about seven items for no more than 20 to 30 seconds at a time. Important information is gradually transferred to long term memory. The more the information is repeated or used, the greater the chance that it will end up in long term memory. Unlike sensory and short term memory, long term memory can store unlimited amounts of information indefinitely. Here are some tips to improve your chances of moving important information to long-term memory.

Be attentive and focused on the information.

Study in a location that is free of distractions and avoid watching TV or listening to music with lyrics while studying.

Recite information aloud.

Ask yourself questions about the material to see if you can recall important facts and details. Pretend you are teaching or explaining the material to someone else. This will help you put the information into your own words.

Associate the information with something you already know.

Think about how you can make the information personally meaningful—how does it relate to your life, your experiences, and your current knowledge? If you can link new information to memories already stored, you create "mental hooks" that help you recall the information. For example, when trying to remember the formula for slope using rise and run, remember that rise would come alphabetically before run, so rise will be in the numerator in the slope fraction and run will be in the denominator.

Use visual images like diagrams, charts, and pictures.

You can make your own pictures and diagrams to help you recall important definitions, theorems, or concepts.

Split larger pieces of information into smaller "chunks."

This is useful when remembering strings of numbers, such as social security numbers and telephone numbers. Instead of remembering a sequence of digits such as 555777213 you can break it into chunks such as 555 777 213.

Group long lists of information into categories that make sense.

For example, instead of remembering all the properties of real numbers individually, try grouping them into shorter lists by operation, such as addition and multiplication.

Use mnemonics or memory techniques to help remember important concepts and facts.

A mnemonic that is commonly used to remember the order of operations is "Please Excuse My Dear Aunt Sally," which uses the first letter of the words Parentheses, Exponents, Multiplication, Division, Addition, and Subtraction to help you remember the correct order to perform basic arithmetic calculations. To make the mnemonic more personal and possibly more memorable, make up one of your own.

Use acronyms to help remember important concepts or procedures.

An acronym is a type of mnemonic device which is a word made up by taking the first letter from each word that you want to remember and making a new word from the letters. For example, the word HOMES is often used to remember the five Great Lakes in North America where each letter in the word represents the first letter of one of the lakes: Huron, Ontario, Michigan, Erie, and Superior.

Questions

1. Create an original mnemonic or acronym for any math topic covered so far in this course.
2. Explain two ways you can incorporate these tips into your study routine.

1 Source: http://science.howstuffworks.com/life/inside-the-mind/human-brain/human-memory2.htm

Strategies for Academic Success

Overcoming Anxiety

People who are anxious about math are often just not good at taking math tests. If you understand the math you are learning but don't do well on math tests, you may be in the same situation. If there are other subject areas in which you also perform poorly on tests, then you may be experiencing test anxiety.

How to Reduce Math Anxiety

- Learn effective math study skills. Sit near the front of your class and take notes. Ask questions when you don't understand the material. Review your notes after class and read new material before it's covered in class. Keep up with your assignments and do a lot of practice problems.
- Don't accept negative self talk such as "I am not good at math" or "I just don't get it and never will." Maintain a positive attitude and set small math achievement goals to keep you positively moving toward bigger goals.
- Visualize yourself doing well in math, whether it's on a quiz or test, or passing a math class. Rehearse how you will feel and perform on an upcoming math test. It may also help to visualize how you will celebrate your success after doing well on the test.
- Form a math study group. Working with others may help you feel more relaxed about math in general and you may find that other people have the same fears.
- If you panic or freeze during a math test, try to work around the panic by finding something on the math test that you can do. Once you gain confidence, work through other problems you know how to do. Then, try completing the harder problems, knowing that you have a large part of the test completed already.
- If you have trouble remembering important concepts during tests, do what is called a "brain drain" and write down all the formulas and important facts that you have studied on your test or scratch paper as soon as you are given the test. Do this before you look at any questions on the test. Having this information available to you should help boost your confidence and reduce your anxiety. Doing practice brain drains while studying can help you remember the concepts when the test time comes.

How to Reduce Test Anxiety

- Be prepared. Knowing you have prepared well will make you more confident and less anxious.
- Get plenty of sleep the night before a big test and be sure to eat nutritious meals on the day of the test. It's helpful to exercise regularly and establish a set routine for test days. For example, your routine might include eating your favorite food, putting on your lucky shirt, and packing a special treat for after the test.
- Talk to your instructor about your anxiety. Your instructor may be able to make accommodations for you when taking tests that may make you feel more relaxed, such as extra time or a more calming testing place.
- Learn how to manage your anxiety by taking deep, slow breaths and thinking about places or people who make you happy and peaceful.
- When you receive a low score on a test, take time to analyze the reasons why you performed poorly. Did you prepare enough? Did you study the right material? Did you get enough rest the night before? Resolve to change those things that may have negatively affected your performance in the past before the next test.
- Learn effective test taking strategies. See the study skill on Tips for Improving Math Test Scores.

Questions

1. Describe your routine for test days. Think of two ways you can improve your routine to reduce stress and anxiety.
2. Research and describe the accommodations that your instructor or school can provide for test taking.

Strategies for Academic Success

Online Resources

With the invention of the internet, there are numerous resources available to students who need help with mathematics. Here are some quality online resources that we recommend.

HawkesTV

tv.hawkeslearning.com

If you are looking for instructional videos on a particular topic, then start with HawkesTV. There are hundreds of videos that can be found by looking under a particular math subject area such as introductory algebra, precalculus, or statistics. You can also find videos on study skills.

YouTube

www.youtube.com

You can also find math instructional videos on YouTube, but you have to search for videos by topic or key words. You may have to use various combinations of key words to find the particular topic you are looking for. Keep in mind that the quality of the videos varies considerably depending on who produces them.

Google Hangouts

plus.google.com/hangouts

You can organize a virtual study group of up to 10 people using Google Hangouts. This is a terrific tool when schedules are hectic and it avoids everyone having to travel to a central location. You do have to set up a Google+ profile to use Hangouts. In addition to video chat, the group members can share documents using Google Docs. This is a great tool for group projects!

Wolfram|Alpha

www.wolframalpha.com

Wolfram|Alpha is a computational knowledge engine developed by Wolfram Research that answers questions posed to it by computing the answer from "curated data." Typical search engines search all of the data on the Internet based on the key words given and then provide a list of documents or web pages that might contain relevant information. The data used by Wolfram|Alpha is said to be "curated" because someone has to verify its integrity before it can be added to the database, therefore ensuring that the data is of high quality. Users can submit questions and request calculations or graphs by typing their request into a text field. Wolfram|Alpha then computes the answers and related graphics from data gathered from both academic and commercial websites such as the CIA's World Factbook, the United States Geological Survey, financial data from Dow Jones, etc. Wolfram|Alpha uses the basic features of Mathematica, which is a computational toolkit designed earlier by Wolfram Research that includes computer algebra, symbol and number computation, graphics, and statistical capabilities.

Questions

1. Describe a situation where you think Wolfram|Alpha might be more helpful than YouTube, and vice versa.
2. What are some pros and cons to using Google Hangouts?

Strategies for Academic Success

Preparing for a Final Math Exam

Since math concepts build on one another, a final exam in math is not one you can study for in a night or even a day or two. To pull all the concepts together for the semester, you should plan to start one or two weeks ahead of time. Being comfortable with the material is key to going into the exam with confidence and lowering your anxiety.

Before You Start Preparing for the Exam

1. What is the date, time, and location of the exam? Check your syllabus for the final exam time and location. If it's not on your syllabus, your instructor should announce this information in class.
2. Is there a time limit on the exam? If you experience test anxiety on timed tests, be sure to speak to your professor about it and see if you can receive accommodations that will help reduce your anxiety, such as extended time or an alternate testing location.
3. Will you be able to use a formula sheet, calculator, and/or scrap paper on the exam? If you are not allowed to use a formula sheet, you should write down important formulas and memorize them. Most of the time, math professors will advise you of the formulas you need to know for an exam. If you cannot use a calculator on the exam, be sure to practice doing calculations by hand when you are preparing for the exam and go back and check them using the calculator.

A Week Before the Exam

1. Decide where to study for the exam and with whom. Make sure it's a comfortable study environment with few outside distractions. If you are studying with others, make sure the group is small and that the people in the group are motivated to study and do well on the exam. Plan to have snacks and water with you for energy and to avoid having to delay studying to go get something to eat or drink. Be sure and take small breaks every hour or two to keep focused and minimize frustration.
2. Organize your class notes and any flash cards with vocabulary, formulas, and theorems. If you haven't used flash cards for vocabulary, go back through your notes and highlight the vocabulary. Create a formula sheet to use on the exam, if the professor allows. If not, then you can use the formula sheet to memorize the formulas that will be on the exam.
3. Start studying for the exam. Studying a week before the exam gives you time to ask your instructor questions as you go over the material. Don't spend a lot of time reviewing material you already know. Go over the most difficult material or material that you don't understand so you can ask questions about it. Be sure to review old exams and work through any questions you missed.

3 Days Before the Exam

1. Make yourself a practice test consisting of the problem types. Don't necessarily put the questions in the order that the professor covered them in class.
2. Ask your instructor or classmates any questions that you have about the practice test so that you have time to go back and review the material you are having difficulty with.

The Night Before the Exam

1. Make sure you have all the supplies you will need to take the exam: formula sheet and calculator, if allowed, scratch paper, plain and colored pencils, highlighter, erasers, graph paper, extra batteries, etc.
2. If you won't be allowed to use your formula sheet, review it to make sure you know all the formulas. Right before going to bed, review your notes and study materials, but do not stay up all night to "cram."
3. Go to bed early and get a good night's sleep. You will do better if you are rested and alert.

The Day of the Exam

1. Get up with plenty of time to get to your exam without rushing. Eat a good breakfast and don't drink too much caffeine, which can make you anxious.
2. Review your notes, flash cards, and formula sheet again, if you have time.
3. Get to class early so you can be organized and mentally prepared.

Checklist for the Exam

Date of the Exam: ____________________ **Time of the Exam:** ____________________

Location of the Exam: ____

Items to bring to the exam:

__ calculator and extra batteries

__ formula sheet

__ scratch paper

__ graph paper

__ pencils

__ eraser

__ colored pencils or highlighter

__ ruler or straightedge

Notes or other things to remember for exam day:

During the Exam

1. Put your name at the top of your exam immediately. If you are not allowed to use a formula sheet, before you even look at the exam, do what is called a "brain drain" or "data dump." Recall as much of the information on your formula sheet as you possibly can and write it either on the scratch paper or in the exam margins if scratch paper is not allowed. You have now transferred over everything on your "mental cheat sheet" to the exam to help yourself as you work through the exam.
2. Read the directions carefully as you go through the exam and make sure you have answered the questions being asked. Also, check your solutions as you go. If you do any work on scratch paper, write down the number of the problem on the paper and highlight or circle your answer. This will save you time when you review the exam. The instructor may also give you partial credit for showing your work. (Don't forget to attach your scratch work to your exam when you turn it in.)
3. Skim the questions on the exam, marking the ones you know how to do immediately. These are the problems you will do first. Also note any questions that have a higher point value. You should try to work these next or be sure to leave yourself plenty of time to do them later.
4. If you get to a problem you don't know how to do, skip it and come back after you finish all the ones you know how to do. A problem you do later may jog your memory on how to do the problem you skipped.
5. For multiple choice questions, be sure to work the problem first before looking at the answer choices. If your answer is not one of the choices, then review your math work. You can also try starting with the answer choices and working backwards to see if any of them work in the problem. If this doesn't work, see if you can eliminate any of the answer choices and make an educated guess from the remaining ones. Mark the problem to come back to later when you review the exam.
6. Once you have an answer for all the problems, review the entire exam. Try working the problems differently and comparing the results or substituting the answers into the equation to verify they are correct. Do not worry about finishing early. You are in control of your own time—and your own success!

Questions

1. Does your syllabus provide any of the information needed for the checklist?
2. Are there any tips or suggestions mentioned here that you haven't thought of before?

Strategies for Academic Success

Managing Your Time Effectively

Have you ever made it to the end of a day and wondered where all of your time went? Sometimes it feels like there aren't enough hours in the day. Managing your time is important because you can never get that time back. Once it's gone, you have to rush and cram the work into your schedule. Not only will you start feeling stressed out, but you may also find yourself turning in late or incomplete work.

Here are three strategies for managing your time more effectively.

Time Budgets

Time budgets help you find the time you need to complete necessary projects and tasks. Just like a financial budget shows you how you spend your money, a time budget shows you how you spend your time. You can then identify "wasted" time that could be used more productively.

To begin budgeting your time, assess how much time each week you spend on different types of activities, like Sleep, Meals, Work, Class, Study, Extracurricular, Exercise, Personal, Other, etc.

- What are some activities you'd like to spend more time doing in the future?
- What are some activities you should spend less time doing in the future?

Based on your answers to the questions above, create a weekly time budget. One week contains only 168 hours. If you want to spend more time on a particular activity, you'll need to find that time somewhere. Use a planner to schedule specific blocks of time for study sessions, meals, travel times, and morning/evening routines. As a general rule, you should set aside at least two hours of study time for every one hour of class time. That means that a three-credit course would require at least six hours of outside work per week.

Breaks

When you are working on an important project or studying for a big exam, you can feel tempted to go as long as possible without taking a break. While staying focused is important, working yourself until you're mentally drained will lower the quality of your work and force you to take even more time recovering. Just like taking breaks helps your physical body recover, it will also help your brain re-energize and refocus. During study sessions, you should plan to take a break at least once an hour. Study and work breaks should usually last around five minutes. The longer the break, the harder it is to start working again. Some courses have a built-in break during the middle of the class period. Stand up and move around, even if you don't feel tired. Even this little bit of physical movement can help you think more clearly.

Avoiding Multitasking

Multitasking is working on more than one task at a time. When you have several assignments that need to be completed, you may be tempted to save time by working on two or three of them at once. While this strategy might seem like a time-saver, you will probably end up using more time than if you had done each task individually. Not only will you have to switch your focus from one task to the next, but you will also make more mistakes that will need to be corrected later. Multitasking usually ends up wasting time instead of saving it.

Instead of trying to do two things at once, schedule yourself time to work on one task at a time. To-do lists can be helpful tools for keeping yourself focused on finishing one item before moving on to another. You'll do better work and save yourself time.

Questions

1. Are there any areas in your day that are taking up too much of your time, making it hard to devote enough time to more important things?
2. Can you think of a time when multitasking has resulted in lower quality outcome in your experience?

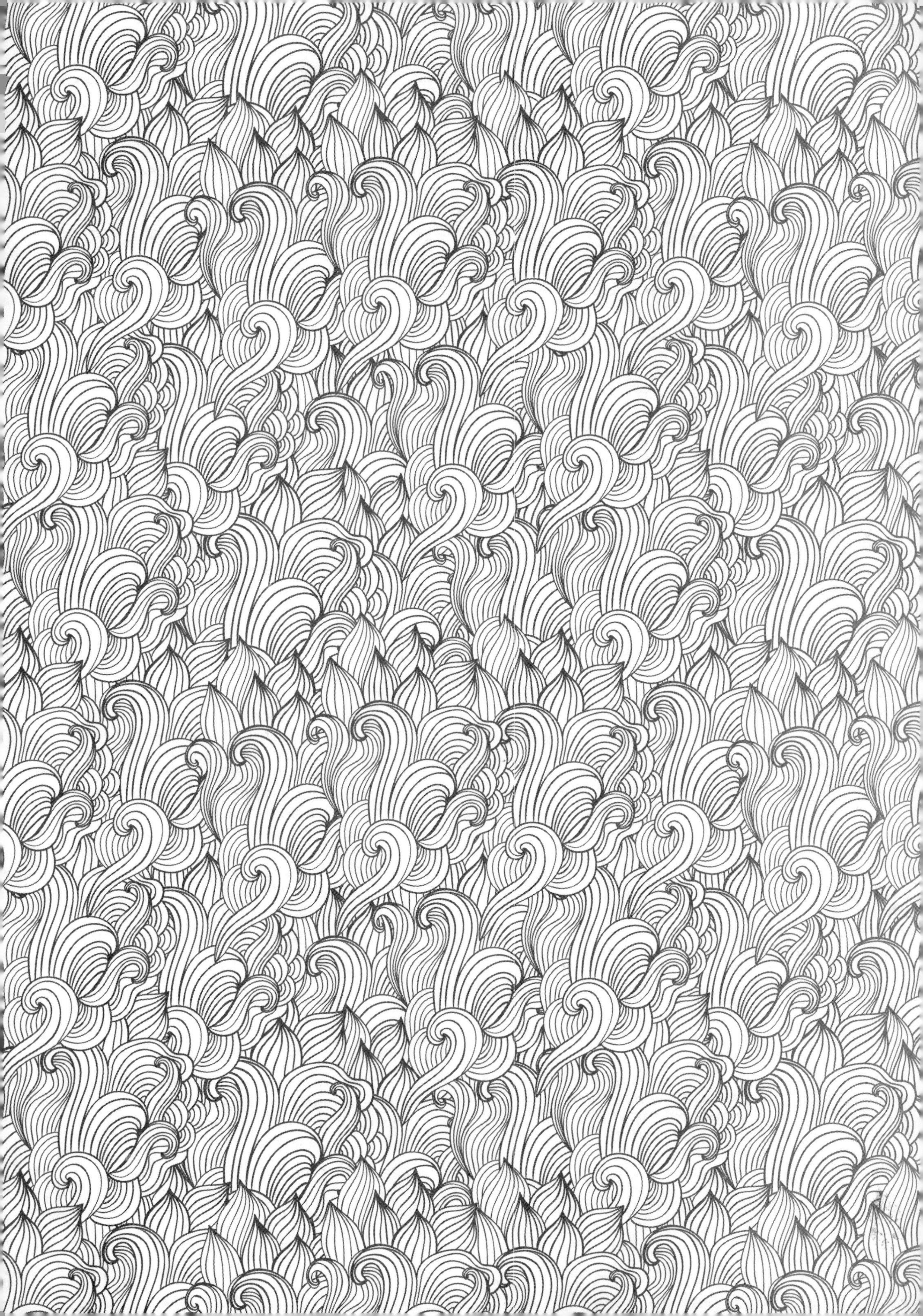

CHAPTER 1

Whole Numbers

Math @ Work

Numbers play a role in any job you may take in the future. If you go into sales, you will need to keep track of the number of sales you make and the profit you've earned for the company. If you go into construction, you will need to be able to accurately take measurements of the buildings and structures you help create. If you go into the medical field, numerical accuracy is important in situations such as measuring out the medicine for a sick child or recording a patient's blood pressure. If you decide to start your own business, you will need to evaluate operation costs to determine how much to charge your clients for your products or services.

During your education, numbers will also play a large role. For example, as a student striving to succeed, you will want to keep track of your progress in the courses you take. One way to do this is to find the average of your test scores throughout the semester. The average of these scores will help you determine if you are doing well in the course or if you need to seek assistance from either the instructor or a tutor. Suppose during the semester you receive the following scores on the first four tests.

78, 85, 94, 83

What is your current test average for the course? Should you seek assistance to improve your grade?

Name: Date:

1.1 Introduction to Whole Numbers

Whole Numbers

The whole numbers are ______________________________ along with ________________________

Natural numbers = N = {____}

Whole numbers = W = {____}

Note that 0 is a ____

DEFINITION

The Decimal System

The **decimal system** (or base ten system) is a place value system that depends on three things.

1. ____

2. the placement of _____________________

3. the value of ______________________

DEFINITION

Reading and Writing Whole Numbers

You should note the following four things when reading or writing whole numbers.

1. Digits are read in ___________________________________

2. Commas are used to ___________________________ if a number has ______________________________

3. The word **and** does ____

And is said only when reading a _________________________

4. Hyphens (-) are used to write words for ____

_________________ For example: ____________________________________

Watch and Work

Watch the video for Example 5 in the software and follow along in the space provided.

Example 5 Reading and Writing Whole Numbers

Each number is written in words. Write it in standard notation.

a. twenty-seven thousand, three hundred thirty-six

b. three hundred forty million, sixty-two thousand, forty-eight

Solution

Now You Try It!

Use the space provided to work out the solution to the next example.

Example A Reading and Writing Whole Numbers

Write each number in standard notation.

a. six thousand forty-one

b. one million, four hundred eighty-three thousand, seven

1.1 Exercises

Concept Check

True/False. Determine whether each statement is true or false. If a statement is false, explain how it can be changed so the statement will be true. (**Note:** There may be more than one acceptable change.)

1. In the number 21,057, the "1" represents 1000.

2. 42,360 can be written as forty-two thousand, three hundred sixty

3. The word "and" is not used when reading or writing whole numbers.

Practice

4. Given the number 284,065 which digit indicates the number of

a. tens?

b. ten thousands?

c. hundreds?

5. Name the place value of each nonzero digit in the following number: 24,608.

6. Write 683,100 in words.

7. Write five hundred thirty-seven thousand, eighty-two in standard notation.

Applications

Write the numbers in each sentence in words.

8. ***Lakes:*** The largest lake in the United States is Lake Superior. It takes up an area of 82,103 square kilometers.

Write the numbers in the sentence in standard form.

9. ***Card Games:*** The largest collection of Joker playing cards consists of eight thousand, five hundred twenty cards amassed by Tony De Santis after inheriting a two thousand piece collection from the magician Fernando Riccardi.

Writing & Thinking

10. How are natural numbers and whole numbers different and how are they the same?

11. When are hyphens used to write numbers in English words?

Name: Date:

1.2 Addition and Subtraction with Whole Numbers

Adding Whole Numbers

1. Write the numbers ____________________ so that the ______________________________________

2. Add only the ____

PROCEDURE

Carrying When Adding Whole Numbers

If the sum of the digits in one column is more than 9,

1. write the ____

2. carry the _________________________ as a number to be added to ______________________________

PROCEDURE

Variable

A variable is a symbol (generally a letter of the alphabet) that is ____________________________________

____

DEFINITION

Commutative Property of Addition

For any whole numbers *a* and *b*, _______________________

____

(The **order** of the numbers in addition can ______________________)

PROPERTIES

Associative Property of Addition

For any whole numbers *a*, *b*, and *c*, ________________________________

For example, __________________________________

(____)

PROPERTIES

Additive Identity Property

For any whole number *a*, ____________________

For example, ____________________

(The sum of a number and 0 is ____________________)

The number 0 is called the ____________________

PROPERTIES

The **perimeter** of a geometric figure is ____________________

Subtraction

Subtraction is the operation of ____________________

The difference is the result of ____________________

____

DEFINITION

Subtracting Whole Numbers

1. Write the numbers ____________ so that the ____________________
2. Subtract only the ____________________
3. Check by ____________________
4. The sum must be the ____________________

PROCEDURE

Watch and Work

Watch the video for Example 13 in the software and follow along in the space provided.

Example 13 Application: Subtracting Whole Numbers

The cost of repairing Ed's used TV is $395. To buy a new TV, he will have to pay $447. How much more would Ed have to pay for a new TV than to have his old TV repaired?

Solution

Now You Try It!

Use the space provided to work out the solution to the next example.

Example A Application: Subtracting Whole Numbers

The cost of repairing Robert's used DVD player will be $165. To buy a new DVD player, Robert will have to pay $129. How much more would it cost to fix the old DVD player than to buy a new one?

1.2 Exercises

Concept Check

True/False. Determine whether each statement is true or false. If a statement is false, explain how it can be changed so the statement will be true. (**Note:** There may be more than one acceptable change.)

1. A polygon is a geometric figure in a plane with two or more sides.

2. To find the perimeter of a rectangle, add the lengths of the four sides.

3. When subtracting, sometimes the digit being subtracted is larger than the digit it is being subtracted from and so "carrying" must occur.

4. If your bank account has a balance of $743 and you want to withdraw $115, you would use subtraction to find that the new balance would be $628.

Practice

Simplify.

5. $\begin{array}{r} 15 \\ +43 \\ \hline \end{array}$

6. $\begin{array}{r} 981 \\ +46 \\ \hline \end{array}$

7. $\begin{array}{r} 275 \\ -131 \\ \hline \end{array}$

8. $\begin{array}{r} 543 \\ -167 \\ \hline \end{array}$

Calculate the perimeter of the geometric figure.

9.

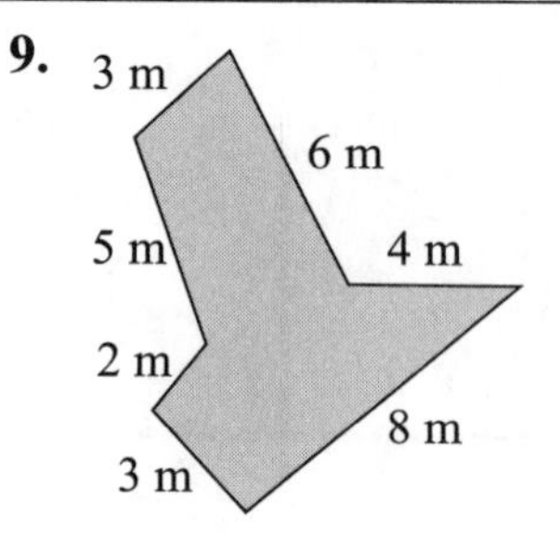

Applications

Solve.

10. ***Budgeting:*** The Magley family has the following monthly budget: $815 mortgage; $69 electric; $47 water; and $122 phone bills (including cell phones). What is the family's budget for each month for these expenses?

11. ***Credit:*** A couple sold their house for $135,000. They paid the realtor $8100, and other expenses of the sale came to $800. If they owed the bank $87,000 for the mortgage, what were their net proceeds from the sale?

Writing & Thinking

12. List three properties of addition and give an example of each.

13. Explain when "carrying" should be used in addition with whole numbers and give an example.

1.3 Multiplication with Whole Numbers

Symbols for Multiplication

Symbol	Example
___ ___________	___________
___ ___________	___________
_ ___________	
	___________

Commutative Property of Multiplication

For any whole numbers a and b, ___________

For example, ___________

(The order of the numbers in multiplication can be ___________)

PROPERTIES

Associative Property of Multiplication

For any whole numbers a, b, and c, ___________

For example, ___________

(The grouping of the numbers in multiplication can ___________)

PROPERTIES

Multiplicative Identity Property

For any whole number a, ___________

____

(The product of any number and 1 is ___________)

The number 1 is called the ___________

Multiplication Property of 0 (or Zero-Factor Law)

For any whole number a, ____________________

For example, ____________________

____)

PROPERTIES

Distributive Property

For any whole numbers a, b, and c, ________________________________

PROPERTIES

Watch and Work

Watch the video for Example 6 in the software and follow along in the space provided.

Example 6 Multiplying Whole Numbers

Multiply: $93 \cdot 46$

Solution

Now You Try It!

Use the space provided to work out the solution to the next example.

Example A Multiplying Whole Numbers

Multiply: $\begin{array}{r} 15 \\ \times\ 32 \\ \hline \end{array}$

Multiplying Whole Numbers by Powers of 10

To multiply a whole number:

by 10, write ____________________
by 100, write ____________________
by 1000, write ____________________
by 10,000, write ____________________
and so on.

PROCEDURE

Area of a Rectangle

The **area** of a rectangle (measured in square units) is found by ____________________

In the form of a formula, $A =$ ____________

DEFINITION

1.3 Exercises

Concept Check

True/False. Determine whether each statement is true or false. If a statement is false, explain how it can be changed so the statement will be true. (**Note:** There may be more than one acceptable change.)

1. The numbers being multiplied are called the divisors.

2. According to the multiplicative identity, $1 \cdot 25 = 52$.

3. According to the distributive property, $4 \cdot (7+2) = 4 \cdot 7 + 4 \cdot 2$.

4. The associative property of multiplication indicates that length can be multiplied by width or width can be multiplied by length to get the same answer.

Practice

Multiply.

5. $\begin{array}{r} 42 \\ \times 56 \\ \hline \end{array}$

6. $20 \cdot 200$

Use your knowledge of the properties of multiplication to find the value of the variable that will make the statement true. State the property illustrated.

7. $(5 \cdot 10) \cdot y = 5 \cdot (10 \cdot 7)$

8. Rewrite $7(8 + 4)$ by using the distributive property then simplify.

Calculate the area of the given rectangle.

9.

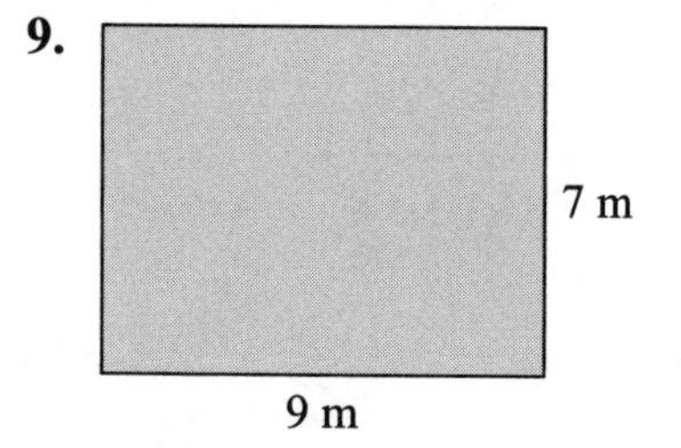

Applications

Solve.

10. ***Dining Out:*** A group of 15 friends are gathering at a restaurant. The restaurant is having a special where each person can order a three-course meal for $35. If all 15 friends order this special, how much will the total bill going be?

11. ***Inventory:*** A sandwich shop buys 372 loaves of bread for the week. If each loaf of bread has 24 slices, how many slices of bread were purchased?

Writing & Thinking

12. Explain, in your own words, what the zero-factor law indicates.

13. Name the property that uses both multiplication and addition and give an example of it.

1.4 Division with Whole Numbers

1. We can use the division sign $(\div)$ to indicate the division procedure as follows.

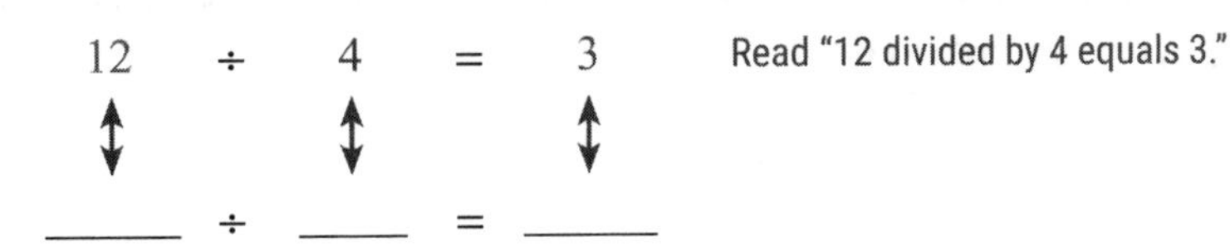

Read "12 divided by 4 equals 3."

2. Two other notations that indicate division are the following.

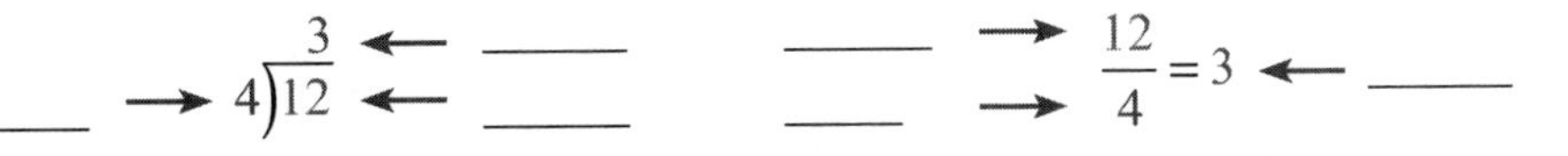

Division by 1

For any number a, ______________

PROPERTIES

Division of a Number by Itself

For any nonzero number a, ______________

PROPERTIES

Division Involving 0

Case 1: If a is any nonzero whole number, then ______________

Example: ______________

Case 2: If a is any whole number, then ______________

Example: ______________

PROPERTIES

3. The long division process can be written in the following format.

$$\begin{array}{r} 4 \\ 6\overline{)27} \\ \underline{-24} \\ 3 \end{array}$$

______ (quotient)

______ (divisor) ______ (dividend)

$6 \cdot 4 = 24$

$27 - 24 = 3$ → ______

Watch and Work

Watch the video for Example 3 in the software and follow along in the space provided.

Example 3 Dividing Whole Numbers

Divide: $683 \div 7$

Solution

Step 1:

Step 2:

Now You Try It!

Use the space provided to work out the solution to the next example.

Example A Dividing Whole Numbers

Divide: $415 \div 6$

1.4 Exercises

Concept Check

True/False. Determine whether each statement is true or false. If a statement is false, explain how it can be changed so the statement will be true. (**Note:** There may be more than one acceptable change.)

1. If a division problem has a nonzero remainder, then the divisor and quotient are factors of the dividend.

2. $13 \div 1 = 13$

3. $12 \div 0 = 12$

4. $\frac{0}{7}$ is undefined.

Practice

Divide.

5. $13\overline{)0}$

6. $0\overline{)51}$

7. $12\overline{)108}$

8. $11\overline{)4406}$

Applications

Solve.

9. ***Nutrition Facts:*** One pint of Ben and Jerry's Crème Brûlée Ice Cream has 64 grams of fat. If there are 4 servings per pint, how many grams of fat are in each serving?

10. ***Space Travel:*** US Astronaut Peggy Whitson orbited the Earth 6032 times during her space flights on the International Space Station. If the International Space Station orbits the Earth 16 times per day, how many days was Petty Whitson in space?

Writing & Thinking

11. Explain how you would check a division problem that has a nonzero remainder.

12. Discuss how division is related to multiplication.

1.5 Rounding and Estimating with Whole Numbers

Rounding Numbers

To **round** a given number means ______________________________

DEFINITION

Rounding Rule for Whole Numbers

1. Look at the single digit just to the right of the digit in the place of desired accuracy.

   a. **If this digit is less than 5**, leave the digit in the place of desired accuracy as it is, and ______________

   ____

   b. **If this digit is 5 or greater**, increase the digit in the desired place of accuracy by ______________

   ______________ All digits to the left remain unchanged unless

   ______________ Then the 9 is replaced by 0 and ______________

   ____

PROCEDURE

To Estimate a Sum or Difference

1. Round each number to ______________________________

2. Perform the ______________________________

PROCEDURE

Watch and Work

Watch the video for Example 4 in the software and follow along in the space provided.

Example 4 Estimating Sums of Whole Numbers

Estimate the sum; then find the actual sum.

$$\begin{array}{r} 68 \\ 925 \\ +487 \\ \hline \end{array}$$

Solution

Now You Try It!

Use the space provided to work out the solution to the next example.

Example A Estimating Sums of Whole Numbers

Estimate the sum; then find the actual sum.

$$\begin{array}{r} 176 \\ 84 \\ +\ 75 \\ \hline \end{array}$$

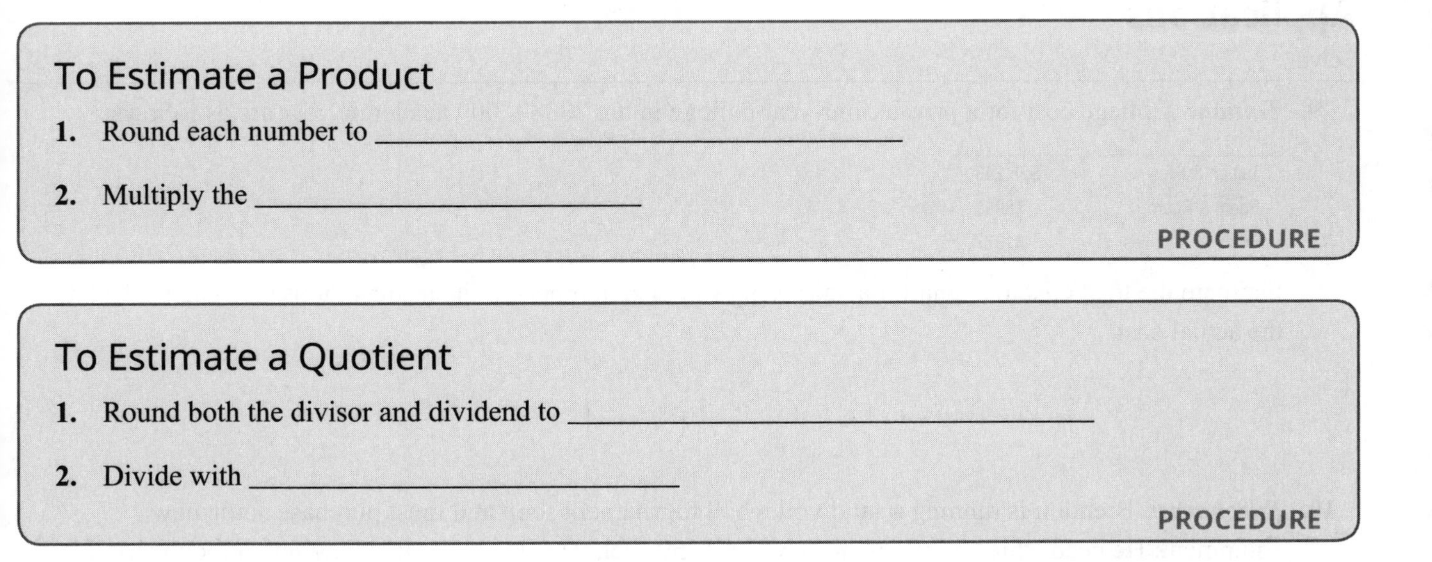
To Estimate a Product

1. Round each number to ______________________________

2. Multiply the ______________________

PROCEDURE

To Estimate a Quotient

1. Round both the divisor and dividend to ______________________________

2. Divide with ______________________

PROCEDURE

1.5 Exercises

Concept Check

Determine whether each statement is true or false. If a statement is false, explain how it can be changed so the statement will be true. (**Note:** There may be more than one acceptable change.)

1. Rounding means finding a number close to the given number, using a specified place of accuracy.

2. When rounded to the ten thousands place, 435,613 becomes 400,000.

3. To estimate the answer for a division problem, begin by rounding both the divisor and dividend.

4. If estimated, $4250 \div 51$ is $4000 \div 50 = 80$.

Practice

Estimate each answer; then find the actual answer.

5. $\begin{array}{r} 83 \\ 62 \\ +\ 78 \\ \hline \end{array}$

6. $\begin{array}{r} 63{,}504 \\ -\ 42{,}700 \\ \hline \end{array}$

7. $\begin{array}{r} 420 \\ \times\ 104 \\ \hline \end{array}$

8. $11\overline{)99}$

Applications

Solve.

9. ***Tuition:*** College cost for a private four-year college in the 2008–2009 academic year are as follows:

Tuition & Fees	$25,243
Room & Board	$8996
Books & Supplies	$1077

Estimate the total cost to attend for a year using rounded numbers to the nearest thousand. Then calculate the actual cost.

10. ***Purchases:*** Brendon is running a sand volleyball tournament soon and must purchase some new equipment. He needs three new nets, which cost $159 each. He also needs five new sets of boundary lines, which cost $86 each. Estimate the total cost of the new equipment. Then calculate the actual cost.

Writing & Thinking

11. In your own words, define estimation.

12. Compare and contrast rounding and estimating.

1.6 Problem Solving with Whole Numbers

Basic Strategy for Solving Word Problems

1. READ: ______________________________
2. SET UP: Draw any type of figure or diagram that might be helpful and ______________________________ ________________
3. SOLVE: ______________________________
4. CHECK: Check your work and ______________________________

PROCEDURE

To Find the Average of a Set of Numbers

1. Find the ______________________________
2. Divide this sum by the ______________________________

PROCEDURE

Watch and Work

Watch the video for Example 7 in the software and follow along in the space provided.

Example 7 Calculating an Average

Find the average of the following set of numbers: 15, 8, 90, 35, 27.

Solution

Now You Try It!

Use the space provided to work out the solution to the next example.

Example A Calculating an Average

Find the average of the following set of numbers: 18, 29, 6, 33, 14, 26.

1.6 Exercises

Concept Check

Determine whether each statement is true or false. If a statement is false, explain how it can be changed so the statement will be true. (**Note:** There may be more than one acceptable change.)

1. Averages are found by performing addition and then division.

2. The sum of 312 and 4 is 1248.

3. The word "quotient" indicates multiplication.

4. After reading a problem carefully, the next step might be to make a diagram or draw a figure.

Applications

Solve.

5. ***Nutrition Facts:*** Steven is calculating how many calories are in his lunch. He has a hamburger that has 354 calories, a medium fry that has 365 calories, and a chocolate milk shake that has 384 calories. How many total calories is his meal?

6. ***Purchases:*** For a class in statistics, Anthony bought a new graphing calculator for $95, special graphing paper for $8, a USB flash drive for $10, a textbook for $105, and a workbook for $37. How much did he spend for this class?

7. ***Area:*** A square that is 10 inches on a side is placed inside a rectangle that has a width of 20 inches and a length of 24 inches. What is the area of the region inside the rectangle that surrounds the square? (Find the area of the shaded region in the figure.)

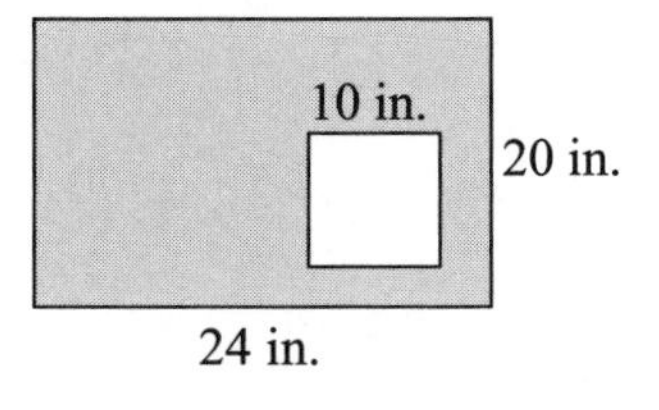

8. ***Purchases:*** The Lee family spent the following amounts for groceries: $338 in June; $307 in July; $318 in August. What was the average amount they spent for groceries in these three months?

Writing & Thinking

9. Make up three word problems that include key words to indicate operations such as addition, subtraction, multiplication and division. Underline the key words.

10. Give an example where you might use average (other than in a class).

1.7 Solving Equations with Whole Numbers ($x + b = c$ and $ax = c$)

Equation, Solution, Solution Set

An **equation** is a statement that ______________________________

A **solution** of an equation is a number that gives a ______________________________

A **solution set** of an equation is the set of ______________________________

DEFINITION

Basic Principles for Solving Equations

1. The addition principle: If A, B, and C are algebraic expressions, then the equations

____

______ ______________

____

2. The subtraction principle: If A, B, and C are algebraic expressions, then the equations

____

______ ______________

____

3. The division principle: If A and B are algebraic expressions and C is a nonzero constant, then the equations

____

______ ______________

____

____

PROPERTIES

1. The objective in solving an equation is to ______________________________

2. That is, we want the ______________________________

Watch and Work

Watch the video for Example 5 in the software and follow along in the space provided.

Example 5 Solving Equations of the Form *ax* = *c*

Solve the equation: $3n = 24$

Solution

Now You Try It!

Use the space provided to work out the solution to the next example.

Example A Solving Equations of the Form *ax* = *c*

Solve the equation: $9y = 36$

1.7 Exercises

Concept Check

True/False. Determine whether each statement is true or false. If a statement is false, explain how it can be changed so the statement will be true. (**Note:** There may be more than one acceptable change.)

1. Evaluating expressions and solving equations are basically two different concepts.

2. Variables may appear only on the left side of an equation.

3. The solution to the equation $n + 3 = 10$ is 13.

4. In solving an equation, we want the variable on one side of the equation by itself with a coefficient of 1.

Practice

For each equation, determine whether the given number is a solution to the equation.

5. $15 = y - 3$ given that $y = 10$

6. $3x = 15$ given that $x = 5$

Solve each equation. Show each step and keep the equal signs aligned vertically.

7. $x + 35 = 65$

8. $15y = 75$

Simplify and then solve the equation.

9. $15 + 9 - 5 = y + 10$

Applications

Solve.

10. ***Purchases:*** Karl purchased 3 pairs of pants for a total of \$84. This situation can be modeled by the equation $3x = \$84$, where x is the cost of each pair of pants. The pants are on sale for \$24 per pair. Did the pants ring up correctly as the sale price? Explain how you know.

11. ***Purchases:*** Apples are on sale for \$2 per pound and Caleb purchases \$18 worth of apples. This situation can be modeled by the equation $\$2x = \$18$, where x is the number of pounds of apples purchased. Solve this equation for x to determine how many pounds of apples Caleb purchased.

Writing & Thinking

12. Consider an equation of the form $x + b = c$, where x is a variable and b and c are constants.

    a. Use the subtraction principle to solve for the variable x.

    b. Keeping in mind that b and c are constants, how many solutions will the original equation, $x + b = c$, have? (**Hint:** Try plugging in different values for b and c to see how many answers you can find each time.)

1.8 Exponents and Order of Operations

1. When looking at $3^5 = 243$, 3 is the ________________, 5 is the ____________________, and 243 is the ____________________. ______________________ are written slightly to the right and above the ________________. The expression 3^5 is __________________________________.

The Exponent 1

For any number *a*, __________________

For example, ______________________________________

DEFINITION

Rules for Order of Operations

1. Simplify within grouping symbols, such as ____
(If there are more than one pair of grouping symbols, start with __________________________________
____

2. Evaluate any __________________________________

3. Moving from left to right, perform any ____
____

4. Moving from left to right, perform any ____
____

PROCEDURE

2. A well-known mnemonic device for remembering the rules for order of operations is the following.

Please	Excuse	My	Dear	Aunt	Sally
↓	↓	↓	↓	↓	↓
________	________	________	________	________	________

Watch and Work

Watch the video for Example 6 in the software and follow along in the space provided.

Example 6 Using the Order of Operations with Whole Numbers

Simplify: $2 \cdot 3^2 + 18 \div 3^2$

Solution

Now You Try It!

Use the space provided to work out the solution to the next example.

Example A Using the Order of Operations with Whole Numbers

Simplify: $6^2 \div 9 + 3 - 14 \div 7$

1.8 Exercises

Concept Check

True/False. Determine whether each statement is true or false. If a statement is false, explain how it can be changed so the statement will be true. (**Note:** There may be more than one acceptable change.)

1. Nine squared is equal to eighteen.

2. $2^7 = 128$

3. 7^0 is undefined.

4. According to the order of operations, multiplication is always performed before division.

Practice

For each exponential expression **a.** identify the base, **b.** identify the exponent, and **c.** evaluate the exponential expression.

5. 2^3

6. 4^0

Simplify.

7. $18 \div 2 - 1 - 3 \cdot 2$

8. $30 \div 2 - 11 + 2(5-1)^3$

Applications

Solve.

9. ***Card Games:*** Neville bought 15 boxes of trading cards. Each box has 10 packs of trading cards. Each pack of trading cards contains 20 cards. He adds 132 cards that he already owns to the newly purchased cards. Then, Neville evenly distributes all of the cards to 6 of his friends. How many trading cards would each person get?

a. If you simplify the expression $15 \cdot 10 \cdot 20 + 132 \div 6$ using the order of operations, will you get the correct answer? If not, explain what is wrong with the expression.

b. What is the answer? If necessary, write the corrected expression to get the correct results when following the order of operations.

10. ***Purchases:*** Robert is purchasing shirts for his weekend soccer team. The shirts he wants to buy are normally \$25 each but are on sale for \$10 off. His team has a total of 11 players. How much will he spend to buy the shirts?

a. If you simplify the expression $\$25 - \$10 \cdot 11$ using the order of operations, will you get the correct answer? If not, explain what is wrong with the expression.

b. What is the answer? If necessary, write the corrected expression to get the correct results when following the order of operations.

Writing & Thinking

11. Give one example where addition should be completed before multiplication.

1.9 Tests for Divisibility

Divisibility

If a number can be divided by another number so that the remainder is 0, then we say

1. the number is ______________________________

2. the divisor ___________________________

DEFINITION

When working with fractions (as we will be shortly), being able to divide quickly and easily by small numbers is a valuable skill. We will be looking for **factors**, so we will want to know if an integer is **divisible by** a number before actually dividing.

Divisibility by 2

A number is divisible by 2 (is an **even number**) if __________________________________

DEFINITION

Divisibility by 3

A number is divisible by 3 if _______________________________________

DEFINITION

Watch and Work

Watch the video for Example 3 in the software and follow along in the space provided.

Example 2 Determining Divisibility by 3

Determine whether each of the following numbers is divisible by 3.

a. 6801

b. 356

Solution

Now You Try It!

Use the space provided to work out the solution to the next example.

Example A Determining Divisibility by 3

Is 7912 divisible by 3? Explain why or why not.

Divisibility by 4

A number is divisible by 4 if ______________________________

DEFINITION

Divisibility by 5

A number is divisible by 5 if ______________________________

DEFINITION

Divisibility by 6

A number is divisible by 6 if ______________________________

DEFINITION

Divisibility by 9

A number is divisible by 9 if ______________________________

DEFINITION

Divisibility by 10

A number is divisible by 10 if ______________________________

DEFINITION

1.9 Exercises

Concept Check

True/False. Determine whether each statement is true or false. If a statement is false, explain how it can be changed so the statement will be true. (**Note:** There may be more than one acceptable change.)

1. A number that is divisible by 10 is also divisible by 2 and 5.

2. 6801 is divisible by 9.

3. 7605 is divisible by 10.

4. 5,187,042 is divisible by 3.

Practice

Using the tests for divisibility, determine which of 2, 3, 4, 5, 6, 9, and 10 (if any) will divide exactly into each given number.

5. 105

6. 150

7. 331

8. 1234

Applications

Solve.

9. ***Fundraising:*** You are on a team that is participating in a charity walk with a goal to raise $12,400. Each team member agrees to raise the same amount of money. If the possible team sizes are 5, 6, 9, or 10 members, which team sizes allow the goal amount to be evenly split between the team members? How much money would each team member raise for each team size that can evenly split the goal amount?

10. ***Time:*** A company is working on a project that will take 440 hours of work to complete. The manager in charge of the project has the option to have 4, 6, or 8 people work on the project. If the manager wants to evenly divide the work between the team members, which team size will evenly split the work hours? How many hours would each team member spend on the project for each team size that evenly splits the work hours?

Writing & Thinking

11. **a.** If a number is divisible by both 3 and 5, then it will be divisible by 15. Give two examples.

    b. However, a number might be divisible by 3 and not by 5. Give two examples.

    c. Also, a number might be divisible by 5 and not 3. Give two examples.

1.10 Prime Numbers and Prime Factorizations

Prime Number

A **prime number** is a counting number ______________________________

____

DEFINITION

Composite Number

A **composite number** is ______________________________

DEFINITION

To Determine Whether a Number is Prime

Divide the number by progressively larger prime numbers (2, 3, 5, 7, 11, and so forth) until one of the following is true.

1. The remainder ____________ This means that the ______________________________

   ____

2. You find a quotient ______________________________ This means that the ____________

   ____

PROCEDURE

The Fundamental Theorem of Arithmetic

Every composite number has ______________________________

DEFINITION

To Find the Prime Factorization of a Composite Number

1. Factor the composite number ______________________
2. Factor each ______________________
3. Continue this process until all factors are prime.

PROCEDURE

Factors of a Composite Number

The only factors (or divisors) of a composite number are

1. ____________________

2. ____________________

3. products formed by ____________________

DEFINITION

Watch and Work

Watch the video for Example 10 in the software and follow along in the space provided.

Example 10 Finding the Factors of a Composite Number

Find all the factors of 60.

Solution

Now You Try It!

Use the space provided to work out the solution to the next example.

Example A Finding the Factors of a Composite Number

Find all the factors of 42.

1.10 Exercises

Concept Check

True/False. Determine whether each statement is true or false. If a statement is false, explain how it can be changed so the statement will be true. (**Note:** There may be more than one acceptable change.)

1. A prime number has exactly 1 factor.

2. A composite number has 2 or more factors.

3. 231 is a prime number.

4. All the factors of 30 are 1, 2, 3, 5, 6, 10, 15 and 30.

Practice

Determine whether each number is prime or composite. If the number is composite, find at least three factors of the number.

5. 47

6. 63

Find the prime factorization of each number. Use the tests for divisibility for 2, 3, 4, 5, 6, 9, and 10 whenever they help to find beginning factors.

7. 125

8. 150

Applications

Solve.

9. ***Inventory:*** Twenty-four pencils are to be distributed evenly between the members of a group. What are the possible group sizes if each person in the group is to receive the same number of pencils?

10. ***Baking:*** A chocolatier makes 72 specialty truffles. She wants to sell packages that each have the same number of truffles. What are her options for the number of truffles that can be in a package?

Writing & Thinking

11. Are all odd numbers also prime numbers? Explain your answer.

12. Explain the difference between factors of a number and multiples of that number.

Chapter 1 Project

Aspiring to New Heights!

An activity to demonstrate the use of whole numbers in real life.

You may have never heard of the Willis Tower, but it once was the tallest building in the United States. This structure was originally named the Sears Tower when it was built in 1973, and it held the title of the tallest building in the world for almost 25 years. The name was changed in 2009 when Willis Group Holdings obtained the right to rename the building as part of their lease for a large portion of the office space in the building.

The Willis Tower, which is 1,451 feet tall and located in Chicago, Illinois, was the tallest building in the Western Hemisphere until May 2, 2013. On this date a 408 foot spire was placed on the top of One World Trade Center in New York to bring its total height to a patriotic 1,776 feet. One World Trade Center now claims the designation of being the tallest building in the United States and the Western Hemisphere.

1. The Willis Tower has an unusual construction. It is comprised of 9 square tubes of equal size, which are really separate buildings, and the tubes extend to different heights. The footprint of the building is a 225 foot by 225 foot square. In answering the questions below, be sure to use the correct units of measurement on your answers.

   a. Since the footprint of the Willis Tower is a square measuring 225 feet on each side and it is comprised of 9 square tubes of equal size, what is the side length of each tube? (It might help to draw a diagram.)

   b. What is the perimeter around the footprint, or base of the Willis Tower?

   c. What is the area of the base of the Willis Tower?

   d. What is the area of the base for one square tube?

   e. Write the values from Parts **c.** and **d.** in words.

2. Suppose there are plans to alter the landscape around Willis Tower. The city engineers have proposed adding a concrete sidewalk 6 feet wide around the base of the building. A drawing of the proposal is shown. (**Note:** This drawing is not to scale.)

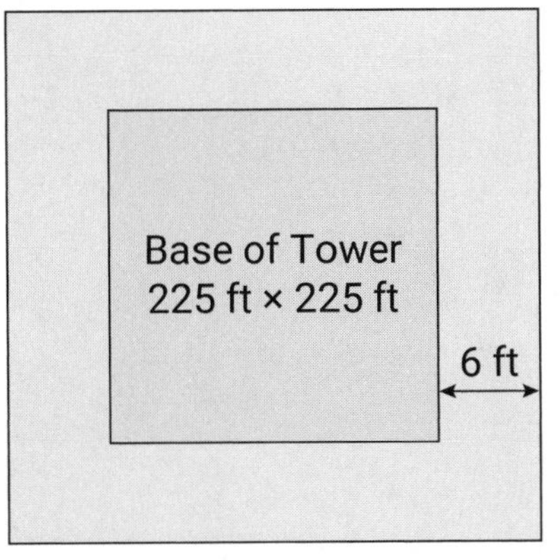

   a. Determine the total area of the base of the tower including the new sidewalk.

   b. Write down the area of just the base of the tower that you determined in Part **1c.**

   c. Determine the area covered by the concrete sidewalk around the building. (**Hint:** You only want the area between the two squares.)

   d. If a border were to be placed around the outside edge of the concrete sidewalk, how many feet of border would be needed?

   e. If the border is only sold by the yard, how many yards of border will be needed? (**Note:** 1 yard = 3 feet.)

   f. Round the value from Part **e.** to the nearest ten.

   g. Round the value from Part **e.** to the nearest one hundred.

CHAPTER 2

Integers

Math @ Work

Algebraic expressions can be used to create mathematical models that generate interesting and useful data about the world around us. These expressions can model nearly anything, such as the profit of a company throughout the year, the population size of a certain species of animal, or the average daily temperature of a city. Knowing what an algebraic expression means and how to evaluate it for different values can help you extract useful information from these models. You can then use this information to make better and well-informed decisions at work and in your everyday life.

Suppose you want to plan a visit to a local amusement park with a group of friends and need to decide which day is the best day to go. While looking at park attendance over the past few years, you find out that the average number of people who attend the amusement park during the summer can be modeled by the expression $-1.25x^2 + 100x + 30{,}000$, where x is the number of days after June 1. The entire group of friends can make it to the amusement park on either June 11 (10 days after June 1) or July 21 (51 days after June 1). If you want to go to the park on the day with the fewest number of people, and therefore shorter lines for the rides, which day should you plan to go?

2.1 Introduction to Integers

Integers

The set of integers is the set of ______________________

Integers $= \mathbb{Z} =$ ______________________

DEFINITION

Opposites of Integers

Note the following facts about signed integers.

1. The opposite of a positive integer is ______________________
For example,

____

2. The opposite of a negative integer is ______________________
For example,

____

3. The opposite of 0 is __________

DEFINITION

Symbols of Inequality

$<$ read __________ $\leq$ read ______________________

$>$ read __________ $\geq$ read ______________________

DEFINITION

Absolute Value

The absolute value of a number is ______________________

The absolute value of a number is ______________________

DEFINITION

Watch and Work

Watch the video for Example 6 in the software and follow along in the space provided.

Example 6 Simplifying Expressions Containing Absolute Value

Simplify.

a. $-(-10)$ **b.** $-|6|$ **c.** $-|-3|$

Solution

Now You Try It!

Use the space provided to work out the solution to the next example.

Example A Simplifying Expressions Containing Absolute Value

Simplify.

a. $-(-2)$ **b.** $-|13|$ **c.** $-|-7|$

2.1 Exercises

Concept Check

True/False. Determine whether each statement is true or false. If a statement is false, explain how it can be changed so the statement will be true. (**Note:** There may be more than one acceptable change.)

1. If -8 lies to the right of a number on a number line, then -8 is less than that number.

2. All whole numbers have an opposite number.

3. All whole numbers are also integers.

4. The absolute value of a positive number is a positive number.

Practice

Graph each set of integers on a number line.

5. $\{-3, -1, 1\}$

6. $\{-5, -4, -3, -2, 0, 1\}$

Fill in each blank with the appropriate symbol that will make the statement true: <, >, or =.

7. 4 ___ 6

8. -20 ___ -19

9. Simplify $|-4|$

10. List the possible values of x for $|x| = 5$

Applications

Represent each quantity with a signed integer.

11. ***Oceans:*** The Alvin is a manned deep-ocean research submersible that has explored the wreck of the Titanic. The operating depth of the Alvin is 4500 meters below sea level.

12. ***Mountains:*** Mount Everest is considered to be the highest mountain on Earth. Its peak reaches to a height of approximately 8844 meters.

Writing and Thinking

13. Explain, in your own words, how an expression such as $-y$ might represent a positive number.

14. Compare and contrast absolute value with opposites.

2.2 Addition with Integers

Rules for Addition with Integers

1. To add two integers with **like signs**,

a. add their ______________________________

b. use the _________________________

2. To add two integers with **unlike** signs,

a. subtract their ______________________ (the smaller from the _____________)

b. use the ____

Additive Inverse

The opposite of an integer is called ______________________________

The sum of any integer and its additive inverse is ____

Symbolically, for any integer a,

____

As an example, ____

DEFINITION

Watch and Work

Watch the video for Example 4 in the software and follow along in the space provided.

Example 4 Checking Solutions in Equations

Determine whether the given integer is a solution to the given equation by substituting for the variable and adding.

a. $x+8=-2;\quad x=-10$

b. $x+(-5)=-6;\quad x=1$

c. $17+y=0;\quad y=-17$

Solution

Now You Try It!

Use the space provided to work out the solution to the next example.

Example A Checking Solutions in Equations

Determine whether $x = -5$ is a solution to the given equation by substituting for the variable and adding.

a. $x + 14 = 9$ **b.** $x + 5 = 0$ **c.** $x + (-25) = -20$

2.2 Exercises

Concept Check

True/False. Determine whether each statement is true or false. If a statement is false, explain how it can be changed so the statement will be true. (**Note:** There may be more than one acceptable change.)

1. When adding integers with unlike signs, the answer will be negative.

2. The sum of two positive numbers can equal zero.

3. The additive inverse of negative seven is -7.

4. When a number substituted for a variable makes a statement true, that number is said to be an equation.

Practice

Find the additive inverse (opposite) of each integer.

5. 15

6. -40

Add.

7. $-3+(-5)$

8. $12+14+(-16)$

Add. Be sure to find the absolute value first.

9. $|-7|+(-7)$

Determine whether the given integer is a solution to the equation by substituting for the variable and then adding.

10. $x+5=7;\quad x=-2$

Applications

Solve.

11. ***Transportation:*** A submarine dives to a depth of 250 feet below the surface. It rises 75 feet before diving an additional 100 feet. What is the final depth of the submarine?

12. ***Weather:*** The temperature at 2 a.m. was -17 °C. By 2 p.m. the temperature increased a total of 15 °C. What was the temperature at 2 p.m.?

Writing & Thinking

Choose the response that correctly completes each statement. In each problem, give two examples that illustrate your reasoning.

13. If x and y are integers, then $x + y$ is (never, sometimes, always) equal to 0.

14. Explain how the sum of the absolute values of two integers might be 0. (Is this possible?)

2.3 Subtraction with Integers

Subtraction with Integers

For any integers a and b,

____

In words, to subtract b from a, ____________________________________

DEFINITION

To calculate the change between two values, including direction (negative for decrease, positive for increase), use the following rule: first find the ____

(Change in value) = ____

Watch and Work

Watch the video for Example 7 in the software and follow along in the space provided.

Example 7 Checking Solutions in Equations

Determine whether the given integer is a solution to the given equation by substituting for the variable and then subtracting.

a. $x-(-6)=-10;$
$x=-14$

b. $7-y=-1;$
$y=8$

c. $a-12=-2;$
$a=-10$

Solution

Now You Try It!

Use the space provided to work out the solution to the next example.

Example A Checking Solutions in Equations

Determine whether $x = 3$ is a solution to the given equation by substituting for the variable and then subtracting.

a. $x-(-2)=5$ **b.** $-16-x=-13$ **c.** $x-6=-3$

Solution

2.3 Exercises

Concept Check

True/False. Determine whether each statement is true or false. If a statement is false, explain how it can be changed so the statement will be true. (**Note:** There may be more than one acceptable change.)

1. Moving to the right on the number line is equivalent to moving in a positive direction.

2. Like addition, subtraction is both commutative and associative.

3. The expression "15 – 7" can be thought of as "fifteen plus negative seven."

4. If an integer is a solution to an equation, it satisfies the equation.

Practice

Perform the indicated operations.

5. $17 - 22$

6. $-17 - 30$

7. $-3-(-3)+(-2)$

Determine whether the given integer is a solution to the equation by substituting for the variable and performing the indicated operation.

8. $a + 5 = -10; \quad a = -15$

Applications

Solve.

9. ***Weather:*** Beginning with a temperature of 8° above zero, the temperature was measured hourly for 4 hours. It rose 3°, dropped 7°, and rose 1°. What was the final temperature recorded?

10. ***Stock Market:*** In a 5-day week, the NASDAQ stock market posted a gain of 145 points, a loss of 100 points, a loss of 82 points, a gain of 50 points, and a gain of 25 points. If the NASDAQ started the week at 6300 points, what was the value of the market at the end of the week?

Writing & Thinking

11. Under what conditions can the difference between two negative numbers be a positive number?

12. Give two examples to illustrate why subtraction is not commutative.

2.4 Multiplication, Division, and Order of Operations with Integers

1. Because the sum of negative integers is negative, we can reason that ______________________________

____

Rules for Multiplication with Integers

If a and b are positive integers, then

1. The product of two positive integers is ______________________________
2. The product of two negative integers is ______________________________
3. The product of a positive integer and a negative integer is ____________________

   ____
4. The product of 0 and any integer is 0.

   ____

In summary:

a. When the signs are alike, ______________________________

b. When the signs are not alike, ______________________________

Meaning of $\frac{a}{b} = x$

For integers a, b, and x (where $b \neq 0$),

____

DEFINITION

Rules for Division with Integers

If a and b are positive integers, then

1. The quotient of two positive integers is ______________________
2. The quotient of two negative integers is ________________________
3. The quotient of a positive integer and a negative integer is _______________ ________________________________
4. 0 divided by any integer is 0. _________________
5. Any integer divided by 0 is _____________________________________

In summary:

a. When the signs are alike, __________________________
b. When the signs are not alike, __________________________

Rules for Order of Operations

1. Simplify within ____ ________________(If there are more than one pair of grouping symbols, start with ___________ __________________________)
2. Evaluate any ______________________________
3. Moving from left to right, perform any ____
4. Moving from left to right, perform any ____

Watch and Work

Watch the video for Example 9 in the software and follow along in the space provided.

Example 9 Using the Order of Operations with Integers

Simplify: $27 \div (-9) \cdot 2 - 5 + 4(-5)$

Solution

Now You Try It!

Use the space provided to work out the solution to the next example.

Example A Using the Order of Operations with Integers

Simplify: $6^2 \div (-9) + 3 - 14 \div 7$

2.4 Exercises

Concept Check

True/False. Determine whether each statement is true or false. If a statement is false, explain how it can be changed so the statement will be true. (**Note:** There may be more than one acceptable change.)

1. The product of zero and an integer is undefined.

2. If a negative integer is divided by a positive integer, the result will be a negative number.

3. When zero is divided by any nonzero integer, the result is zero.

4. If there are no grouping symbols, multiplication should always be performed before division.

Practice

Multiply.

5. $0(-5)$

6. $(-5)(3)(-4)$

Divide.

7. $\frac{-12}{3}$

8. $\frac{35}{0}$

Simplify the expression using the order of operations.

9. $-6^2+7(12)-3^2$

Applications

Solve.

10. ***Stocks:*** Alicia bought shares of two companies on the stock market. She paid $9000 for 90 shares in one company and $6600 for 110 shares in another company. What was the average price per share for the 200 shares?

11. ***Grades:*** In a speech class the students graded each other on a particular assignment. On this speech, three students scored 60, three scored 70, five scored 80, five scored 82, and four scored 85. What was the average score on this speech?

Writing & Thinking

12. If you multiply an odd number of negative numbers together, do you think that the product will be positive or negative? Explain your reasoning.

13. Explain, in your own words, why the following expression cannot be evaluated:
$\left(24-2^4\right)+6\left(3-5\right)\div\left(3^2-9\right)$.

2.5 Simplifying and Evaluating Expressions

Like Terms

Like terms (or similar terms) are terms that are ______________________________

____

Whatever power a variable is raised to in one term, it is ______________________________

DEFINITION

To Evaluate an Expression

1. Combine ______________________________
2. Substitute ______________________________
3. Follow the ______________________________

(**Note:** Terms separated by + and – signs may be evaluated at the ______________________________

______________________________.)

PROCEDURE

Watch and Work

Watch the video for Example 5 in the software and follow along in the space provided.

Example 5 Simplifying and Evaluating Expressions

Simplify and evaluate $x^3 - 5x^2 + 2x^2 + 3x - 4x - 15 + 3$ for $x = -2$.

Solution

Now You Try It!

Use the space provided to work out the solution to the next example.

Example A Simplifying and Evaluating Expressions

Simplify and evaluate $5a^2 - 8a^2 - 2a + 3a - 5a + 6 - 9$ for $a = -3$.

2.5 Exercises

Concept Check

True/False. Determine whether each statement is true or false. If a statement is false, explain how it can be changed so the statement will be true. (**Note:** There may be more than one acceptable change.)

1. A variable that does not appear to have an exponent has an exponent of 1.

2. In the term $-9x$, nine is being subtracted from x.

3. In the term "12*a*," 12 is the constant.

4. Like terms have the same coefficients.

Practice

Identify the like terms in the list of terms.

5. $-5, 3, 7x, 8, 9x, 3y$

Simplify each expression by combining like terms.

6. $2a + 14a - 25a$

7. $5x^2 - 3x^2 + 2x$

8. $3(n+1)+n$

Simplify the expression and then evaluate the expression for $x = -2$.

9. $2x^2 - 3x^2 + 5x - 8 + 1$

Applications

Solve.

10. ***Profit:*** An apartment management company owns a property with 100 units. The company has determined that the profit made per month from the property can be calculated using the equation $P = -10x^2 + 1500x - 6000$, where x is the number of units rented per month. How much profit does the company make when 80 units are rented?

11. ***Physics:*** A ball is thrown upward from an initial height of 96 feet with an initial velocity of 16 feet per second. After t seconds, the height of the ball can be described by the expression $-16t^2 + 16t + 96$. What is the height of the ball after 3 seconds?

Writing & Thinking

12. Discuss like and unlike terms and give an example of each.

13. Explain the difference between -13^2 and $(-13)^2$.

2.6 Translating English Phrases and Algebraic Expressions

Key Words To Look For When Translating Phrases

Addition	Subtraction	Multiplication	Division	Exponent (Powers)

1. Division and subtraction are done with the values in the ________________ that they are given in the problem.

2. An **ambiguous phrase** is one whose meaning is ____________________________________ ______________

Watch and Work

Watch the video for Example 3 in the software and follow along in the space provided.

Example 3 Translating Algebraic Expressions into English Phrases

Change each algebraic expression into an equivalent English phrase. In each case translate the variable as "a number."

a. $5x$

b. $2n+8$

c. $3(a-2)$

Solution

	Algebraic Expression	Possible English Phrase
a.		
b.		
c.		

Now You Try It!

Use the space provided to work out the solution to the next example.

Example A Translating Algebraic Expressions into English Phrases

Change each algebraic expression into an equivalent English phrase.

a. $10x$

b. $4a + 7$

c. $7(n - 5)$

Solution

2.6 Exercises

Concept Check

True/False. Determine whether each statement is true or false. If a statement is false, explain how it can be changed so the statement will be true. (**Note:** There may be more than one acceptable change.)

1. The order in which the values are given is particularly important when working with subtraction and division problems.

2. "More than" and "increased by" are key phrases specifying the operation of subtraction.

3. Division is indicated by the phrase "five less than a number."

4. Key phrases for parentheses can be used to limit ambiguity in English phrases.

Practice

Write the algebraic expressions described by the English phrases. Choose your own variable.

5. six added to a number

6. twenty decreased by the product of four and a number

7. eighteen less than the quotient of a number and two

Translate each pair of English phrases into algebraic expressions. Notice the differences between the algebraic expressions and the corresponding English phrases.

8. a. six less than a number

b. six less a number

9. a. six less than four times a number

b. six less four times a number

Write the algebraic expression described by the English phrases using the given variables.

10. the cost of purchasing a fishing rod and reel if the rod costs x dollars and the reel costs $8 more than twice the cost of the rod

Translate each algebraic expression into an equivalent English phrase. (There may be more than one correct translation.) See Examples 3 and 4.

11. $-9x$

12. $\frac{9}{x+3}$

Writing & Thinking

13. Explain why translating addition and multiplication problems from English into algebra may be easier than changing subtraction or division problems. (Consider the properties previously studied.)

14. Explain the difference between $5(n+3)$ and $5n+3$ when converting from algebra to English.

2.7 Solving Equations with Integers ($ax + b = c$)

Basic Principles for Solving Equations

1. The **addition principle**: If A, B, and C are algebraic expressions, then the equations ____________

   ____________ ____________

   ____

2. The **division principle**: If A and B are algebraic expressions and C is a nonzero constant, then the equations

   ____

   ____________ ____________

   ____

PROPERTIES

To Solve Equations of the Form $ax + b = c$

1. Apply the ____________
2. Combine ____________
3. If a constant is added to a variable, use the ____________

   ____

4. If a variable has a constant coefficient other than 1, use the ____________

   ____

Remember that the object is to ____________

PROCEDURE

Watch and Work

Watch the video for Example 5 in the software and follow along in the space provided.

Example 5 Solving Equations of the Form $ax + b = c$

Solve the equation: $1 - 3x = 19$

Solution

Now You Try It!

Use the space provided to work out the solution to the next example.

Example A Solving Equations of the Form $ax + b = c$

Solve the equation: $6 - 5y = -14$

2.7 Exercises

Concept Check

True/False. Determine whether each statement is true or false. If a statement is false, explain how it can be changed so the statement will be true. (**Note:** There may be more than one acceptable change.)

1. In the addition and division principles, A, B, and C can be represented by positive or negative numbers.

2. The object in solving an equation is to isolate the constant with coefficient 1 on one side of the equation.

3. When solving an equation of the form $ax + b = c$, if a variable has a constant coefficient other than 1, use the addition principle to add the opposite of the coefficient to both sides.

4. When checking the solution to an equation with more than one occurrence of the variable, you can choose which instance of the variable to substitute the solution found into.

Practice

Solve each equation. Combine like terms whenever necessary.

5. $2x + 3 = 13$

6. $23 = 7n - 5$

7. $5y - 2 - 4y = -6$

8. $3 = 13y + 15 - 7y$

Applications

Solve.

9. ***Temperature:*** The temperature decreased by 27 degrees during a 9-hour period. This situation can be modeled by $9t = -27$, where t represents the average change in temperature per hour. Solve the equation for t to determine the average change in temperature per hour.

10. ***Inventory:*** A kitchen has 71 pounds of fresh produce in its inventory. After two days, the kitchen had 45 pounds of fresh produce. This situation can be modeled by the equation $2x + 71 = 45$, where x is the change in the amount of fresh produce per day. Solve the equation for x to determine the change in the amount of fresh produce per day.

Writing & Thinking

Give a brief explanation of what is happening in each step of the solution process.

11. $71y - 62y = -36$ ____________________

$9y = -36$ ____________________

$\frac{9y}{9} = \frac{-36}{9}$ ____________________

$y = -4$ ____________________

12. Explain in your own words why the addition and subtraction principles can be stated as one principle of addition.

Name: Date:

Chapter 2 Project

Going to Extremes!

An activity to demonstrate the use of signed numbers in real life.

When asked what the highest mountain peak in the world is, most people would say Mount Everest. This answer may be correct, depending on what you mean by highest. According to geology.com, there may be other contenders for this important distinction.

The peak of Mount Everest is 8850 meters or 29,035 feet above sea level, giving it the distinction of being the mountain with the highest altitude in the world. However, Mauna Kea is a volcano on the big island of Hawaii whose peak is over 10,000 meters above the nearby ocean floor, which makes it taller than Mount Everest. A third contender for the highest mountain peak is Chimborazo, an inactive volcano in Ecuador. Although Chimborazo only has an altitude of 6310 meters (20,703 feet) above sea level, it is the highest mountain above Earth's center. Most people think that the Earth is a sphere, so how could a mountain that is only 6310 meters tall be higher than a mountain that is 8850 meters tall? Because the Earth is really not a sphere but an "oblate spheroid". It is widest at the equator. Chimborazo is 1° south of the equator which makes it about 2 km farther from the Earth's center than Mount Everest.

What about the other extreme? What is the lowest point on Earth? As you might have guessed, there is more than one candidate for that distinction as well. The lowest exposed area of land on Earth's surface is on the Dead Sea shore at 413 meters below sea level. The Bentley Subglacial Trench in Antarctica is the lowest point on Earth that is not covered by ocean but it is covered by ice. This trench reaches 2555 meters below sea level. The deepest point on the ocean floor occurs 10,916 meters below sea level in the Mariana Trench in the Pacific Ocean.

For the following problems, be sure to show all math work to justify your results.

1. Calculate the **difference** in elevation between Mount Everest and Chimborazo in both meters and feet. What operation does the word **difference** imply?
2. Write an expression to calculate the **difference** in elevation between the peak of Mount Everest and the lowest point on the Dead Sea shore in meters and simplify.
3. If you were to travel from the bottom of the Mariana Trench to the top of Mount Everest, how many meters would you travel?
4. If you were to travel from the bottom of the Dead Sea Shore to the top of Chimborazo, how many meters would you travel?
5. If Mount Everest were magically moved and placed at the bottom of the Mariana Trench, how many meters of water would lie above Mount Everest's peak?
6. How much farther below sea level (in meters) is the Mariana Trench as compared to the Dead Sea shore?
7. How much farther below sea level (in meters) is the Mariana Trench as compared to the Bentley Subglacial Trench?
8. Add the elevations (in meters) together for Mount Everest, Chimborazo, the Dead Sea Shore, the Bentley Subglacial Trench, and the Mariana Trench and show your result. Is this number positive or negative? Would this value represent an elevation above or below sea level?
9. If you calculate the difference in the absolute values of the elevations for Mount Everest and the Dead Sea shore, do you get the same result as in problem 2?
10. Describe how to perform the order of operations in evaluating the expression in problem 9. Be sure to use complete sentences.

CHAPTER 3

Fractions, Mixed Numbers, and Proportions

Math @ Work

To be a well-informed citizen, it's important to pay attention to politics and the political process since they affect everyone in some way. In local, state, and national elections, registered voters make choices about various ballot measures and who will represent them in the government. In major issues at the state and national levels, pollsters use mathematics (in particular, statistics and statistical methods) to indicate attitudes and to predict how the electorate will vote, typically accurate within certain percentage ranges. When there is an important election in your area, read news articles and listen to news reports for mathematically-related statements predicting the outcomes.

Certain states, such as Ohio, are considered swing states. This means that during presidential elections no single candidate has an overwhelming support from the voters in the state. Swing state status is based on elections in recent history. Ohio is an important swing state because, since the 1960s, the candidate that won the electoral votes in Ohio has typically also won the presidential election.

Suppose that to determine funding for a political advertising campaign, a candidate's campaign staff takes a survey of voters in Ohio. They determine there are 8000 registered voters in a certain precinct, and $\frac{3}{8}$ of the voters are undecided about who they will vote for. The survey indicates that $\frac{4}{5}$ of these undecided voters agree with the candidate's platform. From this information, the campaign staff can determine how many people can be influenced to vote for their candidate with a series of campaign advertisements. How many undecided voters agree with the candidate's platform?

3.1 Introduction to Fractions and Mixed Numbers

1. Numbers such as $\frac{2}{3}$ (read "two-thirds") are said to be in ________________. The top number, 2, is called the ________________ and the bottom number, 3, is called the ________________.

Rule for the Placement of Negative Signs

If a and b are integers and $b \neq 0$, then

____

PROCEDURE

Graph the fraction $\frac{2}{3}$ proceed as follows.

1. Divide the interval (distance) from ________________________________

2. Graph (or shade) the ________________________________

3 equal parts

0 $\frac{1}{3}$ $\frac{2}{3}$ 1

A **mixed number** is ________________________________.

Graph the mixed number $1\frac{3}{4}$ proceed as follows.

1. Mark the intervals from ________________________________

2. Graph (or shade) the ________________

0 $\frac{1}{4}$ $\frac{1}{2}$ $\frac{3}{4}$ 1 $1\frac{1}{4}$ $1\frac{1}{2}$ $1\frac{3}{4}$ 2

To Change a Mixed Number to an Improper Fraction

1. Multiply the whole number by ________________________________

2. Add the numerator of the ________________________________

3. Write this sum ________________________________

PROCEDURE

Watch and Work

Watch the video for Example 13 in the software and follow along in the space provided.

Example 13 Changing Mixed Numbers to Improper Fractions

Change $8\frac{9}{10}$ to an improper fraction.

Solution

Step 1: ______________________________

Step 2: ________________

Step 3: ___________________

Now You Try It!

Use the space provided to work out the solution to the next example.

Example A Application: Changing Mixed Numbers to Improper Fractions

Change $10\frac{4}{9}$ to an improper fraction.

To Change an Improper Fraction to a Mixed Number

1. Divide the numerator by ________________
2. The quotient is ______________________________
3. Write the remainder ____

PROCEDURE

3.1 Exercises

Concept Check

True/False. Determine whether each statement is true or false. If a statement is false, explain how it can be changed so the statement will be true. (**Note:** There may be more than one acceptable change.)

1. In $\frac{11}{13}$, the denominator is 11.

2. $\frac{0}{6} = 0$

3. $\frac{17}{0}$ is undefined.

Practice

For the figure, a) write the fraction for the number of days remaining in June (not crossed out) and b) write the fraction for the number of days that have been crossed out for June.

4.

June

S	M	T	W	T	F	S
~~1~~	~~2~~	~~3~~	~~4~~	~~5~~	~~6~~	~~7~~
~~8~~	~~9~~	~~10~~	~~11~~	~~12~~	~~13~~	~~14~~
~~15~~	~~16~~	~~17~~	~~18~~	~~19~~	~~20~~	~~21~~
~~22~~	~~23~~	24	25	26	27	28
29	30					

5. Graph $\frac{3}{5}$ on a number line.

Write the remaining amount as **a.** a mixed number and **b.** an improper fraction.

6. Isabella brought 2 boxes of doughnuts to a meeting. The figure shows the remaining amount of doughnuts.

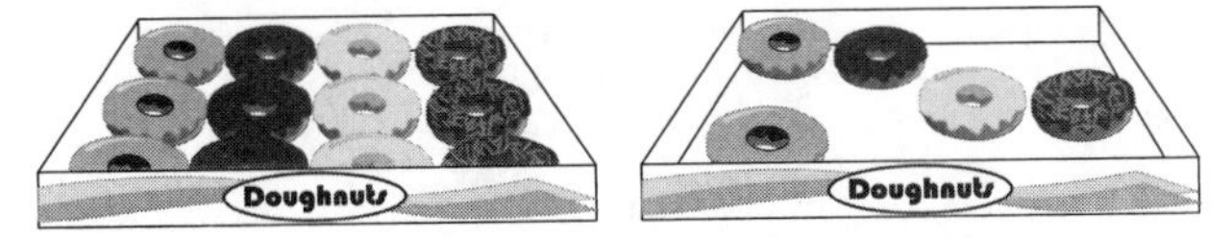

7. Graph $3\frac{1}{4}$ on a number line.

8. Change $1\frac{3}{5}$ to an improper fraction.

Applications

Solve.

9. ***Class Grades:*** In a class of 35 students, 6 students received As on a mathematics exam. What fraction of students received an A? What fraction of students did not receive an A?

10. ***Nutrition:*** A certain brand of plain bagels has 146 calories per bagel. 115 calories come from the carbohydrates in the bagel. What fraction of the calories is from carbohydrates?

Writing & Thinking

11. In your own words, list the parts of a fraction and briefly describe the purpose of each part.

12. Show and explain, using diagrams and words, why $2\frac{3}{5} = \frac{13}{5}$.

3.2 Multiplication with Fractions

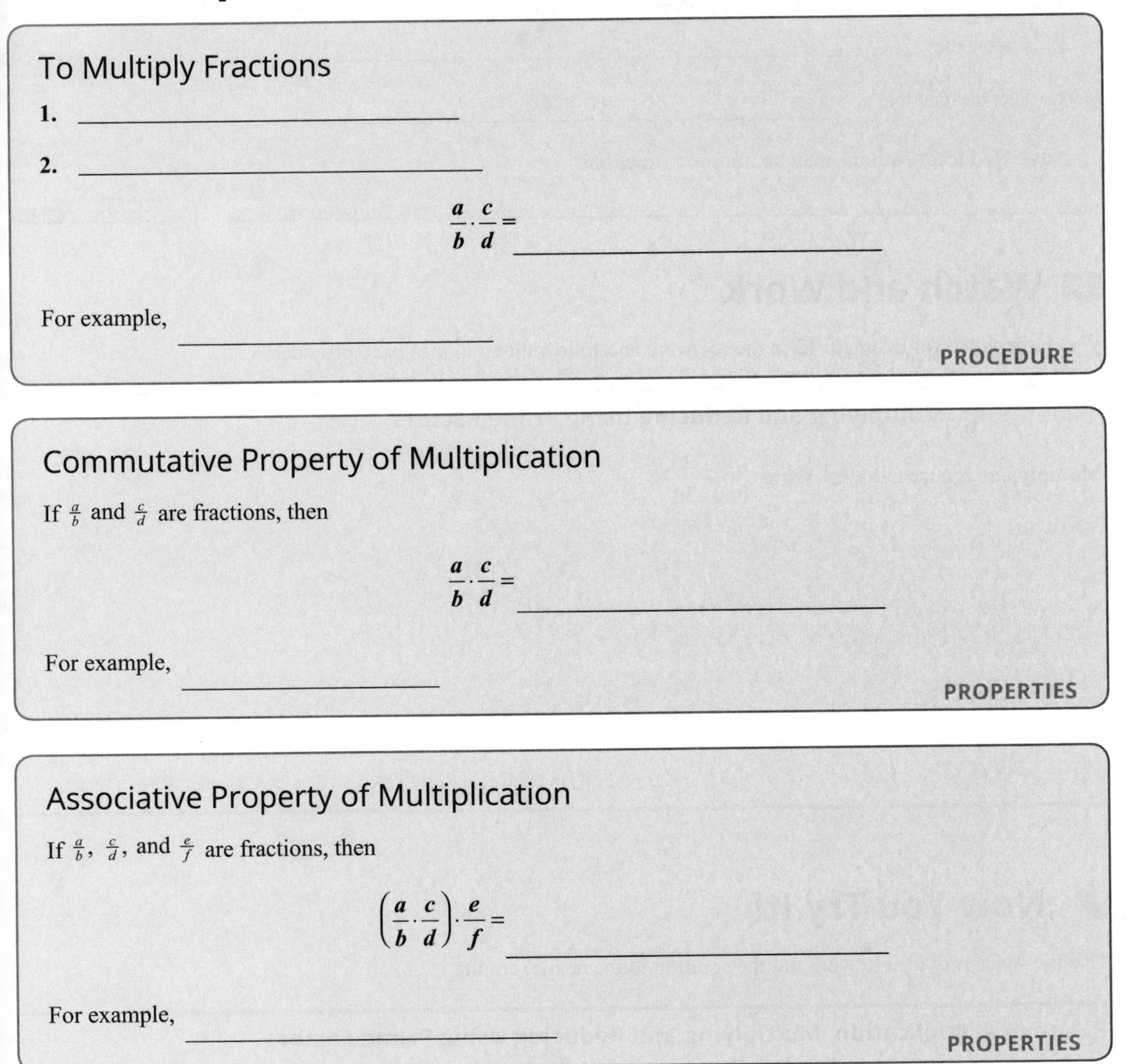

To Multiply Fractions

1. ______________________

2. ______________________

$$\frac{a}{b} \cdot \frac{c}{d} = \_\_\_\_\_\_\_\_\_\_\_\_$$

For example, ______________________

PROCEDURE

Commutative Property of Multiplication

If $\frac{a}{b}$ and $\frac{c}{d}$ are fractions, then

$$\frac{a}{b} \cdot \frac{c}{d} = \_\_\_\_\_\_\_\_\_\_\_\_$$

For example, ______________________

PROPERTIES

Associative Property of Multiplication

If $\frac{a}{b}$, $\frac{c}{d}$, and $\frac{e}{f}$ are fractions, then

$$\left(\frac{a}{b} \cdot \frac{c}{d}\right) \cdot \frac{e}{f} = \_\_\_\_\_\_\_\_\_\_\_\_$$

For example, ______________________

PROPERTIES

1. A fraction is reduced to lowest terms if the numerator and denominator have ______________________

____

To Reduce a Fraction to Lowest Terms

1. Factor the ____________________
2. Use the fact that ____________________

Note: Reduced fractions may be improper fractions.

PROCEDURE

Watch and Work

Watch the video for Example 12 in the software and follow along in the space provided.

Example 12 Multiplying and Reducing Using Prime Factors

Multiply and reduce to lowest terms: $\frac{17}{50}\cdot\frac{25}{34}\cdot 8$

Solution

Now You Try It!

Use the space provided to work out the solution to the next example.

Example A Application: Multiplying and Reducing Using Prime Factors

Multiply and reduce to lowest terms: $16\cdot\frac{12}{100}\cdot\frac{5}{36}$

3.2 Exercises

Concept Check

True/False. Determine whether each statement is true or false. If a statement is false, explain how it can be changed so the statement will be true. (**Note:** There may be more than one acceptable change.)

1. To find $\frac{1}{2}$ of $\frac{2}{9}$ requires multiplication.

2. $-\frac{3}{4} \cdot \frac{9}{10} = -\frac{27}{40}$

3. The statement $\frac{1}{3} \cdot \frac{2}{5} = \frac{2}{5} \cdot \frac{1}{3}$ is an example of the associative property of multiplication.

4. The number 1 is always a factor of the numerator and the denominator.

Practice

Multiply and reduce to lowest terms. (**Hint:** Factor before multiplying.)

5. $\frac{1}{3} \cdot \frac{3}{4}$

6. $\left(-\frac{1}{5}\right)\left(-\frac{4}{7}\right)$

7. $\frac{10}{18} \cdot \frac{9}{5}$

8. $\frac{35a^2b^3c}{20ab^2c^2} \cdot \frac{36ac^3}{14a^2bc^2}$

Applications

Solve.

9. ***Recipes:*** A recipe calls for $\frac{3}{4}$ cups of flour. How much flour should be used if only half of the recipe is to be made?

10. ***Demographics:*** A study showed that $\frac{3}{5}$ of the students in an elementary school were left-handed. If the school had an enrollment of 600 students, how many were left-handed?

Writing &Thinking

11. If two fractions are between 0 and 1, can their product be more than 1? Explain.

12. Explain the process of multiplying two fractions. Give an example of a product that cannot be reduced.

3.3 Division with Fractions

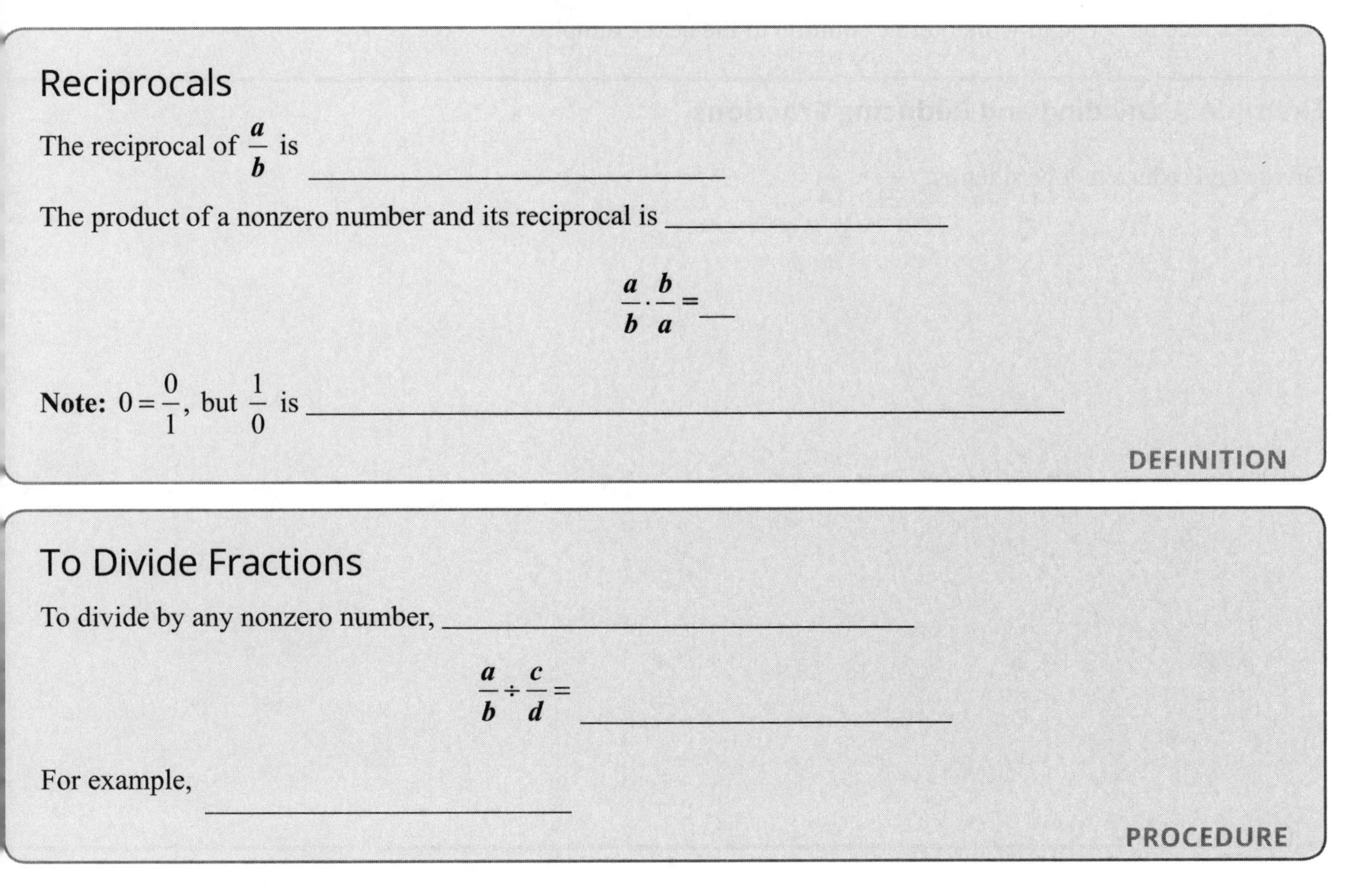

Reciprocals

The reciprocal of $\frac{a}{b}$ is ______________________

The product of a nonzero number and its reciprocal is ______________

$$\frac{a}{b} \cdot \frac{b}{a} = \_\_$$

Note: $0 = \frac{0}{1}$, but $\frac{1}{0}$ is ______________________________________

DEFINITION

To Divide Fractions

To divide by any nonzero number, ________________________

$$\frac{a}{b} \div \frac{c}{d} = \_\_\_\_\_\_\_\_\_\_\_\_\_\_\_\_\_\_\_\_$$

For example, ____________________

PROCEDURE

Watch and Work

Watch the video for Example 5 in the software and follow along in the space provided.

Example 5 Dividing and Reducing Fractions

Divide and reduce to lowest terms: $\frac{16}{27} \div \frac{8}{9}$

Solution

Now You Try It!

Use the space provided to work out the solution to the next example.

Example A Dividing and Reducing Fractions

Divide and reduce to lowest terms: $\frac{10}{21} \div \frac{5}{14}$

3.3 Exercises

Concept Check

True/False. Determine whether each statement is true or false. If a statement is false, explain how it can be changed so the statement will be true. (**Note:** There may be more than one acceptable change.)

1. The reciprocal of 1 is undefined.

2. The product of a nonzero number and its reciprocal is undefined.

3. The reciprocal of -12 is $-\frac{12}{1}$.

4. The result of $\frac{1}{3} \div \frac{1}{6}$ is 2.

Practice

Divide and reduce to lowest terms. See Examples 5 through 7.

5. $\frac{2}{3} \div \frac{3}{4}$

6. $0 \div \frac{5}{6}$

7. $\frac{5}{6} \div 0$

8. $\frac{16}{33} \div \frac{24}{55}$

9. $\frac{5x}{8} \div \frac{-5x}{8}$

Applications

Solve.

10. ***Enrollment:*** A small private college has determined that about $\frac{11}{25}$ of the students that it accepts will actually enroll. If the college wants 255 freshmen to enroll, how many should it accept?

11. ***Airplane Capacity:*** An airplane is carrying 180 passengers. This is $\frac{9}{10}$ of the capacity of the airplane.

a. Is the capacity of the airplane more or less than 180?

b. If you were to multiply 180 times $\frac{9}{10}$, would the product be more or less than 180?

c. What is the capacity of the airplane?

Writing & Thinking

12. Explain why the number 0 has no reciprocal.

13. Is division a commutative operation? Explain briefly and give three examples using fractions to help justify your answer.

3.4 Multiplication and Division with Mixed Numbers

To Multiply Mixed Numbers

1. Change each mixed number to ______________________________
2. Factor the numerator and denominator of ____
3. Change the answer to a ____

PROCEDURE

1. To find the area of a triangle, use the formula ________________

To Divide with Mixed Numbers

1. Change each mixed number to ______________________________
2. ____
3. ______________________________

PROCEDURE

Watch and Work

Watch the video for Example 10 in the software and follow along in the space provided.

Example 10 Dividing and Reducing Mixed Numbers

Divide and reduce to lowest terms: $3\frac{1}{4} \div 7\frac{4}{5}$

Solution

Now You Try It!

Use the space provided to work out the solution to the next example.

Example A Dividing and Reducing Mixed Numbers

Divide and reduce to lowest terms: $2\frac{2}{5} \div 3\frac{3}{4}$

3.4 Exercises

Concept Check

True/False. Determine whether each statement is true or false. If a statement is false, explain how it can be changed so the statement will be true. (**Note:** There may be more than one acceptable change.)

1. When multiplying or dividing with mixed numbers, the answer should always be simplified, if possible.

2. Multiplication or division with mixed numbers can be accomplished by changing the mixed numbers to improper fractions.

3. The mixed number $4\frac{1}{5}$ is equal to $\frac{9}{5}$.

4. The reciprocal of $7\frac{2}{5}$ is $\frac{5}{37}$.

Practice

Multiply and reduce to lowest terms. Write your answer in mixed number form.

5. $\frac{2}{3} \cdot 3\frac{1}{4}$

6. $\left(-12\frac{1}{2}\right)\left(-3\frac{1}{3}\right)$

Divide and reduce to lowest terms. Write your answer in mixed number form.

7. $3\frac{1}{2} \div \frac{7}{8}$

8. $7\frac{1}{5} \div 3$

Applications

Solve.

9. ***Traveling:*** You are planning a trip of 615 miles (round trip), and you know that your car gets an average of $27\frac{1}{3}$ miles per gallon of gas. You also know that your gas tank holds $15\frac{1}{2}$ gallons of gas.

a. How many gallons of gas will you use on this trip?

b. If the gas you buy costs $2 per gallon, how much should you plan to spend on this trip for gas?

10. ***Triangles:*** A right triangle is a triangle with one right angle (measure 90°). The two sides that form the right angle are called legs, and they are perpendicular to each other. The longest side is called the hypotenuse. The legs can be treated as the base and height of the triangle. Find the area of the right triangle shown here.

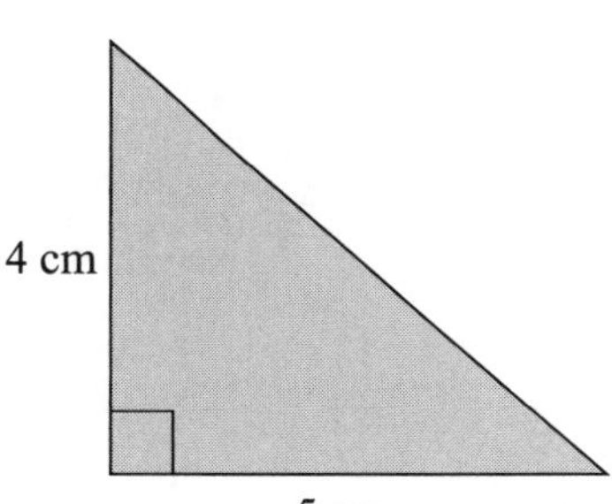

Writing & Thinking

11. Suppose that a fraction between 0 and 1, such as $\frac{1}{2}$ or $\frac{2}{3}$, is multiplied by some other number. Give brief discussions and several examples in answering each of the following questions.

a. If the other number is a positive fraction, will this product always be smaller than the other number?

b. If the other number is a positive whole number, will this product always be smaller than the other number?

c. If the other number is a negative number (integer or fraction), will this product ever be smaller than the other number?

12. Compare and contrast multiplying two mixed numbers and dividing two mixed numbers.

3.5 Least Common Multiple (LCM)

1. The **multiples** of an integer are ______________________________

Least Common Multiple (LCM)

The **least common multiple (LCM)** of two (or more) counting numbers is ______________________________

____

DEFINITION

The following method, involving prime factorizations, is one way to find the least common multiple of a set of counting numbers.

To Find the LCM of a Set of Counting Numbers

1. Find the ______________________________
2. List the ______________________________
3. Find the product of these primes using each ______________________________

   ____

PROCEDURE

Watch and Work

Watch the video for Example 4 in the software and follow along in the space provided.

Example 4 Finding the Least Common Multiple (LCM)

Find the LCM of 27, 30, and 42.

Solution

Now You Try It!

Use the space provided to work out the solution to the next example.

Example A Finding the Least Common Multiple (LCM)

Find the LCM of 36, 45, and 60.

Finding the LCM of a Set of Algebraic Terms

1. Find the prime factorization of each term in the set and ____________________

____

2. Find the largest power of each prime factor present in ____________________
3. The LCM is the ____

PROCEDURE

Finding Equivalent Fractions

To find a fraction equivalent to $\frac{a}{b}$, multiply the numerator and denominator by ____________

____

$$\frac{a}{b} = \underline{\hspace{6cm}}$$

Example:

PROCEDURE

3.5 Exercises

Concept Check

True/False. Determine whether each statement is true or false. If a statement is false, explain how it can be changed so the statement will be true. (**Note:** There may be more than one acceptable change.)

1. The LCM of 15 and 25 is 50.

2. The first five multiples of 9 are 9, 18, 27, 36, and 45.

3. The first five multiples of 4 are 4, 8, 12, 20, and 24.

4. When given larger numbers, the most efficient way to find the LCM is to use the prime factorization method.

Practice

Find the LCM of each set of numbers. See Examples 1 through 5.

5. 6, 10

6. 3, 4, 8

7. For 14, 35, and 49, **a.** find the LCM and **b.** state how many times each number divides into the LCM.

8. Find the LCM of $24xyz$, $16xy^2$, and $6x^2z$

For each equation, find the missing numerator that will make the fractions equivalent.

9. $\frac{5}{8} = \frac{?}{24}$

10. $\frac{-3x}{16y} = \frac{?}{80y}$

Applications

Solve.

11. ***Security:*** Three security guards meet at the front gate for coffee before they walk around inspecting buildings at a manufacturing plant. The guards take 15, 20, and 30 minutes, respectively, for the inspection trip.

a. If they start at the same time, in how many minutes will they meet again at the front gate for coffee?

b. How many trips will each guard have made?

12. ***Fruit:*** A fruit production company has three packaging facilities, each of which uses different-sized boxes as follows: 24 pieces/box, 36 pieces/box, and 45 pieces/box.

a. Assuming that the truck provides the same quantity of uniformly-sized pieces of fruit to all three packaging facilities, what is the minimum number of pieces of fruit that will be delivered so that no fruit will be left over?

b. How many boxes will each facility package?

Writing & Thinking

13. Explain, in your own words, why each number in a set divides evenly into the LCM of that set of numbers.

14. Explain why simply multiplying two numbers together will not necessarily find the LCM of those numbers. Give an example of when it would find the LCM and an example when it would not.

Name: Date:

3.6 Addition and Subtraction with Fractions

To Add Fractions with the Same Denominator

1. ______________________

2. ______________________ $\frac{a}{b}+\frac{c}{b}=$ ______________________

3. ______________________

For example,

____

PROCEDURE

To Add Fractions with Different Denominators

1. Find the ______________________

2. Change each fraction into ______________________

3. ______________________

4. ______________________

PROCEDURE

Commutative Property of Addition

If $\frac{a}{b}$ and $\frac{c}{d}$ are fractions, then

$$\frac{a}{b}+\frac{c}{d}= \_\_\_\_\_\_\_\_\_\_\_\_\_\_\_\_\_\_\_\_$$

For example,

____

PROPERTIES

Associative Property of Addition

If $\frac{a}{b}$, $\frac{c}{d}$, and $\frac{e}{f}$ are fractions, then

$$\frac{a}{b}+\left(\frac{c}{d}+\frac{e}{f}\right)= \_\_\_\_\_\_\_\_\_\_\_\_\_\_\_\_\_\_\_\_$$

For example,

____

PROPERTIES

To Subtract Fractions with the Same Denominator

1. ______________________
2. ______________________ $\frac{a}{b}-\frac{c}{b}=$ ____________
3. ______________________

For example,

PROCEDURE

To Subtract Fractions with Different Denominators

1. Find the ______________________
2. Change each fraction into ______________________
3. ______________________
4. ______________________

PROCEDURE

Watch and Work

Watch the video for Example 10 in the software and follow along in the space provided.

Example 10 Subtracting Fractions with Different Denominators

Subtract: $1-\frac{5}{8}$

Solution

Now You Try It!

Use the space provided to work out the solution to the next example.

Example A Subtracting Fractions with Different Denominators

Subtract: $3-\frac{5}{12}$

3.6 Exercises

Concept Check

True/False. Determine whether each statement is true or false. If a statement is false, explain how it can be changed so the statement will be true. (**Note:** There may be more than one acceptable change.)

1. The final step in adding fractions is to reduce, if possible.

2. The process for finding the LCD is the same as the process for finding the LCM.

3. LCD represents the Least Common Digit.

4. When subtracting fractions, simply subtract the numerators and the denominators.

5. Subtraction of fractions requires that the fractions have the same denominators.

Practice

Add and reduce to lowest terms.

6. $\frac{3}{25}+\frac{12}{25}$

7. $\frac{2}{7}+\frac{4}{21}+\frac{1}{3}$

Subtract and reduce to lowest terms.

8. $\frac{1}{15}-\frac{4}{15}$

9. $-\frac{31}{40}-\left(-\frac{5}{8}\right)$

10. $\frac{5}{6}-\frac{4}{5y}$

Applications

Solve.

11. ***Postage:*** Three pieces of mail weigh $\frac{1}{2}$ ounce, $\frac{1}{5}$ ounce, and $\frac{3}{10}$ ounce. What is the total weight of the letters?

12. ***Cooking:*** A recipe calls for the following spices: $\frac{1}{2}$ teaspoon of turmeric, $\frac{1}{4}$ teaspoon of ginger, and $\frac{1}{8}$ teaspoon of cumin. What is the total quantity of these three spices?

Writing & Thinking

13. Explain how finding the LCM relates to LCDs.

3.7 Addition and Subtraction with Mixed Numbers

To Add Mixed Numbers

1. ______________________________
2. ______________________________
3. Write the answer as ______________________________

PROCEDURE

To Subtract Mixed Numbers

1. ______________________________
2. ______________________________

PROCEDURE

If the Fraction Part Being Subtracted is Larger than the First Fraction

1. "Borrow" 1 from ______________________________
2. Add this 1 to ______________________________

   (This will always result in ______________________________)______
3. ____________________

PROCEDURE

Watch and Work

Watch the video for Example 8 in the software and follow along in the space provided.

Example 8 Subtracting Mixed Numbers by Borrowing

Subtract: $4\frac{2}{9}-1\frac{5}{9}$

Solution

Now You Try It!

Use the space provided to work out the solution to the next example.

Example A Subtracting Mixed Numbers by Borrowing

Subtract: $10\frac{1}{8}-3\frac{5}{8}$

3.7 Exercises

Concept Check

True/False. Determine whether each statement is true or false. If a statement is false, explain how it can be changed so the statement will be true. (**Note:** There may be more than one acceptable change.)

1. $3\frac{1}{5}+5\frac{1}{2}=8\frac{7}{10}$

2. When adding (or subtracting) mixed numbers, the final answer should be written as a mixed number.

3. LCDs are not required when adding or subtracting mixed numbers.

4. $12-5\frac{1}{3}=7\frac{1}{3}$

Practice

Add and reduce to lowest terms. Write your answer in mixed number form.

5. $3\frac{1}{4}+7\frac{1}{8}$

6. $2\frac{5}{8}+6\frac{5}{6}$

Subtract and reduce to lowest terms. Write your answer in mixed number form.

7. $\begin{array}{r} 7\frac{9}{10} \\ -3\frac{3}{10} \\ \hline \end{array}$

8. $\begin{array}{r} 14\dfrac{7}{10} \\ -3\dfrac{4}{5} \\ \hline \end{array}$

9. $-6\dfrac{1}{2}-\left(-10\dfrac{3}{4}\right)$

Applications

Solve.

10. ***Travel:*** A bus trip is made in three parts. The first part takes $2\frac{1}{3}$ hours, the second part takes $2\frac{1}{2}$ hours, and the third part takes $3\frac{3}{4}$ hours. How long does the entire trip take?

11. ***Physiology:*** On average, the air that we inhale includes $1\frac{1}{4}$ parts water and the air we exhale includes $5\frac{9}{10}$ parts water. How many more parts water are in the exhaled air?

Writing & Thinking

12. In subtracting with mixed numbers explain why fractional parts should be subtracted before the whole numbers.

Name: Date:

3.8 Comparisons and Order of Operations with Fractions

1. For any two numbers on a number line, the smaller number is to the __________ of the larger number.

To Compare Two Fractions

1. Find the ______________________________
2. Change each fraction into ____________________________________
3. __________________________

PROCEDURE

Rules for Order of Operations

1. Simplify within grouping symbols, such as ______________________________
   (If there are more than one pair of grouping symbols, start with ______________________)
2. Evaluate any ________________________
3. Moving from left to right, perform any ______________________________________
4. Moving from left to right, perform any ______________________________________

PROPERTIES

Watch and Work

Watch the video for Example 6 in the software and follow along in the space provided.

Example 6 Using the Order of Operations with Fractions

Simplify: $3\frac{2}{5} \div \left(\frac{1}{4} + \frac{3}{5}\right)$

Solution

Now You Try It!

Use the space provided to work out the solution to the next example.

Example A Finding the Sample Space Using a Tree Diagram

Simplify: $\left(\frac{5}{12} + \frac{2}{3}\right) \div 4\frac{1}{3}$

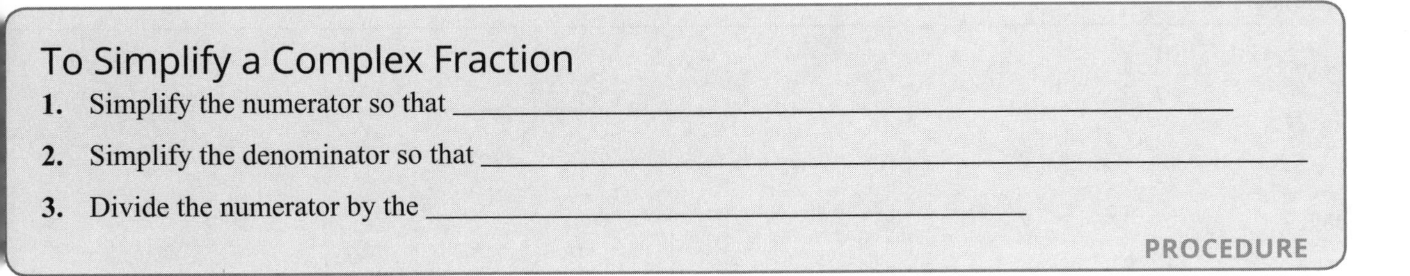

To Simplify a Complex Fraction

1. Simplify the numerator so that ____________________
2. Simplify the denominator so that ____________________
3. Divide the numerator by the ____________________

PROCEDURE

3.8 Exercises

Concept Check

True/False. Determine whether each statement is true or false. If a statement is false, explain how it can be changed so the statement will be true. (**Note:** There may be more than one acceptable change.)

1. The rules for order of operations are the same for fractions and mixed numbers as they are for whole numbers.

2. According to the rules for order of operations, multiplication occurs before division.

3. An average is found by adding all the numbers in the set and then dividing by the number of numbers in the set.

4. To simplify a complex fraction, first divide the numerator by the denominator.

Practice

5. For $\frac{4}{10}$, and $\frac{3}{8}$, determine which fraction is larger and by how much it is larger.

6. Arrange $\frac{2}{3}, \frac{3}{4},$ and $\frac{5}{8}$ in order from smallest to largest. Then find the difference between the largest and smallest fractions.

Simplify.

7. $\frac{1}{2} \div \frac{7}{8} + \frac{1}{7} \cdot \frac{2}{3}$

8. $-\frac{1}{2} \div \left|-\frac{2}{3}\right| - \left(\frac{1}{3}\right)^2$

9. $\dfrac{\frac{5}{8} - \frac{1}{2}}{\frac{1}{8} - \frac{3}{16}}$

Find each average.

10. Find the average of the numbers $\frac{5}{6}, \frac{1}{15}$, and $\frac{17}{30}$.

Applications

Solve.

11. ***Painting:*** Two painters paint $76\frac{1}{2}$ feet of fencing in one day. The first painter contributes $2\frac{1}{2}$ hours and the second works for $5\frac{3}{5}$ hours. How many feet of fencing are painted in each hour? (**Hint:** Add the number of hours then divide the length of fencing by this sum.)

12. ***Books:*** An art book has 40 two-sided pages of pictures, each of which is $\frac{1}{32}$ inch thick. Each two-sided page is protected by a $\frac{1}{80}$ inch thick piece of paper. Each side of the book is bound by a $\frac{1}{6}$ inch cover. What is the total thickness of the book?

Writing & Thinking

13. **a.** If two fractions are between 0 and 1, can their sum be more than 1? Explain.

b. If two fractions are between 0 and 1, can their product be more than 1? Explain.

3.9 Solving Equations with Fractions

Basic Principles for Solving Equations

1. **The addition principle** If A, B, and C are algebraic expressions, then the equations

   ______

   ____ ________

   ____

2. **The multiplication principle** If A and B are algebraic expressions and C is a nonzero constant, then the equations

   ______

   ______

   ______

   ____

PROPERTIES

Watch and Work

Watch the video for Example 1 in the software and follow along in the space provided.

Example 1 Solving Equations with Fractions

Solve the equation $\frac{2}{3}x = 14$ by multiplying both sides of the equation by the reciprocal of the coefficient.

Solution

Now You Try It!

Use the space provided to work out the solution to the next example.

Example A Solving Equations with Fractions

Solve the equation $\frac{5}{8}y = -25$

3.9 Exercises

Concept Check

True/False. Determine whether each statement is true or false. If a statement is false, explain how it can be changed so the statement will be true. (**Note:** There may be more than one acceptable change.)

1. The equations $x + 3 = 4$ and $16(x+3) = 16(4)$ have the same solutions.

2. The goal of solving equations is to have all of the constants on one side of the equation and the variable on the other side of the equation with a coefficient of 0.

3. The equation $\frac{3}{4}x = \frac{8}{15}$ can be solved by multiplying both sides of the equation by $\frac{15}{8}$.

4. The first step to solve $\frac{1}{3}x + 5 = \frac{1}{7}$ is to multiply both sides of the equation by 28.

Practice

Solve each equation.

5. $16x + 23x - 5 = 8$

6. $\frac{7}{10}y = -28$

7. $\frac{1}{5}x + 10 = -32$

8. $\frac{n}{5} - \frac{1}{5} = \frac{1}{5}$

Applications

Solve.

9. ***Voting:*** Thomas Brown won the election for class president by earning 96 votes, which was $\frac{2}{3}$ of the total votes. This situation can be modeled by the equation $\frac{2}{3}x = 96$, where x is the number of people who voted. Solve the equation for x to determine the number of students who voted in the election for class president.

10. ***Nutrition Facts:*** One serving of Ritz Bits Peanut Butter Sandwiches has $2\frac{1}{2}$ g of saturated fat. This represents $\frac{5}{22}$ of the total fat content. This situation can be modeled by the equation $\frac{5}{22}x = 2\frac{1}{2}$, where x represents the total number of grams of fat in the serving. Solve the equation for x to determine the total number of grams of fat are in one serving of Ritz Bits Peanut Butter Sandwiches.

Writing & Thinking

11. In your own words, explain why multiplying an equation containing fractions by the LCD will result in an equation with only integer coefficients.

3.10 Ratios and Unit Rates

Ratios

A **ratio** is ____

The ratio of a to b can be written as

____

DEFINITION

Ratios have the following characteristics.

1. Ratios can be ____
2. The common units in a ratio can be ____
3. Generally, ratios are written with whole numbers in the numerator and denominator, and the ________

   ____

Watch and Work

Watch the video for Example 2 in the software and follow along in the space provided.

Example 2 Writing Ratios that Compare Mixed Numbers

Write each ratio as a fraction in lowest terms.

a. $2\frac{1}{2}$ to $5\frac{1}{2}$

b. 3 days to $1\frac{1}{2}$ days

Solution

Now You Try It!

Use the space provided to work out the solution to the next example.

Example A Writing Ratios that Compare Mixed Numbers

Write each ratio as a fraction in lowest terms.

a. $1\frac{1}{3}$ to $4\frac{2}{3}$

b. 7 months to $1\frac{3}{4}$ months

B Rates

4. A **rate** is ______________________________

Changing Rates to Unit Rates

To make a rate a unit rate, divide the ______________________________

PROCEDURE

3.10 Exercises

Concept Check

True/False. Determine whether each statement is true or false. If a statement is false, explain how it can be changed so the statement will be true. (**Note:** There may be more than one acceptable change.)

1. The units in the numerator and denominator of a ratio must be the same, or need to be able to be converted to the same units.

2. The order of the numbers in a ratio or a rate is irrelevant as long as the numbers are reduced.

3. The ratio 8:2 can be reduced to the ratio 4.

4. To make a unit rate, divide the numerator by the denominator.

Practice

Write each ratio as a fraction in lowest terms. See Examples 1 through 3.

5. Write 18 to 28 as a fraction in lowest terms.

6. Write 5 nickels to 3 quarters as a fraction in lowest terms by first finding common unites in the numerator and denominator.

7. Write the rate $200 in profit to $500 invested as a fraction in lowest terms.

8. Write the rate 50 miles to 2 gallons of gas as a unit rate.

Find the unit price (to the nearest dollar) of each set of items and tell which one is the better (or best) purchase.

9. ***T-shirts:*** 2 shirts for $12, 5 shirts for $20

Applications

Solve.

10. ***Nutrition:*** A serving of four home-baked chocolate chip cookies weighs 40 grams and contains 12 grams of fat. What is the ratio, in lowest terms, of fat grams to total grams?

11. ***Standardized Testing:*** In recent years, 18 out of every 100 students taking the SAT (Scholastic Aptitude Test) at a local school have scored 600 or above on the mathematics portion of the test. Write the ratio, in lowest terms, of the number of scores 600 or above to the number of scores below 600.

Writing & Thinking

12. Demonstrate three different ways the ratio comparing 5 apples to 3 apples can be written. Choose one form and explain why it is the preferred form when using ratios in a math course.

13. When finding price per unit, will monetary units be located in the numerator or the denominator of the rate?

3.11 Proportions

Proportions

A **proportion** is a statement that ________________________

In symbols, ________________________

A proportion is true if ________________________

DEFINITION

To Solve a Proportion

1. Find the cross products (or cross multiply) and then ________________________
2. Divide both sides of the equation by the ________________________
3. ____________

PROCEDURE

Watch and Work

Watch the video for Example 3 in the software and follow along in the space provided.

Example 3 Solving Proportions

Find the value of x if $\frac{4}{8}=\frac{5}{x}$.

Solution

Now You Try It!

Use the space provided to work out the solution to the next example.

Example A Solving Proportions

Find the value of x if $\frac{12}{x}=\frac{9}{15}$.

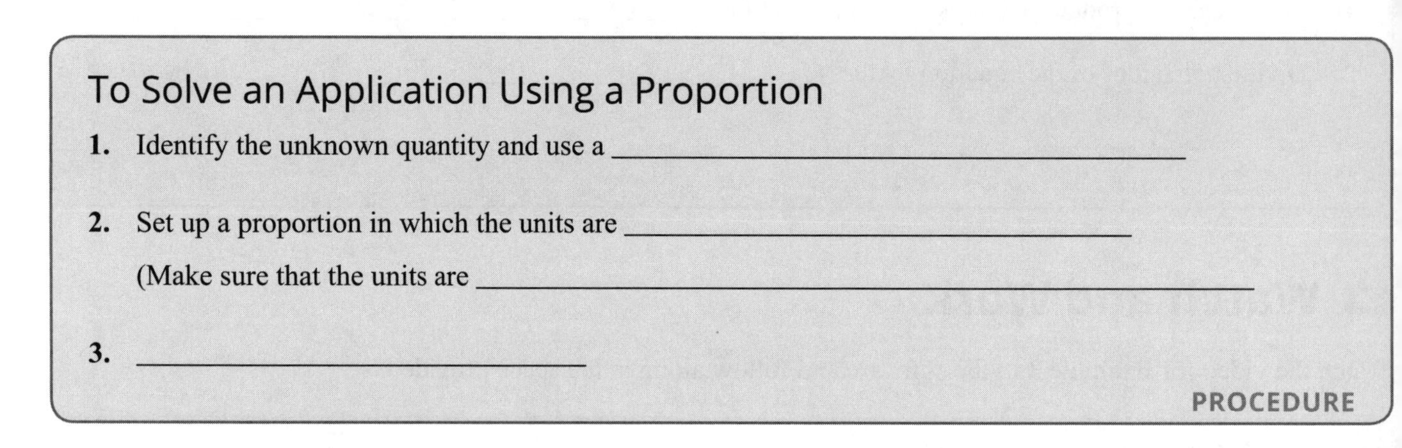

To Solve an Application Using a Proportion

1. Identify the unknown quantity and use a ______________________________
2. Set up a proportion in which the units are ______________________________

   (Make sure that the units are ______________________________
3. ______________________________

PROCEDURE

3.11 Exercises

Concept Check

True/False. Determine whether each statement is true or false. If a statement is false, explain how it can be changed so the statement will be true. (**Note:** There may be more than one acceptable change.)

1. A proportion is a statement that two ratios are being multiplied.

2. Cross canceling is used to determine if a proportion is true.

3. When using proportions to solve a word problem, there is only one correct way to set up the proportion.

4. The proportions $\frac{36 \text{ tickets}}{\$540} = \frac{x \text{ tickets}}{\$75}$ and $\frac{x \text{ tickets}}{36 \text{ tickets}} = \frac{\$75}{\$540}$ will yield the same answer.

Practice

Determine whether each proportion is true or false.

5. $\frac{3}{6} = \frac{4}{8}$

6. $\frac{1}{3} = \frac{33}{100}$

Solve each proportion.

7. $\frac{5}{4} = \frac{x}{8}$

8. $\frac{1}{12} = \frac{1\frac{2}{3}}{x}$

Applications

Solve.

9. ***Concrete:*** The quality of concrete is based on the ratio of bags of cement to cubic yards of gravel. One batch of concrete consists of 27 bags of cement mixed into 9 cubic yards of gravel, while a second has 15 bags of cement mixed with 5 cubic yards of gravel. Determine whether the ratio of cement to gravel is the same for both batches.

10. ***Grading:*** An English teacher must read and grade 27 essays. If the teacher takes 20 minutes to read and grade 3 essays, how much time will he need to grade all 27 essays?

Writing & Thinking

11. In your own words, clarify how you can know that a proportion is set up correctly or not.

12. List the steps to solve a word problem using a proportion.

3.12 Probability

1. Activities involving chance such as tossing a coin, rolling a die, spinning a wheel in a game, and predicting weather are called ________________. The likelihood of a particular result is called its ______________

Terms Related to Probability

________________	An individual result of an experiment.
____________________	The set of all possible outcomes of an experiment.
______________	Some (or all) of the outcomes from the sample space.

DEFINITION

2. A **tree diagram** can be used to____. Each branch of the tree diagram shows________________________.

Watch and Work

Watch the video for Example 3 in the software and follow along in the space provided.

Example 3 Finding the Sample Space Using a Tree Diagram

A coin is tossed and then one of the numbers (1, 2, and 3) is chosen at random from a box. Draw a tree diagram illustrating the possible outcomes of the experiment and list the outcomes in the sample space.

Solution

Now You Try It!

Use the space provided to work out the solution to the next example.

Example A Finding the Sample Space Using a Tree Diagram

A coin is tossed and then a six-sided die is rolled. Draw a tree diagram illustrating the outcomes of the experiment and list the outcomes in the sample space.

Probability of an Event

probability of an event = $\frac{\rule{5cm}{0.4pt}}{}$

DEFINITION

Basic Characteristics of Probabilities.

1. Probabilities are between 0 and 1, inclusive.

   If an event can never occur, ______________________________

   If an event will always occur, ______________________________

2. The sum of the probabilities of the outcomes in a sample space is 1.

PROPERTIES

3.12 Exercises

Concept Check

True/False. Determine whether each statement is true or false. If a statement is false, explain how it can be changed so the statement will be true. (**Note:** There may be more than one acceptable change.)

1. The individual result of an experiment is a probability.

2. An event is some or all of the outcomes from the sample space.

3. A single result of an experiment is an outcome.

4. The probability of a tossed coin showing either heads or tails is 1.

Applications

For each experiment, draw a tree diagram illustrating the possible outcomes and list the outcomes in the sample space.

5. Four marbles are in a box: one red, one white, one blue, and one purple. One ball is chosen.

6. There are three flavors of potato chips to choose from: original, BBQ, and cheddar. One flavor is chosen.

For each problem, calculate the probability described.

7. ***Probability:*** A box contains 5 marbles: two red, one white, two blue. What is the probability of choosing a blue marble from the box?

8. ***Candy:*** A machine contains only 5 gumballs: three yellow, one white, one green. What is the probability of getting a yellow gumball when you put a coin in the machine?

Writing & Thinking

9. List at least three activities that are experiments of chance.

10. Explain the benefit/s of using a tree diagram to determine probability.

Chapter 3 Project

On a Budget

An activity to demonstrate the use of fractions in real life.

Samantha's friend Meghan is always traveling to exotic locations when she goes on vacation, so one day Samantha asked her how she was able to afford it. Meghan told her it was simple. She makes a budget and sets aside a portion of her salary each month for a vacation. Meghan told Samantha that they could meet at her house for dinner on Saturday and she would be glad to show her how to make a budget.

The table below shows the fraction of Meghan's salary that she spends for each category in her budget. Assume that Meghan's salary is \$4000 a month. Answer the questions below, reducing any fractional answers to lowest terms.

Meghan's Budget

Category	Fraction	Category	Fraction
Rent	$\frac{1}{4}$	Savings	$\frac{1}{10}$
Utilities	$\frac{1}{20}$	Gas	$\frac{1}{25}$
Phone	$\frac{1}{25}$	Vacation	$\frac{3}{40}$
Car payment	$\frac{1}{10}$	Car Insurance	$\frac{1}{25}$
IRA	$\frac{1}{8}$	Eating Out	$\frac{1}{16}$

1. What fraction of Meghan's salary does she spend on rent, utilities, and her phone?
   - **a.** First, find the LCM of all the denominators for these budget categories.
   - **b.** Write an equivalent fraction for each of these three budget categories using the value found in Part **a.** as the LCD.
   - **c.** Add the equivalent fractions from Part **b.** together to answer the question.
2. Meghan sets aside $\frac{3}{40}$ of her salary each month for her yearly vacation and $\frac{1}{16}$ of her budget for eating out. Which of these two fractions is larger?
3. Meghan saves $\frac{1}{10}$ of her salary and puts $\frac{1}{8}$ of her salary in an IRA for her retirement fund.
   - **a.** Find the sum of these two fractions.
   - **b.** How much of Meghan's salary is budgeted for investment and savings each month?
4. There are three budget categories that are car-related.
   - **a.** Determine the **average** fraction of Meghan's salary that she budgets for car-related expenses.
   - **b.** Determine the **average** amount spent for each car-related expense.
5. Determine the fractional portion of Meghan's salary remaining to spend on miscellaneous expenses.
   - **a.** What is the LCM of all the denominators in the table?
   - **b.** Write an equivalent fraction for each budget category using the value found in Part **a.** as the LCD.
   - **c.** Add the equivalent fractions from Part **b.** together to get the total fractional portion of Meghan's monthly salary that is budgeted for the categories in the table.
   - **d.** What fractional portion of Meghan's monthly salary is not budgeted?
   - **e.** How much of Meghan's monthly salary is unbudgeted?

6. Let's make some changes to Meghan's current budget using the new table below.

Meghan's Budget

Category	Fraction	Category	Fraction
Rent		Savings	
Utilities		Gas	
Phone		Vacation	
Car payment		Car Insurance	
IRA		Eating Out	

a. Double the fractions spent in the first Category column in the original budget and place in the new table above.

b. Halve the fractions spent in the second Category column in the original budget and place them in the table.

c. Sum the fractions in the second column of the table to find the total fractional portion budgeted for the new budget plan.

d. Would Meghan be able to afford this new budget on her current salary? In other words, do the new fractions in column two add up to a fraction less than 1?

e. If the new budget in Part **d.** is over budget, what fraction is it over? If the new budget is under budget, what fractional portion is left over?

CHAPTER 4

Decimal Numbers

Math @ Work

Many people use decimal numbers on a daily basis to discuss sports. Batting averages in baseball and save percentages in hockey are calculated to the nearest thousandth. Quarterback ratings in football and average points per game in basketball are calculated to the nearest tenth. Decimal numbers are also used to record many of the world's sport records. For example, for the 100-meter freestyle, the men's swimming record is 46.91 seconds by César Cielo of Brazil, and the women's swimming record is 51.71 seconds by Sarah Sjöström of Sweden.

During the NBA 2016–2017 regular season, the top three players based on points per game were Russell Westbrook of the Oklahoma City Thunder, James Harden of the Houston Rockets, and Isaiah Thomas of the Boston Celtics. Westbrook led with an average of 31.6 points per game, followed by Harden with 29.1, and Thomas with 28.9. During the season, Westbrook played 81 games, Harden played 81 games, and Thomas played 76 games. How would you determine how many total points each player scored during the 2016–2017 regular season? Which player scored the most points during the season?

(Source: stats.nba.com)

4.1 Introduction to Decimal Numbers

To Read or Write a Decimal Number

1. Read (or write) the ____________________
2. Read (or write) the word ________________________________
3. Read (or write) the fraction part as a ____ ________________________________

PROCEDURE

To Compare Two Positive Decimal Numbers

1. Moving **left to right**, compare digits with the ____________________
2. When one compared digit is ____

PROCEDURE

To Compare Two Negative Decimal Numbers

1. Moving **left to right**, compare digits with the ________________________________ ____________________________
2. When one compared digit is ____________________________________

PROCEDURE

Watch and Work

Watch the video for Example 6 in the software and follow along in the space provided.

Example 6 Comparing Negative Decimal Numbers

Which number is larger: –4.7 or –4.78?

Solution

Now You Try It!

Use the space provided to work out the solution to the next example.

Example A Comparing Negative Decimal Numbers

Which number is larger: −5.3 or −5.32?

Rounding Rule for Decimal Numbers

1. Look at the single digit one place value to the right of the digit in the place of desired accuracy.

   a. **If this digit is less than 5,** ________________ and replace all digits to the ________ All digits to the ________________

   b. **If this digit is 5 or greater,** ________________ and replace all digits to the ________________ All digits to the ________________
   ____

2. Zeros to the right of the place of accuracy that are also to the right of the ________________
   ____

   In this way, the place of accuracy is ________________

   If a rounded number has a 0 in the desiredplace of accuracy, then ________________

PROCEDURE

4.1 Exercises

True/False. Determine whether each statement is true or false. If a statement is false, explain how it can be changed so the statement will be true. (**Note:** There may be more than one acceptable change.)

1. Two hundred thousand, four hundred six and twelve hundredths can be written as 200,406.12.

2. 92.586 is greater than 92.6.

3. On a number line, any number to the left of another number is larger than that other number.

4. When a decimal number is rounded, all numbers to the right of the place of accuracy become zeros in the final answer.

5. Write $2\frac{57}{100}$ in decimal notation.

6. Write 20.7 in words.

7. Write six and twenty-eight thousandths in decimal notation.

8. Arrange 0.2, 0.26, and 0.17 in order from smallest to largest. Then, graph the numbers on a number line.

Fill in the blanks to correctly complete each statement.

9. Round 3.00652 to the nearest ten-thousandth.

a. The digit in the ten-thousandths position is ___.

b. The next digit to the right is ___.

c. Since ___ is less than 5, leave ___ as it is and replace ___ with 0.

d. So 3.00652 rounds to _____ to the nearest ten-thousandth.

Applications

In each exercise, write the decimal numbers that are not whole numbers in words.

10. ***Unicycles:*** The tallest unicycle ever ridden was 114.8 feet tall, and was ridden by Sam Abrahams (with a safety wire suspended from an overhead crane) for a distance of 28 feet in Pontiac, Michigan, on January 29, 2004.

11. ***Water Weight:*** One quart of water weighs approximately 2.0825 pounds.

Writing & Thinking

12. Discuss situations where you think it is particularly appropriate (or necessary) to write numbers in English word form.

13. With **a.** and **b.** as examples, explain in your own words how you can tell quickly when one decimal number is larger (or smaller) than another decimal number.

a. The decimal number 2.765274 is larger than the decimal number 2.763895.

b. The decimal number 17.345678 is larger than the decimal number 17.345578.

4.2 Addition and Subtraction with Decimal Numbers

To Add Decimal Numbers

1. Write the ______________________________
2. Keep the ______________________________
3. Keep digits with the ______________________________
4. Add, just as with whole numbers, keeping the ______________________________
____

PROCEDURE

Watch and Work

Watch the video for Example 3 in the software and follow along in the space provided.

Example 3 Adding Decimal Numbers

Add: $9 + 4.86 + 37.479 + 0.6$

Solution

Now You Try It!

Use the space provided to work out the solution to the next example.

Example A Adding Decimal Numbers

Add: $23.8 + 4.2567 + 11 + 3.01$

To Subtract Decimal Numbers

1. Write the ____________________
2. Keep the ____________________
3. Keep digits with the ____________________
4. Subtract, just as with whole numbers, keeping the ____________________ ____________________

PROCEDURE

4.2 Exercises

Concept Check

True/False. Determine whether each statement is true or false. If a statement is false, explain how it can be changed so the statement will be true. (**Note:** There may be more than one acceptable change.)

1. Vertical alignment is the preferred method of adding decimal numbers because digits with the same place value are easily lined up.

2. It is important to align the decimal points vertically when adding decimal numbers.

3. In subtracting decimal numbers, line up all the last digits vertically.

4. Once decimal points and corresponding digits have been aligned vertically, add or subtract from left to right.

Practice

5. Add $\begin{array}{r} 42.08 \\ +8.005 \\ \hline \end{array}$

6. Subtract $\begin{array}{r} 39.542 \\ -28.411 \\ \hline \end{array}$

7. Add or Subtract $-88.6-(-91.9)$

8. Simplify by combining like terms $8.3x+x-22.7x$

Applications

Solve.

9. ***Goods and Services:*** Mr. Johnson bought the following items at a department store: slacks, \$32.50; shoes, \$43.75; shirt, \$18.60.

   a. How much did he spend?

   b. What was his change if he gave the clerk a \$100 bill? (Tax was included in the prices.)

10. ***Gardening:*** David is preparing a four-sided garden plot with unequal sides of 7.5 feet, 26.34 feet, 36.92 feet, and 12.07 feet. How many feet of edging material must he use? (This is the same as finding the perimeter of the plot.)

Writing & Thinking

11. Why is it important that the decimal points and numbers be aligned vertically when adding or subtracting decimals?

12. Suppose that you are given two decimal numbers with 0 as their whole number part.

    a. Explain how the sum might be more than 1.

    b. Explain why the sum cannot be more than 2.

4.3 Multiplication and Division with Decimal Numbers

To Multiply Decimal Numbers

1. Multiply the two numbers as if ______

2. Count the total number of ______

3. Place the decimal point in the product so that the ______

______

PROCEDURE

To Multiply by Powers of 10 (10, 100, 1000, and so on)

1. Count the number of ______

2. Move the decimal point to the ______

Multiplication by **10** moves the decimal point ______

Multiplication by **100** moves the decimal point ______

Multiplication by **1000** moves the decimal point ______

And so on.

PROCEDURE

To Divide Decimal Numbers

1. Move the decimal point in the divisor to the ______

2. Move the decimal point in the dividend the ______

3. Place the decimal point in the quotient directly ______

4. Divide just as with ______

PROCEDURE

When the Remainder is Not 0

1. Decide how many ______________________________
2. Divide until the quotient has been calculated to ______________________________
3. Using this last digit, ______________________________

PROCEDURE

Terminating and Nonterminating Decimal Numbers

If the remainder is eventually 0, the decimal number is said ____________________

For example, ______________________________

If the remainder is not eventually 0, the decimal number is said ____________________

For example, ______________________________

DEFINITION

Watch and Work

Watch the video for Example 9 in the software and follow along in the space provided.

Example 9 Dividing Decimal Numbers

Divide (to the nearest tenth): $82.3 \div 2.9$

Solution

Now You Try It!

Use the space provided to work out the solution to the next example.

Example A Dividing Decimal Numbers

Divide (to the nearest tenth): $83.5 \div 5.6$

To Divide a Decimal Number by a Power of 10 (10, 100, 1000, and so on)

1. Count the number of ____________________
2. Move the decimal point to the ____________________

Division by **10** moves the decimal point ____________________

Division by **100** moves the decimal point ____________________

Division by **1000** moves the decimal point ____________________

And so on.

PROCEDURE

4.3 Exercises

Concept Check

True/False. Determine whether each statement is true or false. If a statement is false, explain how it can be changed so the statement will be true. (**Note:** There may be more than one acceptable change.)

1. The decimal points should be aligned vertically when multiplying decimal numbers.

2. When multiplying decimal numbers, the answer should have the same number of decimal places as the total number of decimal places in the numbers being multiplied.

3. Multiplying by 100 requires that the decimal point be moved 100 places to the right.

4. Moving the decimal point in a divisor requires that the decimal point also be moved in the dividend.

Practice

Multiply. See Examples 1 through 4.

5. Multiply $(5.6)(-0.02)$

6. Multiply $10(-45.6)$

7. Divide $-1.62 \div 9$

8. Divide $\frac{167}{10}$

Applications

Solve.

9. ***Financing:*** To buy a car, you can pay \$2036.50 in cash, or you can put down \$400 and make 18 monthly payments of \$104.30. How much would you save by paying cash?

10. ***Test Average:*** A professor has graded a test of five students, and their scores were 76.4, 100, 84.7, 10.2, and 68.3. What is the average of these five scores?

Writing & Thinking

11. In your own words, discuss the similarities and differences between multiplication with whole numbers and multiplication with decimal numbers.

4.4 Estimating and Order of Operations with Decimal Numbers

1. We can estimate a sum (or difference) by rounding each number to the place of the ____________________

____

Watch and Work

Watch the video for Example 1 in the software and follow along in the space provided.

Example 1 Estimating Sums of Decimal Numbers

Estimate the sum; then find the actual sum.

$$74 + 3.529 + 52.61$$

Solution

Now You Try It!

Use the space provided to work out the solution to the next example.

Example A Estimating Sums of Decimal Numbers

Estimate the sum; then find the actual sum.

$6.68 + 103 + 21.94$

---

2. Estimating products can be done by rounding each number to the place of the ______________________

____

3. In order to estimate with division, round both the ____

____

4.4 Exercises

Concept Check

True/False. Determine whether each statement is true or false. If a statement is false, explain how it can be changed so the statement will be true. (**Note:** There may be more than one acceptable change.)

1. An estimate of the sum $71.369 + 49.1$ is 120.

2. One way to estimate the product of decimal numbers is to round the numbers to the rightmost nonzero digit before performing the multiplication.

3. An estimate of the quotient $16.469 \div 3.87$ would be 4.

4. Experience and understanding are needed to decide whether or not a particular answer is reasonably close to an estimate.

5. According to the rules for order of operations, addition and subtraction should be performed before multiplication and division.

Practice

6. Estimate each answer, then find the actual answer

$$\begin{array}{r} 29.03 \\ +\,3.79 \\ \hline \end{array}$$

7. Estimate each answer, then find the actual answer

$$\begin{array}{r} 51.21 \\ -\,25.13 \\ \hline \end{array}$$

8. Estimate each answer, then find the actual answer $(6.3)(1.6)$

9. Estimate each answer, then find the actual answer $3.1\overline{)6.36}$

10. Simplify $3.1(50-25.8)-12.9$

Applications

Solve.

11. ***Shipping:*** Jim is packing three sculptures in a box for shipping. The weights of the sculptures are 5.63 pounds, 12.4 pounds, and 3 pounds. The shipping materials weigh 17.4 pounds.

a. Estimate the total weight.

b. Find the actual weight.

12. ***Bicycling:*** Peter Sagan rode 125.09 miles in 5.35 hours.

a. Estimate how fast he was riding per hour.

b. What was his average speed per hour (to the nearest hundredth)?

4.5 Statistics: Mean, Median, Mode, and Range

1. **Statistics** is the study of how to ____

   ______________.

2. A **statistic** is a particular measure or characteristic of a part, or ____________, of a larger collection of items called the ________________.

Terms Used in the Study of Statistics

__________	Value(s) measuring some characteristic of interest such as income, height, weight, grade point averages, scores on tests, and so on. (We will consider only numerical data.)
__________	A single number describing some characteristic of the data.
__________	The sum of all the data divided by the number of data items. (The arithmetic average of the data.)
__________	The middle of the data after the data have been arranged in order (smallest to largest or vice versa). (The median may or may not be one of the data items.)
__________	The data item(s) that appears the most number of times. (A set of data may have more than one mode.)
__________	The difference between the largest and smallest data items.

DEFINITION

Watch and Work

Watch the video for Example 1 in the software and follow along in the space provided.

Example 1 Application: Finding the Mean

Find the mean annual income for the families in Group A.

Group A

Annual Income for 8 Families							
\$28,000	\$45,000	\$22,000	\$80,000	\$25,000	\$25,000	\$27,000	\$30,000

Table 1

Solution

Now You Try It!

Use the space provided to work out the solution to the next example.

Example A Application: Finding the Mean

Find the mean movie time for the movies in Group B, rounded to the nearest minute.

Group B

The Time (in Minutes) of 11 Movies					
100 min	90 min	113 min	110 min	88 min	90 min
155 min	88 min	105 min	93 min	90 min	

Table 2

To Find the Median

1. Arrange the data in ______________________________
2. a. If there is an odd number of items, ______________________

   b. If there is an even number of items, ______________________________ ______________

   (**Note:** This value ______________________.)

PROCEDURE

4.5 Exercises

Concept Check

True/False. Determine whether each statement is true or false. If a statement is false, explain how it can be changed so the statement will be true. (**Note:** There may be more than one acceptable change.)

1. Data is the value(s) of a particular characteristic of interest such as number of people, weight, temperatures, or innings pitched.

2. The range is the difference between the first number listed in the data and the last number listed.

3. The number that appears the greatest number of times in a set of data is the sample.

4. The mean and median are never the same number.

Practice

For each set of data, find **a.** the mean, **b.** the median, **c.** the mode (if any), and **d.** the range.

5. ***Presidents:*** The ages of the first five US presidents on the date of their inaugurations were as follows. (The presidents were Washington, Adams, Jefferson, Madison, and Monroe.)

   57, 61, 57, 57, 58

6. ***Grades:*** Dr. Wright recorded the following nine test scores for students in his statistics course.

   95, 82, 85, 71, 65, 85, 62, 77, 98

7. ***Finance:*** Family incomes in a survey of eight students are as follows.

   \$35,000, \$63,000, \$28,000, \$36,000, \$42,000, \$51,000, \$71,000, \$63,000

8. ***Auto Repair:*** Stacey went to six different auto repair shops to get the following estimates to repair her car.

   \$425, \$525, \$325, \$300, \$500, \$325

Applications

Solve.

9. ***Grades:*** Suppose that you are to take four hourly exams and a final exam in your chemistry class. Each exam has a maximum of 100 points and you must average between 75 and 82 points to receive a passing grade of C. If you have scores of 83, 65, 70, and 78 on the hourly exams, what is the minimum score you can make on the final exam and receive a grade of C? (First, explain your strategy in solving this problem. Then, solve the problem.)

10. ***Agriculture:*** The number of farms in the United States is decreasing. The following graph shows the number of farms (in millions) for each decade since 1940. As you can tell from the graph, the number of farms seems to be leveling off somewhat.

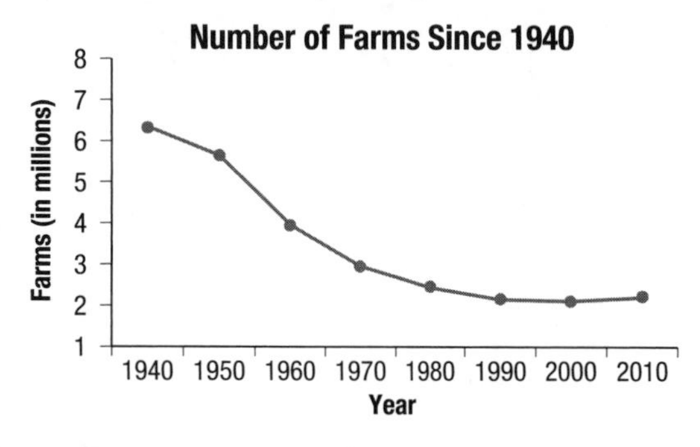

1940: 6.35	**1950:** 5.65
1960: 3.96	**1970:** 2.95
1980: 2.44	**1990:** 2.15
2000: 2.11	**2010:** 2.2

a. Find the mean number of farms from 1940 to 2010.

b. Find the mean number of farms from 1980 to 2010.

Writing & Thinking

11. Determine whether or not the mean and median represent the same number. Give examples to justify your answer.

12. Give three different and specific examples where some statistical measure is used (outside of a class).

4.6 Decimal Numbers and Fractions

To Change from Decimal Numbers to Fractions

A decimal number less than 1 (digits are to the right of the decimal point) can be written in fraction form by writing a fraction with:

1. A numerator that consists of the ____ ______________________________
2. A denominator that is the ____

PROCEDURE

1. ____________________ numbers can be **repeating** or **nonrepeating**. A **repeating decimal** has ________________________ Every fraction with a whole number numerator and nonzero denominator is ________________________ Such numbers are called ________________

Watch and Work

Watch the video for Example 8 in the software and follow along in the space provided.

Example 8 Simplifying Expressions with Decimals and Fractions

Find the sum $10\frac{1}{2}+7.32+5\frac{3}{5}$ in decimal form.

Solution

Now You Try It!

Use the space provided to work out the solution to the next example.

Example A Simplifying Expressions with Decimals and Fractions

Find the sum $2.88 + \frac{1}{4} + 13\frac{9}{10}$ in decimal form.

4.6 Exercises

Concept Check

True/False. Determine whether each statement is true or false. If a statement is false, explain how it can be changed so the statement will be true. (**Note:** There may be more than one acceptable change.)

1. When a decimal number is changed to a fraction, the denominator will be the power of 10 that names the rightmost digit of the decimal number.

2. When a decimal number is changed to a fraction, the numerator can be determined by using the whole number that is formed by all the digits of the decimal number.

3. Fractions can always be converted to decimal form without losing accuracy.

4. In decimal form, $\frac{1}{3}$ is repeating and nonterminating.

Practice

Change each decimal number to a fraction or mixed number in lowest terms.

5. -0.17

6. 4.31

Change each fraction to a decimal number rounded to the nearest hundredth.

7. $\frac{20}{3}$

8. $\frac{40}{9}$

Simplify expression by writing all of the numbers in decimal form and performing the indicated operations. Round to the nearest hundredth, if necessary.

9. $\frac{5}{8}+\frac{3}{5}+0.41$

10. Arrange $0.76, \frac{3}{4}, \frac{7}{10}$ in order from smallest to largest.

Applications

Solve.

11. ***Rectangles:*** A rectangle measures 6.4 inches in length, and has a width that measures $\frac{2}{5}$ of the length. Find the perimeter of the rectangle.

12. ***Groceries:*** A loaf of bread weighs 21.6 ounces. Mauricio cut off a third of the loaf to save for later and then cut the remaining portion into 16 equal slices. What was the weight of each slice of the 16 slices he cut?

Writing & Thinking

13. Describe the process used to change a terminating decimal number to a fraction.

14. List 2 different ways to solve this problem: $\frac{1}{2}+3.67-\frac{1}{8}$. State which method you prefer and why.

4.7 Solving Equations with Decimal Numbers

Solving Equations that Simplify to the Form $ax + b = cx + d$

1. Simplify by removing any ____________________

2. Use the **addition principle** and add the ____________________
____

3. Use the **multiplication** (or **division**) **principle** to multiply both sides by the ____________________
____

4. The coefficient of the ____________________

PROCEDURE

Watch and Work

Watch the video for Example 1 in the software and follow along in the space provided.

Example 1 Solve Equations of the Form $ax + b = c$

Solve the equation: $5(x+3.7)=27.05$

Solution

Check

Now You Try It!

Use the space provided to work out the solution to the next example.

Example A Solve Equations of the Form *ax* + *b* = *c*

Solve the equation: $3(1.7y+3.6)=12.33$

4.7 Exercises

Concept Check

True/False. Determine whether each statement is true or false. If a statement is false, explain how it can be changed so the statement will be true. (**Note:** There may be more than one acceptable change.)

1. When an equation is solved, the variable should always be on the left side.

2. In solving an equation, do not divide both sides of the equation by a negative number.

Practice

Solve each equation.

3. $0.2(y+6.3)=17.5$

4. $\frac{1}{10}y-4.56=27.8$

5. $1.3-0.4z+z=-1.2$

6. $8x-5+9x=-x+12-8$

Applications

Solve.

7. ***Video Games:*** The cost of a video game and game guide was \$72.75. If the game guide was \$38.15 less than the price of the game, what was the price of each item?

8. ***Clothes:*** A pair of tennis shoes and a shirt cost a total of \$53.74. If the price of the tennis shoes was \$20.50 more than the price of the shirt, what was the price of each item?

Writing & Thinking

9. Why is checking your answer in the original equation especially important when working with decimal numbers?

Chapter 4 Project

What Would You Weigh on the Moon?

An activity to demonstrate the use of decimal numbers in real life.

The table below contains the surface gravity of each of the planets in the same solar system as the Earth, as wells Earth's moon, and the sun. The acceleration due to gravity g at the surface of a planet is given by the formula

$$g = \frac{GM}{R^2},$$

where M is the mass of the planet, R is its radius, and G is the gravitational constant. From the formula you can see that a planet with a larger mass M will have a greater value for surface gravity. Also the larger the radius R of the planet, the smaller the surface gravity.

If you look at different sources, you may find that surface gravity varies slightly from one source to another due to different values for the radius of some planets, especially the gas giants: Jupiter, Saturn, Uranus, and Neptune.

Planet	Surface Gravity (m/s^2)	Relative Surface Gravity	Fractional Equivalent
Earth	9.78	1.00	
Jupiter	23.10	2.36	
Mars	3.72		
Mercury	3.78		
Moon	1.62		
Neptune	11.15		
Saturn	9.05		
Sun	274.00		
Uranus	8.69		
Venus	9.07		

1. Compare the surface gravity of each planet or celestial body to the surface gravity of the Earth by dividing each planet's surface gravity by that of the Earth's, as listed in the table above. (This is referred to as **relative surface gravity**.) Round your answer to the nearest hundredth and place your results in the third column of the table. The values for Earth and Jupiter have been done for you. (**Note:** Comparing Earth to itself results in a value of 1.)

2. For Jupiter, the relative surface gravity value of 2.36 means that the gravity on Jupiter is 2.36 times that of Earth, therefore your weight on Jupiter would be approximately 2.36 times your weight on Earth (Although mass is a constant and doesn't change regardless of what planet you are on, your weight depends on the pull of gravity). Explain what the relative surface gravity value means for Mars.

3. Calculate your weight on the Moon by taking your present weight (in kg or pounds) and multiplying it by the Moon's relative surface gravity.

4. Approximately how many times larger is the surface gravity of the Sun compared to that of Mars? Round to the nearest whole number.

5. Convert each value in column three to a mixed number and place the result in column four. Be sure to reduce all fractions to lowest terms.

   a. Which is larger, the relative surface gravity of Mercury or $\frac{2}{5}$?

   b. Which is smaller, the relative surface gravity of the Moon or $\frac{4}{25}$?

   c. Write the fractional equivalent of Jupiter's relative surface gravity as an improper fraction in lowest terms.

CHAPTER 5

Percents

Math @ Work

Homeowners often need to purchase home-improvement supplies such as paint, fertilizer, and grass seed for repairs and maintenance. In many cases, a homeowner must purchase supplies in quantities greater than what is needed for a particular job because the manufacturer's or distributor's packaging sizes do not fit the homeowner's exact needs. When this happens, the store will typically not sell part of a can of paint or a part of a bag of fertilizer. Fortunately, a little application of mathematics can help stretch any homeowner's dollar by minimizing any excess amounts that must be bought.

Suppose that during the late spring, you decide to treat your lawn with a fertilizer and weed killer combination. One bag contains 18 pounds with a recommended coverage of 5000 square feet. If your lawn is in the shape of a rectangle that is 150 feet long by 220 feet wide, how many pounds of the fertilizer and weed killer combination do you need to cover the lawn? How many bags will you need to purchase?

5.1 Basics of Percent

1. The word percent comes from the Latin *per centum*, meaning ____________________. So **percent means** ____________________, or **the ratio of a number to** ______________.

2. The symbol % is called the ____________________. This sign has the same meaning as the fraction $\frac{1}{100}$.

To Change a Decimal Number to a Percent

1. Move the ____
2. Write the ____________________

PROCEDURE

To Change a Percent to a Decimal Number

1. Move the ____
2. Delete the ____________________

PROCEDURE

To Change a Fraction to a Percent

1. Change the fraction to ____________________
2. Change the decimal number to ____________________

PROCEDURE

Watch and Work

Watch the video for Example 6 in the software and follow along in the space provided.

Example 6 Changing Mixed Numbers to Percents

Change $2\frac{1}{4}$ to a percent.

Solution

Now You Try It!

Use the space provided to work out the solution to the next example.

Example A Changing Percents to Decimal Numbers

Change $1\frac{1}{2}$ to a percent.

To Change a Percent to a Fraction or a Mixed Number

1. Write the percent as a fraction with ______________________
2. ______________________

PROCEDURE

5.1 Exercises

Concept Check

True/False. Determine whether each statement is true or false. If a statement is false, explain how it can be changed so the statement will be true. (**Note:** There may be more than one acceptable change.)

1. It is not possible to have a percent greater than 100%.

2. A decimal number that is between 0.01 and 0.10 is between 10% and 100%.

3. To change from a percent to a decimal, simply omit the percent sign.

4. Fractions that have denominators other than 100 cannot be changed to a percent.

Practice

5. Write $\frac{20}{100}$ as a percent.

6. Write 1.12 as a percent.

7. Write 60% as a decimal number.

8. Write $\frac{4}{5}$ as a percent.

9. Write 150% as a fraction or mixed number and reduce, if possible.

Applications

Solve.

10. ***Interest:*** A savings account is offering an interest rate of 0.04 for the first year after opening the account. Change 0.04 to a percent.

11. ***Sales Tax:*** Suppose that sales tax is figured at 7.25%. Change 7.25% to a decimal.

Writing & Thinking

12. Describe a situation where more than 100% is possible. Describe a situation where it is impossible to have more than 100%.

13. Justify why mixed numbers are a larger percentage than proper fractions alone. (Consider the value of 100%.)

5.2 Solving Percent Problems Using Proportions

The Percent Proportion $\frac{P}{100}=\frac{A}{B}$

For the proportion $\frac{P}{100}=\frac{A}{B}$

$P\% =$ ____________ (written ____________________).

$B =$ __________ (number that ______________________________).

$A =$ ____________ (a part of ________________).

FORMULA

Three Basic Types of Percent Problems and the Proportion $\frac{P}{100}=\frac{A}{B}$

Type 1: Find the amount given the ________________________

What is ________________

For example, what is ____________________________

Type 2: Find the base given the __________________________

$P\%$ of what __________________

For example ____

Type 3: Find the percent given the __________________________

What percent of ______________

For example, what __________________________________

FORMULA

Watch and Work

Watch the video for Example 1 in the software and follow along in the space provided.

Example 1 Finding the Amount

What is 65% of 500?

Solution

Now You Try It!

Use the space provided to work out the solution to the next example.

Example A Finding the Amount

What is 15% of 80?

5.2 Exercises

Concept Check

True/False. Determine whether each statement is true or false. If a statement is false, explain how it can be changed so the statement will be true. (**Note:** There may be more than one acceptable change.)

1. Percent problems can be solved with a proportion if two of the three parts P, A, and B are known.

2. In the proportion $\frac{P}{100} = \frac{65}{200}$, the base is 65.

3. In the problem "What is 26% of 720?" the missing number is the base.

4. Because the base represents the whole, it is always larger than the amount.

Practice

Use the proportion $\frac{P}{100} = \frac{A}{B}$ to find each unknown quantity. Round percents to the nearest tenth of a percent. All other answers should be rounded to the nearest hundredth, if necessary.

5. Find 15% of 50.

6. What is 85% of 60?

7. 25% of 60 is ______.

8. What percent of 48 is 12?

9. ____% of 56 is 140.

Applications

Solve.

10. ***Baseball Attendance:*** In 2016 the Los Angeles Dodgers led the major leagues in home attendance, drawing an average of 45,720 fans to their home games. This figure represented 81.64% of the capacity of Dodger Stadium. Estimate how many fans the stadium can hold (to the nearest ten) when it is filled to capacity.[1]

11. ***Real Estate:*** You want to purchase a new home for $122,000. The bank will loan you 80% of the purchase price. How much will the bank loan you? (This amount is called your mortgage and you will pay it off over several years with interest. For example, a 30-year loan will probably cost you a total of more than 3 times the original loan amount.)

Writing & Thinking

12. List the four parts of the proportion equation and give a brief definition of each one.

13. Can a mixed number be used in a proportion? Justify your answer.

1 Source: espn.go.com/mlb/attendance

Name: Date:

5.3 Solving Percent Problems Using Equations

Terms Related to the Basic Equation $R \cdot B = A$

For the basic equation $R \cdot B = A$,

$R =$ ______________ (as a ______________________________).

$B =$ __________ (number that we ______________________________).

$A =$ ____________ (a part of ______________).

DEFINITION

Three Basic Types of Percent Problems and the Formula $R \cdot B = A$

Type 1: Find the amount given the ______________________________

For example, what is ____________________

____

Type 2: Find the base given the ______________________________

For example, 42% of what ______________________

____

Type 3: Find the percent (rate) given the __________________________

For example, what __________________________

____

FORMULA

Watch and Work

Watch the video for Example 4 in the software and follow along in the space provided.

Example 4 Finding the Amount

Find 75% of 56.

Solution

Now You Try It!

Use the space provided to work out the solution to the next example.

Example A Finding the Amount

Find 150% of 60.

5.3 Exercises

Concept Check

True/False. Determine whether each statement is true or false. If a statement is false, explain how it can be changed so the statement will be true. (**Note:** There may be more than one acceptable change.)

1. In order to solve the equation $0.56 \cdot B = 12$ for the base B one would multiply 12 by 0.56.

2. In the problem "126% of 720 is what number?" the missing number is the amount.

3. The solution to the problem "50% of what number is 352?" could be found by solving the equation $50 \cdot B = 352$.

4. If the base is 120 and the rate is greater than 100%, then the amount will be greater than 120.

Practice

Use the equation $R \cdot B = A$ to find each unknown quantity. Round percents to the nearest tenth of a percent. All other answers should be rounded to the nearest hundredth, if necessary.

5. 10% of 70 is what number?

6. Find 75% of 12.

7. 150% of ____ is 63.

8. What percent of 75 is 15?

9. ____% of 30 is 6.

Applications

Solve.

10. ***Presidential Vetoes:*** During his presidency, from 1945 to 1953, Harry S. Truman vetoed 250 congressional bills, and 12 of those vetoes were overridden. What percent of Truman's vetoes were overridden?

11. ***Mortgages:*** The minimum down payment to obtain the best financing rate on a house is 20%. Assuming that John has set aside $35,000 and wants to take advantage of the best financing rate, what is the most expensive house he can purchase? [1]

Writing & Thinking

12. Explain the connection between the proportion $\frac{P}{100} = \frac{A}{B}$ and the equation $R \cdot B = A$.

13. Explain how to determine which number is the rate (percent), which one is the amount, and which one is the base.

1 Source: http://www.kiplinger.com/magazine/archives/what-it-takes-to-get-a-mortgage.html

5.4 Applications of Percent

Basic Steps for Solving Word Problems

1. ______________________

2. ______________________

3. ______________________

4. ______________________

PROCEDURE

Terms Related to Discount

Discount: difference between the ______________________

Sale price: original price minus ______________________

Rate of discount percent of ______________________

DEFINITION

Terms Related to Sales Tax

Sales tax: ______________________

Rate of sales tax: ______________________

DEFINITION

Watch and Work

Watch the video for Example 3 in the software and follow along in the space provided.

Example 3 Application: Solving Sales Tax Problems

If the rate of sales tax is 6%, what would be the final cost of a laptop priced at $899?

Solution

Now You Try It!

Use the space provided to work out the solution to the next example.

Example A Application: Solving Sales Tax Problems

Assuming a 7% sales tax rate, what would be the final cost of a discounted pair of shoes priced at $39?

1. A **commission** is a ______________________________.

2. At times, it is helpful to know by what percent the value changed. This is called finding the __________ __________ (or the ____________________).

Profit and Percent of Profit

Profit: the difference between ____________________

Profit = ____________________

Percent of Profit: There are two types; both are ______________________________

1. Percent of profit **based on** ______________________________

$$\frac{\textbf{Profit}}{\textbf{Cost}} = \% \text{ of profit based on } \_\_\_\_\_\_\_\_\_\_$$

2. Percent of profit **based on** ______________________________

$$\frac{\textbf{Profit}}{\textbf{Selling Price}} = \% \text{ of profit based on } \_\_\_\_\_\_\_\_$$

FORMULA

5.4 Exercises

Concept Check

True/False. Determine whether each statement is true or false. If a statement is false, explain how it can be changed so the statement will be true. (**Note:** There may be more than one acceptable change.)

1. If an item is selling for a 35% discount, the customer will pay 65% of the original price.

2. If you must pay 7% sales tax on a purchase, the total cost you will pay is 170% of the total before tax.

3. A car was purchased in 1965 for $3800. It sold for $1200 in 2011. This is an example of depreciation.

4. Profit is determined by subtracting selling price from the cost.

Applications

Solve.

5. ***Office Supplies:*** A new briefcase was priced at $275. If it were to be marked down 30%:

   a. What would be the amount of the discount?

   b. What would be the new price?

6. ***Sales Tax:*** If sales tax is figured at 7.25%, how much tax will be added to the total purchase price of three textbooks priced at $25.00, $35.00, and $52.00?

7. ***Real Estate:*** A realtor works on 6% commission. What is his commission on a house he sold for $195,000?

8. ***Electronics:*** The cost of a 20-inch television set to a store owner was $450, and she sold the set for $630.

   a. What was her profit?

   b. What was her percent of profit based on cost?

   c. What as her percent of profit based on selling price?

Writing & Thinking

9. Determine how to calculate sales tax when eating out and relate this process to either a proportion and/or using the amount/base/rate equation. Give an example.

5.5 Simple and Compound Interest

Simple Interest Formula

Interest = Principal · rate · time

Writing the formula using letters, we have $I = P \cdot r \cdot t$, where

I = ______________________________

P = ______________________________

r = ______________________ in decimal or fraction form, and

t = ______________________________

FORMULA

Watch and Work

Watch the video for Example 1 in the software and follow along in the space provided.

Example 1 Application: Calculating Simple Interest

You want to borrow $2000 from your bank for one year. If the interest rate is 5.5%, how much interest would you pay?

Solution

Now You Try It!

Use the space provided to work out the solution to the next example.

Example A Application: Calculating Simple Interest

If you were to borrow $1500 at 8.5% for one year, how much interest would you pay?

To Calculate Compound Interest

1. Use the formula ______________________________________

   Let $t = \frac{1}{n}$ where n ____
2. Add this interest to the ____
3. Repeat steps ____

PROCEDURE

Compound Interest Formula

When interest is compounded, the total **amount** A accumulated (including principal and interest) is given by the formula

$$A = P\left(1 + \frac{r}{n}\right)^{nt},$$

where

$P =$ ____

$r =$ ____

$t =$ _____________________

$n =$ ____

FORMULA

Total Interest Earned

To find the total interest earned on an investment that has earned interest by compounding, _____________ ______________________________________

$$I = \_\_\_\_\_\_\_\_\_\_$$

FORMULA

Inflation

The adjusted amount A due to **inflation** is

$$A = P(1 + r)^t,$$

where

$P =$ ______________________________,

$r =$ ____

$t =$ _____________________.

FORMULA

Depreciation

The current value V of an item due to **depreciation** is

$$V = P(1-r)^t,$$

where

$P =$ ____________________,

$r =$ ______________________________

$t =$ ______________.

FORMULA

5.5 Exercises

Concept Check

True/False. Determine whether each statement is true or false. If a statement is false, explain how it can be changed so the statement will be true. (**Note:** There may be more than one acceptable change.)

1. In the simple interest formula, the rate can be written as a decimal number or a fraction.

2. Simple interest can be compounded monthly or quarterly.

3. Interest cannot be earned on interest, only the principal.

4. Inflation can be treated in the same manner as simple interest.

Applications

Solve. Round to the nearest cent, if necessary.

5. ***Loans:*** How much interest would be paid on a loan of $3000 at 5% for 9 months?

Solve each problem by repeatedly using the formula for calculating simple interest. Round your answer to the nearest cent, if necessary.

6. ***Loans:*** You loan your cousin \$2000 at 5% compounded annually for 3 years. How much interest will your cousin owe you?

   a. First year: $I = 2000 \cdot 0.05 \cdot 1 =$ ________

   b. Second year: $I =$ ________ $\cdot\ 0.05 \cdot 1 =$ ________

   c. Third year: $I =$ ________ $\cdot\ 0.05 \cdot 1 =$ ________

   d. The total interest is ________.

Solve each problem by using the compound interest formula. Round your answer to the nearest cent, if necessary.

7. ***Finance:*** You deposit \$1500 at 4% to be compounded semiannually. How much interest will you earn in 3 years?

Solve. Round your answer to the nearest cent, if necessary.

8. ***Truck:*** Stan bought a new truck last year for \$29,900. This year, he decided that he wants to trade it in for a smaller car. He can resell the truck for 26,500. What was the rate of depreciation for the year?

Writing & Thinking

9. List the four parts involved in the simple interest formula. In your own words, define each one.

10. Compare and contrast simple interest with compound interest.

5.6 Reading Graphs

Four Types of Graphs and Their Purposes

________________	To emphasize comparative amounts
__________________	To help in understanding percents or parts of a whole (__________ __________ are also called pie charts.)
_________________	To indicate tendencies or trends over a period of time
________________	To indicate data in classes (a range or interval of numbers)

DEFINITION

Properties of Graphs

Every graph should:

1. ____
2. ____
3. ____

PROPERTIES

Watch and Work

Watch the video for Example 2 in the software and follow along in the space provided.

Example 2 Reading a Circle Graph

Examine the circle graph. This graph shows the percent of a household's annual income they plan to budget for various expenses. Suppose the household has an annual income of $45,000. Use the information in the graph to calculate how much money will be budgeted for each expense.

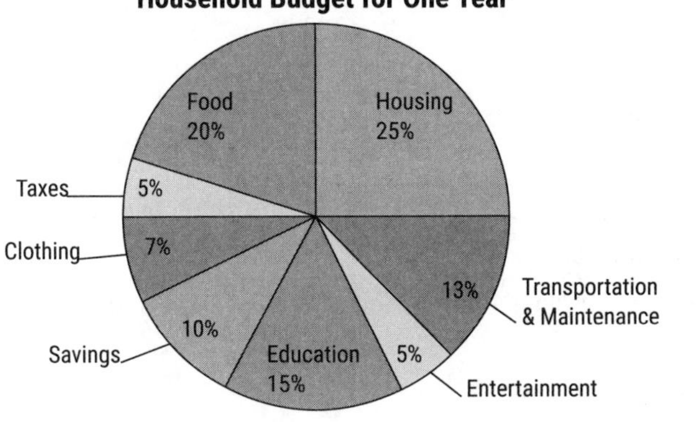

Solution

Now You Try It!

Use the space provided to work out the solution to the next example.

Example A Reading a Circle Graph

Using the circle graph in Example 2, how much will the family spend on each of the following expenses if the family income increases to $55,000 and the percent budgeted for each expense does not change?

a. Housing

b. Savings

c. Clothing

d. Food

Terms Related to Histograms

Term	Definition
____________	A range (or interval) of numbers that contains data items.
____________________	The smallest whole number that belongs to a class.
____________________	The largest whole number that belongs to a class.
____________________	Numbers that are halfway between the upper limit of one class and the lower limit of the next class.
________________	The difference between the class boundaries of a class (the width of each bar).
_______________	The number of data items in a class.

DEFINITION

5.6 Exercises

Concept Check

True/False. Determine whether each statement is true or false. If a statement is false, explain how it can be changed so the statement will be true. (**Note:** There may be more than one acceptable change.)

1. Graphs should always be clearly labeled, easy to read, and have appropriate titles.

2. Circle graphs show trends over a period of time.

3. The frequency is the number of data items in a class.

4. Numbers that are halfway between the upper limit of one class and the lower limit of the next class are the class boundaries.

Applications

Answer the questions using the given graphs.

5. ***Education:*** The following bar graph shows the number of students in five fields of study at a university.

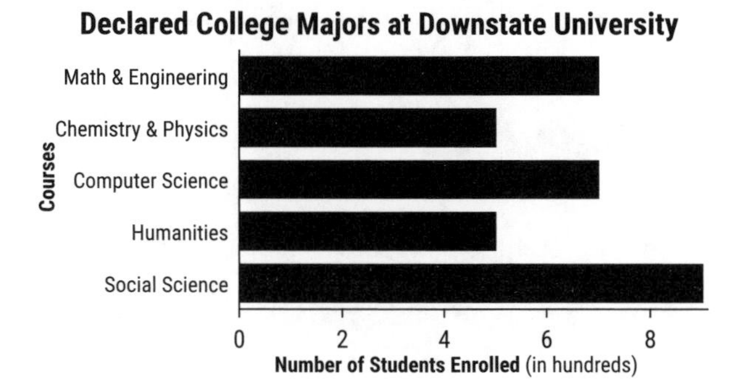

a. Which field(s) of study has the largest number of declared majors?

b. Which field(s) of study has the smallest number of declared majors?

c. How many declared majors are indicated in the entire graph?

d. What percent are computer science majors? Round your answer to the nearest tenth of a percent.

6. ***Budgeting:*** The following circle graph represents the various areas of spending for a school with a total budget of $34,500,000.

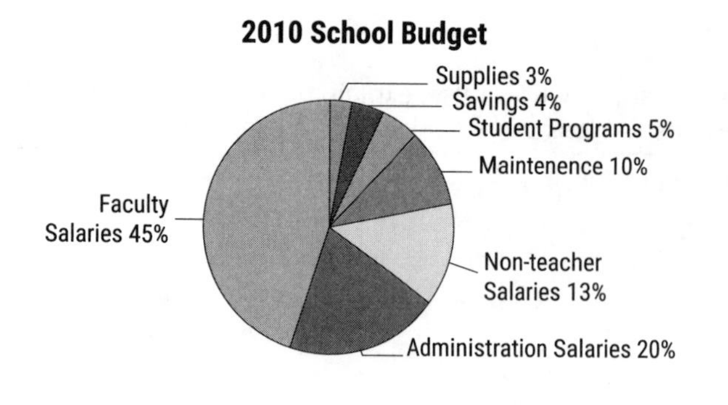

a. What amount will be allocated to each category?

b. What percent will be for expenditures other than salaries?

c. How much will be spent on maintenance and supplies?

d. How much more will be spent on teachers' salaries than on administration salaries?

7. ***Weather:*** The following line graph shows the total monthly rainfall in Tampa, Florida for the first 5 months of 2010.[1]

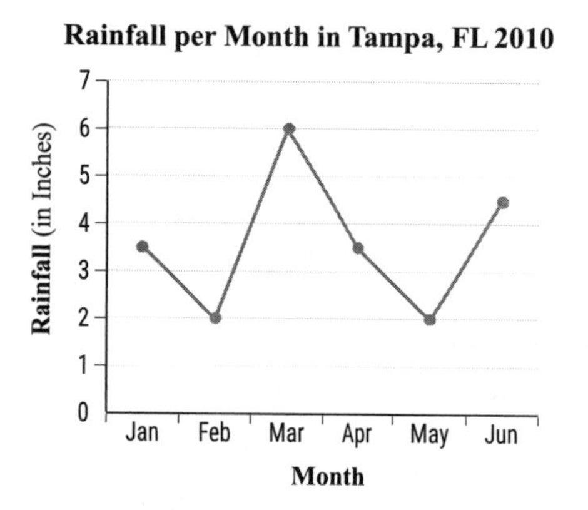

a. Which months had the least rainfall?

b. What was the most rainfall in a month?

c. What month had the most rainfall?

d. What was the mean rainfall over the six-month period (to the nearest hundredth)?

1 Source: weather.gov

8. ***Tires:*** The following histogram summarizes the tread life for 100 types of new tires.

a. How many classes are represented?

b. What is the width of each class?

c. Which class has the highest frequency?

d.

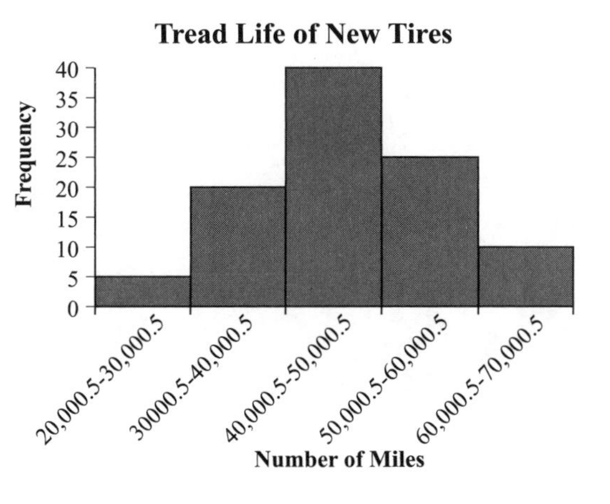

Writing & Thinking

9. State three properties or characteristics that should be true of all graphs so that they can communicate numerical data quickly and easily.

10. Compare and contrast a bar graph and a histogram.

Chapter 5 Project

Take Me Out to the Ball Game!

An activity to demonstrate the use of percents and percent increase or decrease in real life.

The Atlanta Braves baseball team has been one of the most popular baseball teams for fans, not only from Georgia, but throughout the Carolinas and the southeastern United States. The Braves franchise started playing at the Atlanta-Fulton County Stadium in 1966 and this continued to be their home field for 30 years. In 1996, the Centennial Olympic Stadium that was built for the 1996 Summer Olympics was converted to a new ballpark for the Atlanta Braves. The ballpark was named Turner Field and was opened for play in 1997. In 2017, the Braves moved to a new stadium named SunTrust Park.

Round all percents to the nearest whole percent.

1. The Atlanta-Fulton County Stadium had a seating capacity of 52,769 fans. Turner Field had a seating capacity of 50,096. SunTrust Park has a seating capacity of 41,149.
   - **a.** Determine the decrease in seating capacity between Turner Field and the original Braves stadium.
   - **b.** Determine the percent decrease in seating capacity between SunTrust Park and Turner field.
2. The Centennial Olympic Stadium had approximately 85,000 seats. Some of the seating was removed in order to convert it to the Turner Field ballpark. Rounding the number of seats in Turner Field to the nearest thousand, what is the approximate percent decrease in seating capacity from the original Olympic stadium?
3. When Turner Field opened in 1997, the average attendance at an Atlanta Braves game was 42,771. In 2016 the average attendance was 24,950. What is the percent decrease in attendance from 1997 to 2016? [1]
4. The highest average attendance for the Braves was 47,960 in 1993 at the Atlanta-Fulton County Stadium. The lowest average attendance was 6642 in 1975 at the Atlanta-Fulton County Stadium. What is the percent increase from the lowest attendance to the highest? [2]
5. Chipper Jones, a popular Braves third baseman, retired in July 2013. He started his career with the Braves in 1993 at the age of 21. [3]
   - **a.** In 2001, Chipper had 189 hits in 572 at-bats. Calculate Chipper's batting average for the season by dividing the number of hits by the number of at-bats. Round to the nearest thousandth.
   - **b.** In 2008, Chipper had 160 hits in 439 at-bats. Calculate Chipper's batting average for the season by dividing the number of hits by the number of at-bats. Round to the nearest thousandth.
   - **c.** Calculate the percent change in Chipper's batting average from 2001 to 2008.
   - **d.** Does this represent a percent increase or decrease?
6. In 2001, Chipper had 102 RBIs (runs batted in). In 2008, Chipper had only 75 RBIs.
   - **a.** Calculate the percent change in RBIs from 2001 to 2008.
   - **b.** Does this represent a percent increase or decrease?

[1] Source: baseball-almanac.com
[2] Source: baseball-almanac.com
[3] Source: espn.go.com

CHAPTER 6

Measurement and Geometry

Math @ Work

Geometric figures and their properties are a part of our daily lives. Geometry is an integral part of art concepts such as vanishing points, which are used to create three-dimensional perspectives on a two-dimensional canvas. Architects and engineers use geometric shapes when designing buildings and other infrastructures to optimize beauty and structural strength. Geometric patterns are also evident throughout nature. Examples of these patterns can be found in the development of snowflakes, crystal formations, leaves, and pine cones as well as in the camouflage of snakes, fish, and tigers.

The manhole covers in the streets are an example of the use of geometric concepts in the human-created world. You may have wondered why these covers are typically circular in shape and not squares or rectangles. One reason is that a circle cannot "fall through" a slightly smaller circle, whereas a square or rectangle can be rotated so that either shape can fall through a slightly smaller version of itself. As a result, circular manhole covers are considered to be safer than other shapes because they avoid the risk of the cover accidentally falling down into the hole.

To understand this idea, draw a square, a rectangle, a triangle, and a hexagon (a six-sided figure) on a piece of paper. Cut out these figures and notice how easily they can be made to pass through the holes left from the corresponding cutouts. Next follow the same procedure with a circle. You'll notice that you cannot fit the circle through the hole left from the cutout. What properties of the shapes prevents the circle from falling through but allows the other shapes to fall through?

6.1 US Measurements

Using Multiplication and Division to Convert Measurements

1. Multiply to convert to ____________________ (There will be ____________________.)
2. Divide to convert to ____________________. (There will be ____________________.)

PROCEDURE

Using Unit Fractions to Convert Measurements

1. The numerator should be in the ______________________________
2. The denominator should be in the ______________________________

PROCEDURE

Watch and Work

Watch the video for Example 6 in the software and follow along in the space provided.

Example 6 Application: Converting US Units of Measure

Determine how many seconds are in a 5-day work week assuming an 8 hr work day.

Solution

Now You Try It!

Use the space provided to work out the solution to the next example.

Example A Application: Finding the Mean

How many fluid ounces are in 8 gallons of apple juice?

6.1 Exercises

Concept Check

True/False. Determine whether each statement is true or false. If a statement is false, explain how it can be changed so the statement will be true. (**Note:** There may be more than one acceptable change.)

1. Capacity can be measured using ounces, quarts, and gallons.

2. One mile is equivalent to 2000 feet.

3. To convert from smaller units to larger units, division will be required.

4. Multiplication by a unit fraction does not change the value of the expressions being converted.

Practice

Convert each measurement.

5. 4 pt = ____ c

6. 10 mi = ____ ft

7. 39 ft = ____ yd

8. 150 min = ____ hr

Applications

Solve.

9. ***Interior Decorating:*** Sheer fabric costs $7.99 per yard. If it will take 35 feet of fabric to make drapes for the entire house, how much must you spend on fabric for the drapes, to the nearest cent?

10. ***Publishing:*** The author of this textbook spent 1 year, 23 weeks, 5 days, and 14 hours writing it. How many seconds is this? (**Hint:** There are 52 weeks in a year.)

Writing & Thinking

11. Colby needs to find out how many yards are in one mile. What two sets of equivalent units would he need to make that determination?

12. In your own words, explain when you would multiply and when you would divide when converting between units.

6.2 The Metric System: Length and Area

Writing Metric Units of Measure

In the metric system,

1. A 0 is written to the left of the decimal point if ____________________

   For example, ____________

2. No commas are used in writing numbers. If a number has more than four digits (to the left or right of the decimal point), the digits are ____________________

   ____

   For example, ____________________

PROCEDURE

1. There are two basic methods of converting units of measurement in the metric system:

   1. multiplying by____________,
   2. moving the ____________________

Using Unit Fractions to Convert Measures

1. The numerator should be in the ____________________
2. The denominator should be in the ____________________

PROCEDURE

Watch and Work

Watch the video for Example 6 in the software and follow along in the space provided.

Example 6 Converting Metric Units of Area

Convert each measurement using unit fractions.

a. $5\text{ cm}^2 =$ _____ mm^2

b. $4600\text{ mm}^2 =$ _____ m^2

Solution

Now You Try It!

Use the space provided to work out the solution to the next example.

Example A Converting Metric Units of Area

Convert each measurement using unit fractions.

a. $86 \text{ m}^2 = \_\_\_\_\_ \text{ cm}^2$

b. $0.06 \text{ mm}^2 = \_\_\_\_\_ \text{ dm}^2$

6.2 Exercises

Concept Check

True/False. Determine whether each statement is true or false. If a statement is false, explain how it can be changed so the statement will be true. (**Note:** There may be more than one acceptable change.)

1. To change from smaller units to larger units, multiplication must be used.

2. Units of length in the metric system are named by putting a prefix in front of the basic unit meter, for example, centimeter.

3. In metric units, a square that is 1 centimeter long on each side is said to have an area of 1 centimeter.

Practice

Convert each measurement.

4. 3 m = ___ cm

5. 19.77 m = ___km

6. 6 500 000 hertz = ___ megahertz

7. 13 dm^2 = __________ cm^2 = __________ mm^2

Applications

Solve.

8. ***Geometry:*** A triangle has a base measuring 4 cm and a height measuring 16 mm. Determine the area of the triangle in cm^2.

9. ***Transportation:*** A section of railroad track measuring 2.1 km in length needs to be replaced. Each railroad tie is 4 decimeters wide and they are to be spaced 0.8 m apart. How many railroad ties will be needed to complete this section of track?

Writing & Thinking

10. Compare and contrast ease of converting units in the US customary system and the metric system.

11. Discuss the meaning of prefixes like milli-, centi-, and kilo- in metric units. Give examples.

6.3 The Metric System: Capacity and Weight

1. In the metric system, capacity (liquid volume) is measured in ____________ (abbreviated ________).
2. A liter is the volume enclosed in a cube that is ______________ on each edge.
3. Mass is ____________________________ in an object.
4. The basic unit of mass in the metric system is the ________________,

Watch and Work

Watch the video for Example 7 in the software and follow along in the space provided.

Example 7 Converting Metric Units of Weight

Convert 34 g to milligrams **a.** using a unit fraction and **b.** using a metric conversion line.

Solution

Now You Try It!

Use the space provided to work out the solution to the next example.

Example A Application: Finding the Mean

Convert 14.9 kg to grams using a unit fraction or a metric conversion line.

6.3 Exercises

Concept Check

True/False. Determine whether each statement is true or false. If a statement is false, explain how it can be changed so the statement will be true. (**Note:** There may be more than one acceptable change.)

1. One milliliter is equivalent to one cubic centimeter.

2. Volume is measured in square units.

3. In 1 liter there are 100 milliliters.

4. A metric ton and a US customary ton are equal (a metric ton weighs about 2000 US pounds).

5. A dekagram contains 10 grams.

ractice

Convert each measurement.

6. 2 L = ___ mL

7. 6.3 kL = ___ L

8. 2 g = ___ mg

9. 2000 g = ___ kg

Applications

Solve.

10. ***Medicine:*** How many 5-mL doses of liquid medication can be given from a vial containing 3 deciliters?

11. ***Cooking:*** One cup of flour is approximately 120 grams. How many cups of flour can you get out of a bag of flour weighing 2.4 kg?

Writing & Thinking

12. In the metric system, the common unit of capacity is the liter. Discuss how you would change from a measure of liters to milliliters.

6.4 US and Metric Equivalents

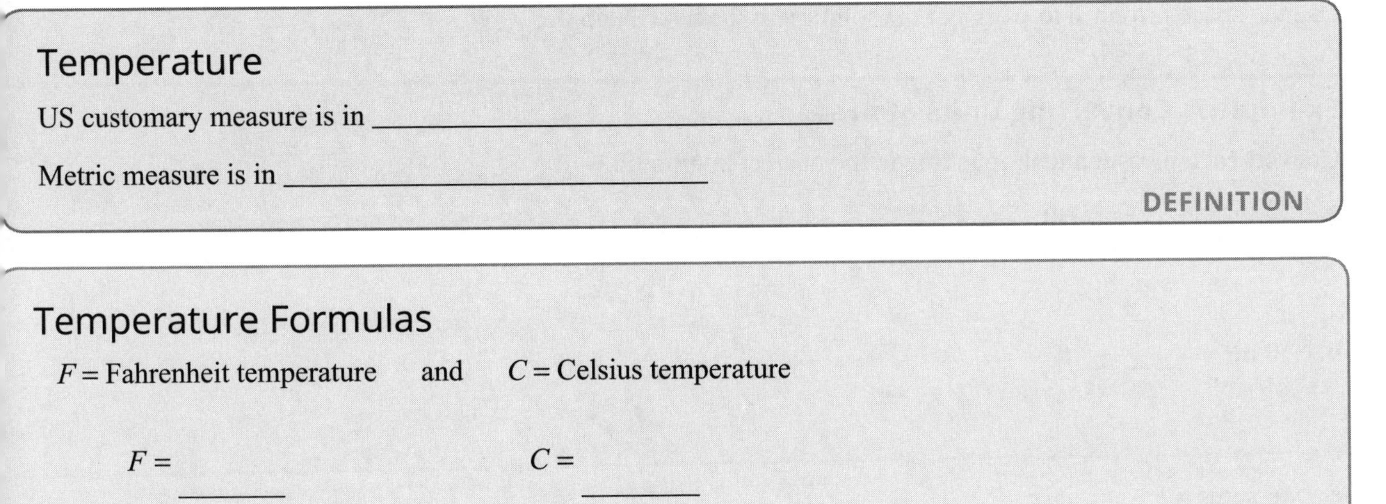

Temperature

US customary measure is in ____________________

Metric measure is in ____________________

DEFINITION

Temperature Formulas

F = Fahrenheit temperature and C = Celsius temperature

$F =$ ______ $C =$ ______

FORMULA

Watch and Work

Watch the video for Example 5 in the software and follow along in the space provided.

Example 5 Converting Units of Area

Convert each measurement, rounding to the nearest hundredth.

a. $40 \text{ yd}^2 =$ ______ m^2

b. $100 \text{ cm}^2 =$ ______ in.^2

c. 6 acres = ______ ha

d. 5 ha = ______ acres

Solution

Now You Try It!

Use the space provided to work out the solution to the next example.

Example A Converting Units of Area

Convert each measurement, rounding to the nearest hundredth.

a. 53 in.^{2} = _______ cm^2

b. 50 m^2 = _______ ft^2

c. 16 acres = _______ ha

d. 3 ha = _______ acres

6.4 Exercises

Concept Check

True/False. Determine whether each statement is true or false. If a statement is false, explain how it can be changed so the statement will be true. (**Note:** There may be more than one acceptable change.)

1. Water freezes at 32 degrees Celsius.

2. When converting between US customary and metric units, often the results will be approximations.

3. A 5K (km) run is longer than a 5 mile run.

4. One square meter covers more area than one square yard.

Practice

Convert each measurement. Round to the nearest hundredth if necessary.

5. 25 °C = ___ °F

6. 9 ft = ___ m

7. 3 in.^{2} = ________ cm^2

8. 4 qt = ___ L

9. 33 kg = ___ lb

Applications

Solve.

10. ***Baking:*** While visiting her aunt in Germany, Helga wants to surprise her aunt with a cake. She brought her mom's cake recipe with her from Georgia. The recipe says to bake the cake at 350 degrees Fahrenheit but the temperature gauge on her aunt's oven is in degrees Celsius. To what temperature should Helga set her aunt's oven in order to bake the cake at the correct temperature? Round the temperature to the nearest degree.

11. ***Sports:*** The Ironman Triathlon championship in Hawaii consists of a swim of 3.86 km, a bike ride of 180.25 km, and finishes with a run equal to the length of a standard marathon. A marathon is typically 26.2 miles. What is the total length of the Ironman Triathlon in kilometers? Round the length to the nearest tenth of a km.

Writing & Thinking

12. Most conversions between the US customary system of measure and metric system are not exact. Explain why this is true and give any exceptions.

6.5 Angles and Triangles

Point, Line, Plane

Undefined Term	Representation	Discussion
Point	______________	A point is represented by ______________ Points are labeled with ______________
Line	______________	A line has no ______________ Lines are labeled with ______________ ______________
Plane	______________	Flat surfaces, such as a table top or wall, represent ______________ ______________ Planes are labeled with ______________

DEFINITION

Ray and Angle

Term	Definition	Illustrations with Notation
Ray	A ray consists of ______________ ______________ ______________ ______________ ______________	______________
Angle	An angle consists of ______________ ______________ ______________ ______________ ______________ ______________	______________

DEFINITION

Labeling Angles

There are three common ways of labeling angles:

A
O
B
1
C

A. ∠_____
Using three __________
____
____

B. ∠___
Using single
____
____

C. ∠___
Using the single
____
____
____

PROCEDURE

1. The base unit when measuring angles is ____________________ (symbolized __________).

Angles Classified by Measure

Name	Measure	Illustrations with Notation
Acute	__________	__________
Right	__________	__________
Obtuse	__________	__________
Straight	__________	__________

DEFINITION

Complementary and Supplementary Angles

1. Two angles are **complementary** if ______________________________

2. Two angles are **supplementary** if ______________________________

DEFINITION

2. If two angles have the same measure, they are said to be ____________________ (symbolized as ≅).

Vertical Angles

Vertical angles ____________________

That is, vertical angles have ______________________

DEFINITION

Adjacent Angles

Two angles are adjacent if ___________________________

DEFINITION

Parallel Lines and Perpendicular Lines

Term	Definition	Illustrations with Notation
Parallel Lines	Two lines are parallel (symbolized ⊠) if ____________________ ____________________ ____________________	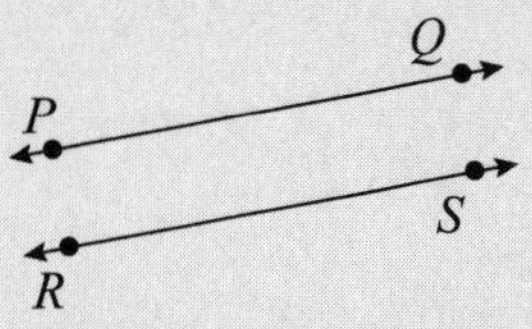 $\overleftrightarrow{PQ}$ is parallel to ($\overleftrightarrow{PQ}$ ________)
Perpendicular Lines	Two lines are perpendicular (symbolized ⊥) if ________ ____________________ ____________________	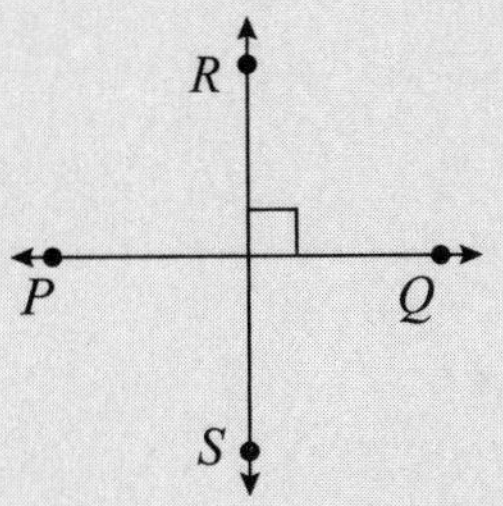 $\overleftrightarrow{PQ}$ is perpendicular to ($\overleftrightarrow{PQ}$ ________)

DEFINITION

Parallel Lines and a Transversal

If two parallel lines are cut by a transversal, then the following two statements are true.

1. ______________________________

2. ______________________________

PROCEDURE

Watch and Work

Watch the video for Example 8 in the software and follow along in the space provided.

Example 8 Calculating Measures of Angles

In the figure below, lines k and l are parallel, t is a transversal, and $m\angle 1 = 50°$. Find the measures of the other 7 angles.

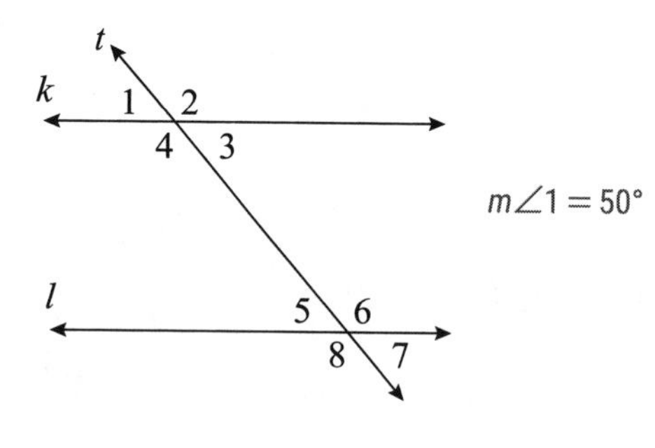

Solution

Now You Try It!

Use the space provided to work out the solution to the next example.

Example A Calculating Measures of Angles

In the figure below, lines l and m are parallel, t is a transversal, and $m\angle 2 = 80°$. Find $m\angle 4$, $m\angle 5$, and $m\angle 6$.

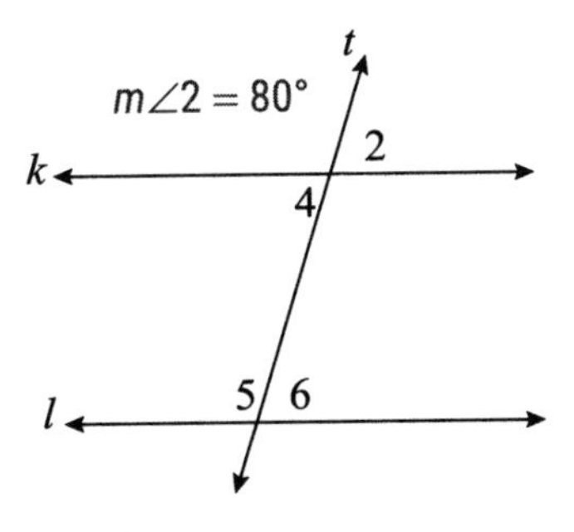

3. A **line segment** consists of ____________________________________

4. A **triangle** consists of ____________________________________.

Triangles Classified by Sides

(**Note:** In the figures, sides with equal length are indicated by the same number of tic marks.)

Name	Property	Example
Scalene	________ ________ ________	ΔABC is scalene since ________ ____________ ____________
Isosceles	________ ________ ________	ΔPQR is isosceles since ________ ______
Equilateral	________ ________ ________	ΔXYZ is equilateral since ________ ________

DEFINITION

Triangles Classified by Angles

Name	Property	Example	
Acute	____________ ____________ ____________	____________	ΔABC is acute since ____________ ____________
Right	____________ ____________	____________	ΔPRQ is a right triangle since ____________
Obtuse	____________ ____________	____________	ΔXYZ is an obtuse triangle since ____________

DEFINITION

Three Properties of Triangles

In a triangle:

1. The sum of the measures ____________________

2. The sum of the lengths of ____________________

3. Longer sides are ____________________

PROPERTIES

6.5 Exercises

Concept Check

True/False. Determine whether each statement is true or false. If a statement is false, explain how it can be changed so the statement will be true. (**Note:** There may be more than one acceptable change.)

1. The sum of the measures of two complementary angles is equal to the measure of one right angle.

2. The sum of the measures of complementary angles is greater than the sum of the measures of supplementary angles.

3. Adjacent angles are two angles that share a side.

4. If two lines in a plane are not parallel, then they are perpendicular.

5. A triangle with sides of 4 inches, 4 inches, and 3 inches is an isosceles triangle.

6. A triangle with three angles that each measure less than 90 degrees is an acute triangle.

Practice

7. Name the type of angle formed by the hands on a clock.

a. at six o'clock

b. at three o'clock

c. at one o'clock

d. at five o'clock

8. Assume that $\angle 1$ and $\angle 2$ are complementary.

a. If $m\angle 1 = 15°$, what is $m\angle 2$?

b. If $m\angle 1 = 3°$, what is $m\angle 2$?

c. If $m\angle 1 = 45°$, what is $m\angle 2$?

d. If $m\angle 1 = 75°$, what is $m\angle 2$?

9. The figure shows two intersecting lines.

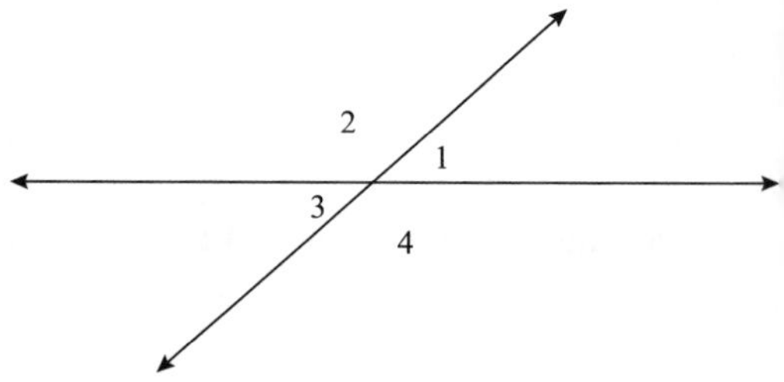

a. If $m\angle 1 = 30°$, what is $m\angle 2$?

b. Is $m\angle 3 = 30°$? Give a reason for your answer other than the fact that $\angle 1$ and $\angle 3$ are vertical angles.

c. Name two pairs of congruent angles.

d. Name four pairs of adjacent angles.

Classify each triangle in the most precise way possible, given the indicated lengths of its sides and/or measures of its angles.

10.

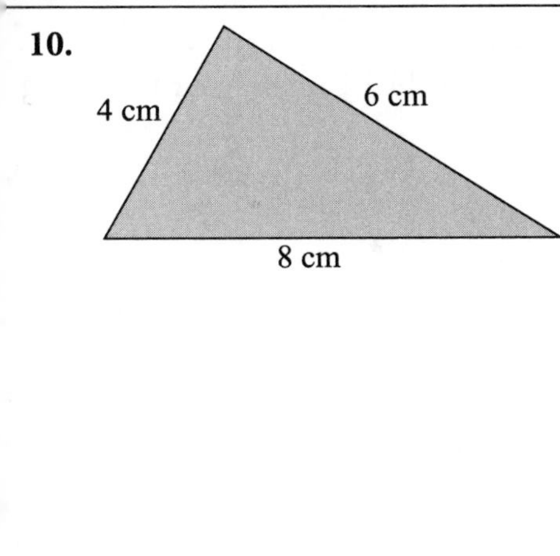

11.

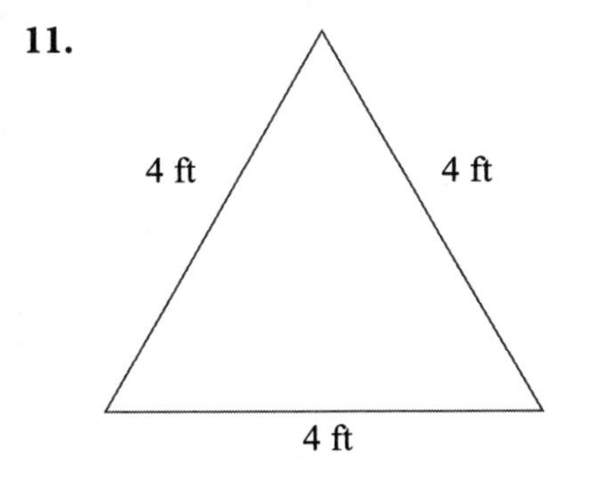

Applications

Solve.

12. Suppose the lengths of the sides of $\triangle DEF$ are as shown in the figure. Is this possible? Explain your reasoning.

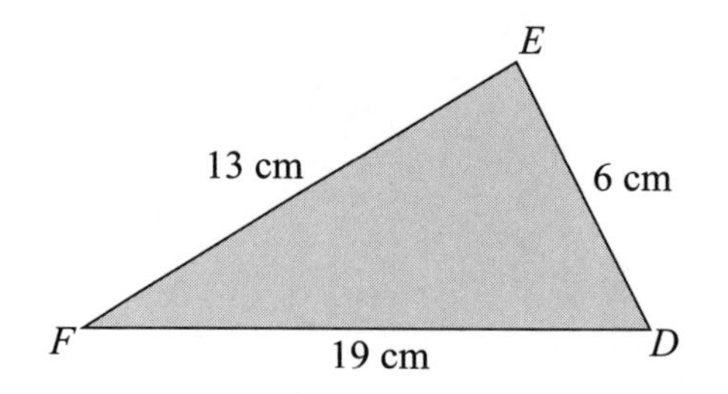

13. In the triangle shown, $m\angle X = 30°$ and $m\angle Y = 70°$.

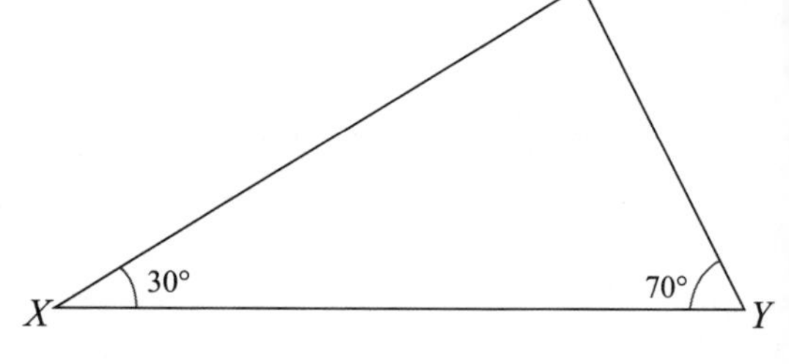

a. What is $m\angle Z$?

b. What kind of triangle is $\triangle XYZ$?

c. Which side is opposite $\angle X$?

d. Which sides include $\angle X$?

e. Is $\triangle XYZ$ a right triangle?

Writing & Thinking

14. Explain, in your own words, the relationships between vertex, ray, angle, and line.

6.6 Perimeter

Polygon

A **polygon** is a closed plane figure, with ______________________________

Each point where two sides meet is ______________

Note: A **closed figure** begins and ends at the same point.

DEFINITION

Perimeter

The **perimeter** P of a polygon is the ______________________

DEFINITION

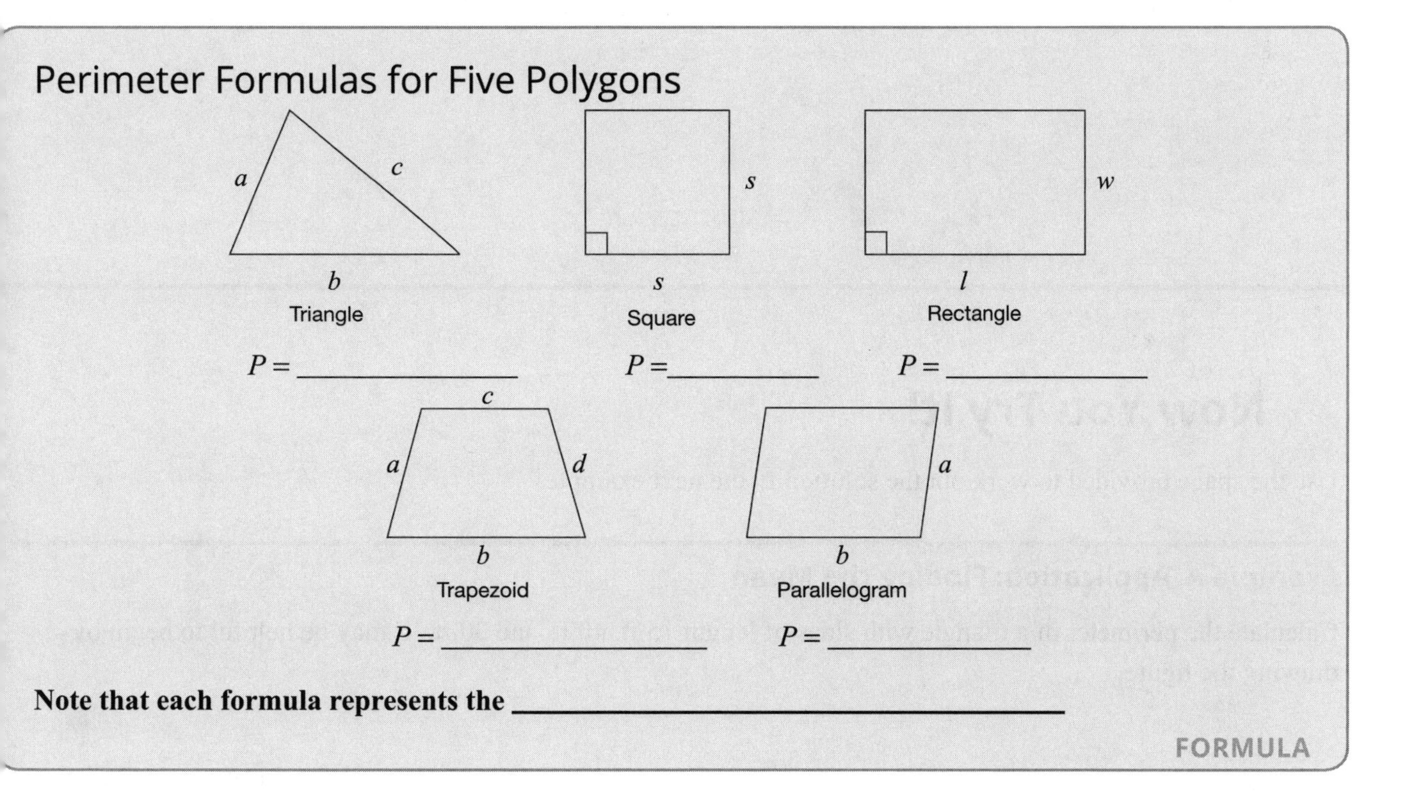

Watch and Work

Watch the video for Example 2 in the software and follow along in the space provided.

Example 2 Calculating the Perimeter of a Triangle

Calculate the perimeter of a triangle with sides of length 40 mm, 70 mm, and 80 mm.

Solution

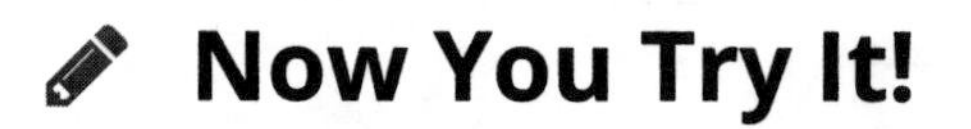

Now You Try It!

Use the space provided to work out the solution to the next example.

Example A Application: Finding the Mean

Calculate the perimeter of a triangle with sides of length 15 ft, 40 ft, and 30 ft. It may be helpful to begin by drawing the figure.

Circles

Circle: The set of all points in a plane that are ______________________________

____

Radius: The distance from the center of a circle to ______________________________
(The letter r is used to represent ______________________________________.)

Diameter: The distance from one point on a circle to ________________________

____

(The letter d is used to represent ______________________________________.)

Circumference: ____

Center

Diameter (d)

Radius (r)

DEFINITION

The Circumference of a Circle

To find the circumference C of a circle, use one of the following formulas,

$C =$ __________ **and** $C =$ __________

where r is the ________________ and d is the ______________________________

FORMULA

6.6 Exercises

Concept Check

True/False. Determine whether each statement is true or false. If a statement is false, explain how it can be changed so the statement will be true. (**Note:** There may be more than one acceptable change.)

1. a. Every square is a rectangle.

b. Every rectangle is a square.

2. a. Every parallelogram is a rectangle.

b. Every rectangle is a parallelogram.

3. A trapezoid has only one pair of parallel lines.

4. The length of the diameter of a circle is half of the length of the radius.

Match each formula for perimeter to its corresponding geometric figure.

5. a. Square **A.** $P = 2l + 2w$

b. Parallelogram **B.** $P = 4s$

c. Rectangle **C.** $P = 2b + 2a$

d. Trapezoid **D.** $P = a + b + c$

e. Triangle **E.** $P = a + b + c + d$

Practice

Calculate the perimeter of each figure described. See Examples 1 through 3.

6. A parallelogram with sides of length 15 cm and 7 cm.

7. A circle with diameter 60 cm.

Calculate the perimeter of each figure.

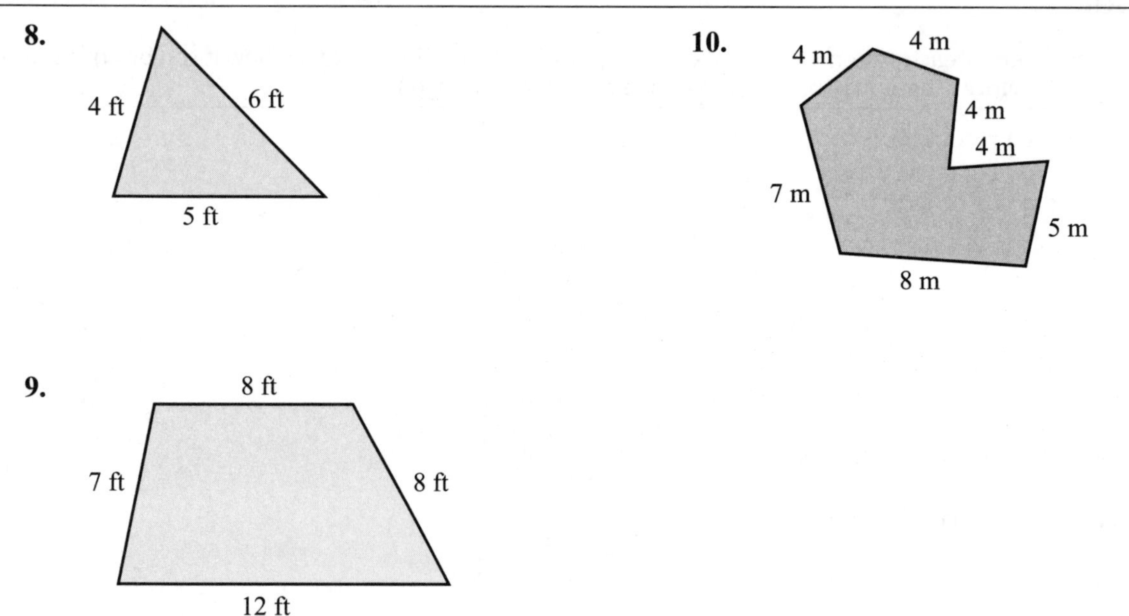

Applications

Solve.

11. ***Construction:*** The Pentagon near Washington, D.C., is a five-sided building where each outside wall is 921 feet. [1]

a. What is the perimeter of the building?

b. If it takes a person 0.00341 minutes to walk 1 foot, how long will it take the person to walk completely around the building? Round your answer to the nearest tenth of a minute.

12. ***Home Improvement:*** Jessica wants to add a decorative fringe to a throw rug. The rug is a rectangle with length 8 feet and width 5 feet. If Jessica wants to buy 1 foot more than the perimeter of the rug, how many feet of fringe must she buy?

1 Source: http://www.infoplease.com/spot/pentagon1.html

Writing & Thinking

13. Name as many polygons as you can and include the number of sides for each one.

14. Explain, briefly, the meaning of perimeter. Write the formula for the perimeter of each of the five types of polygons discussed in this section.

6.7 Area

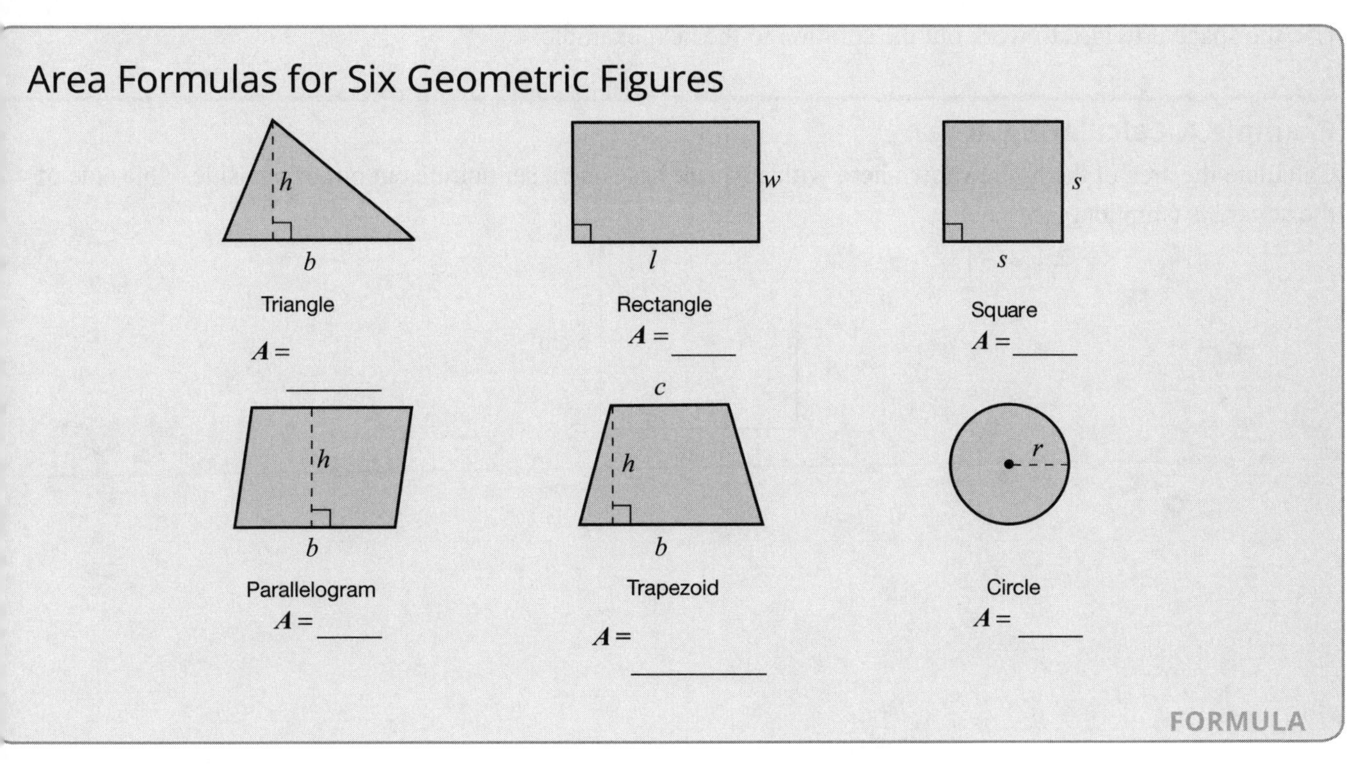

Watch and Work

Watch the video for Example 7 in the software and follow along in the space provided.

Example 7 Calculating Area

Calculate the area of the figure shown here with a square base and a semicircle cut out of one side. One side of the square is 10 in. long.

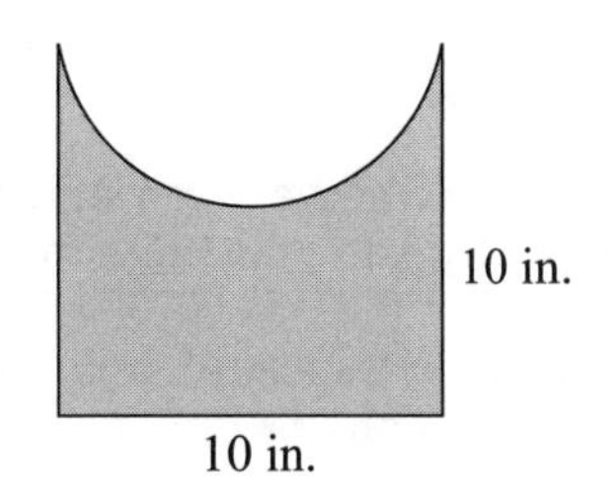

Solution

Now You Try It!

Use the space provided to work out the solution to the next example.

Example A Calculating Area

Calculate the area of the figure shown here with a square base and a semicircle cut out of one side. One side of the square is 6 cm long.

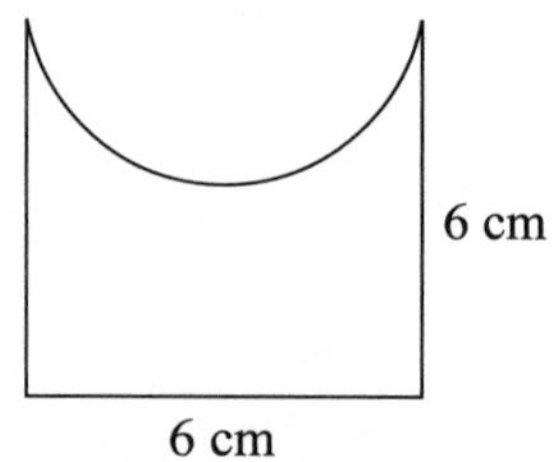

6.7 Exercises

Concept Check

True/False. Determine whether each statement is true of false. If a statement is false, explain how it can be changed so the statement will be true. (**Note:** There may be more than one acceptable change.)

1. The $(b+c)$ in the trapezoid area formula represents the sum of the lengths of the base and the corners.

2. The height of a triangle is the distance between the base and the vertex opposite the base.

3. The area formula for a triangle is $A = a + b + c$.

4. The area formula for a trapezoid is $A = \frac{1}{2}h(b+c)$.

ractice

alculate the area of each figure described. Use π = 3.14.

5. A square with sides of length 9 ft.

6. A triangle with height $\frac{8}{9}$ in. and base $\frac{5}{12}$ in.

7.

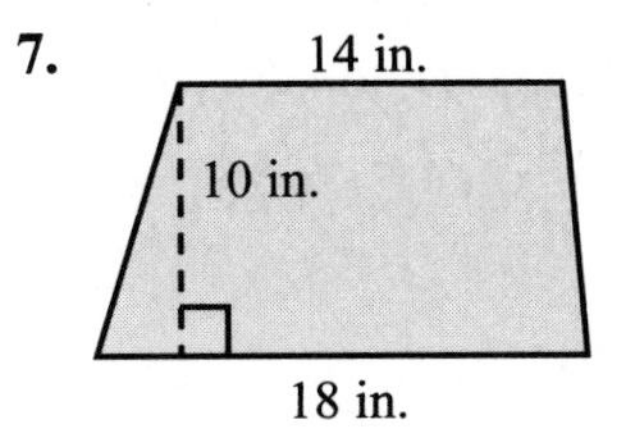

8.

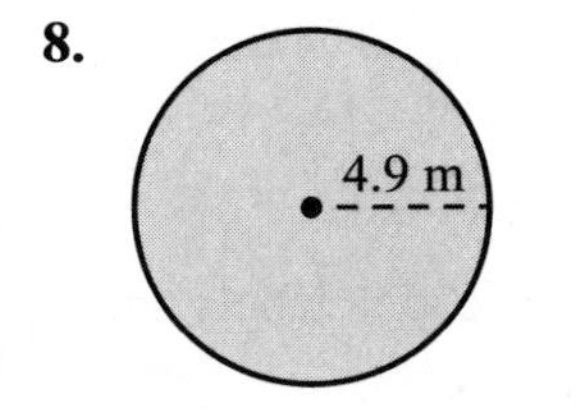

9.

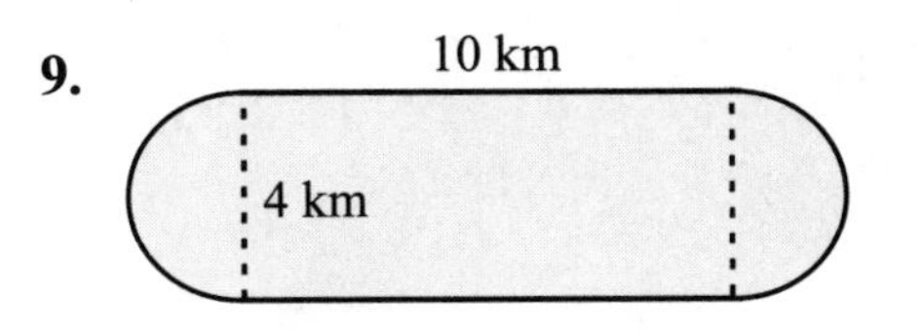

Applications

Solve. Use $\pi = 3.14$.

10. ***Geography:*** The boundaries of a certain small town form a parallelogram with a length of 4.5 miles and a height of 2.6 miles. What is the area within the town limits?

11. ***Construction:*** Vinyl tile is to be laid on the floor of a rectangular room which is 17 feet long and 12 feet wide. How many square feet of tile must be put down?

Writing & Thinking

12. Explain why square units are used for labeling areas. Give two examples each of metric area labels and US customary area labels.

5.8 Volume and Surface Area

1. **Volume** is a measure of the ________________________________.

2. Volume is measured in ____________________

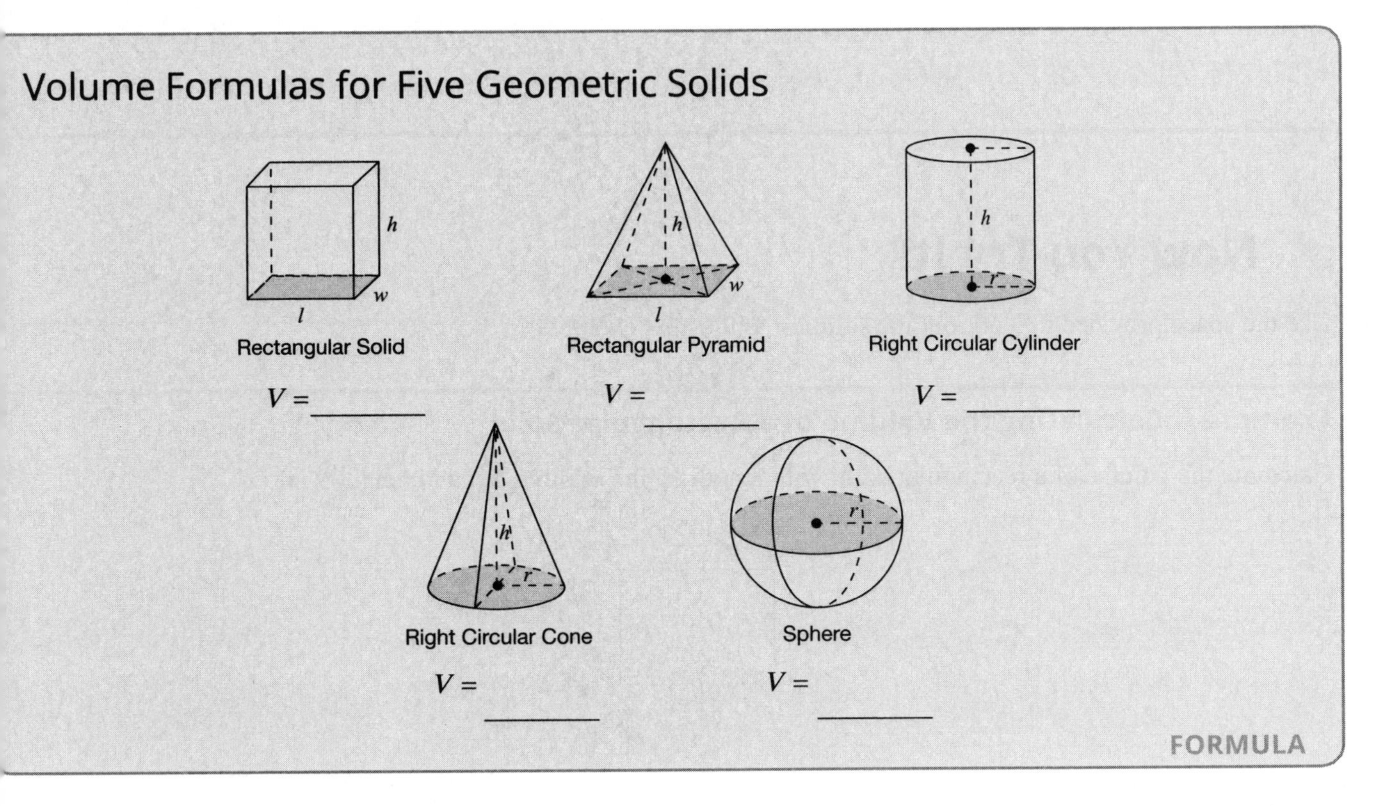

Watch and Work

Watch the video for Example 1 in the software and follow along in the space provided.

Example 1 Calculating the Volume of a Rectangular Solid

Calculate the volume of a rectangular solid with length 8 in., width 4 in., and height 12 in.

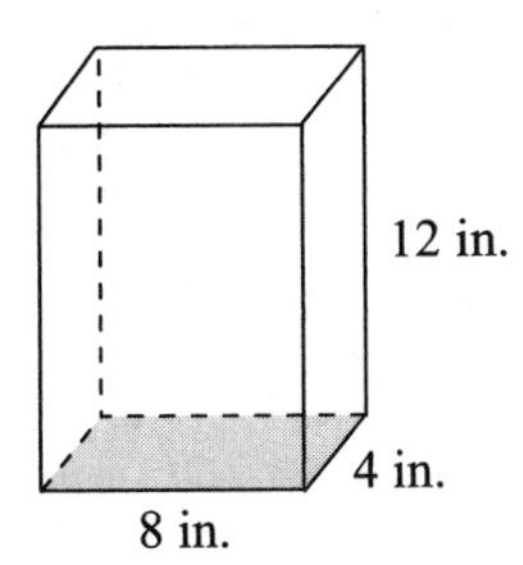

Solution

Now You Try It!

Use the space provided to work out the solution to the next example.

Example A Calculating the Volume of a Rectangular Solid

Calculate the volume of a rectangular solid with length 15 in., width 6 in., and height 9 in.

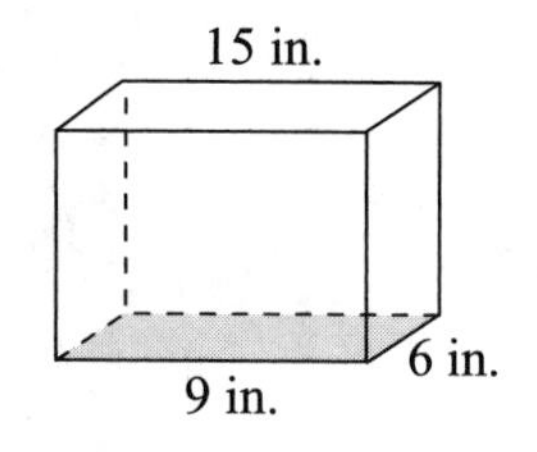

2. The **surface area** (*SA*) of a geometric solid is ______________________________

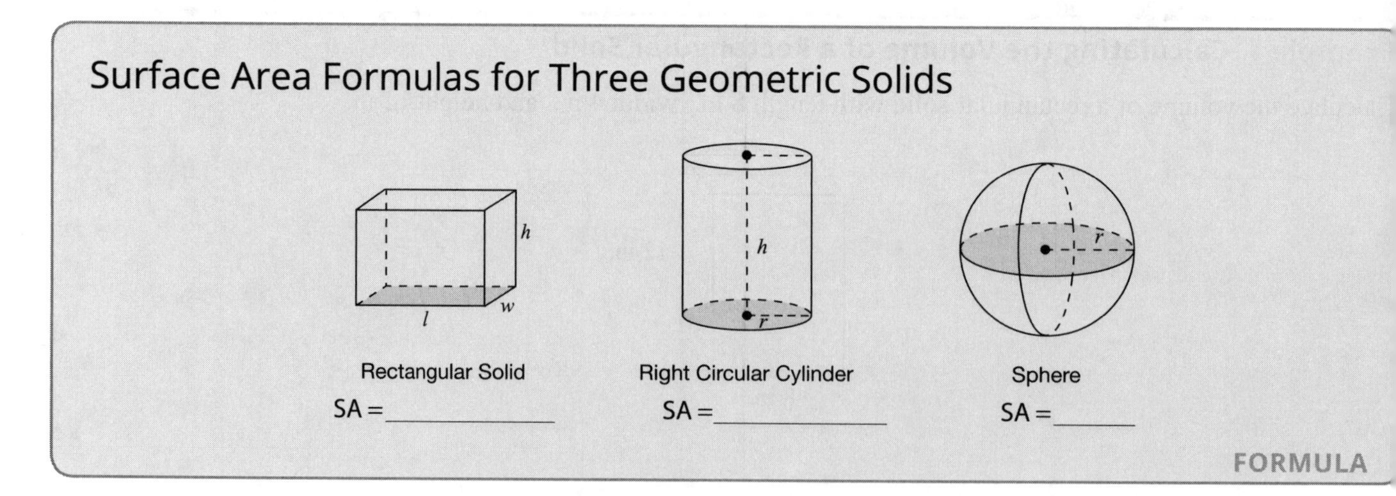

6.8 Exercises

Concept Check

True/False. Determine whether each statement is true or false. If a statement is false, explain how it can be changed so the statement will be true. (**Note:** There may be more than one acceptable change.)

1. To find the volume of a can of corn, the formula $V = \pi r^2 h$ would be used.

2. $V = lwh$ is the formula for the surface area of a rectangular solid.

3. The area of the paper label on a can of peaches is an example of surface area.

4. To find the volume of a rectangular solid, the areas of each surface are added together.

Match each formula for volume to its corresponding geometric figure.

5. a. Rectangular solid **A.** $V = \frac{4}{3}\pi r^3$

b. Rectangular pyramid **B.** $V = \frac{1}{3}\pi r^2 h$

c. Right circular cylinder **C.** $V = lwh$

d. Right circular cone **D.** $V = \pi r^2 h$

e. Sphere **E.** $V = \frac{1}{3}lwh$

Practice

Calculate the volume of each solid. Use $\pi \approx 3.14$.

6. A rectangular solid with length 5 in., width 2 in., and height 7 in.

7. A right circular cone 3 mm high with a 2 mm radius.

8.

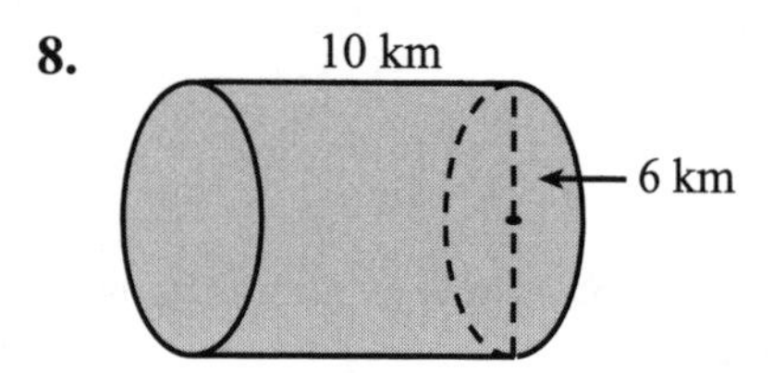

Calculate the surface area of each solid. Use $\pi \approx 3.14$.

9.

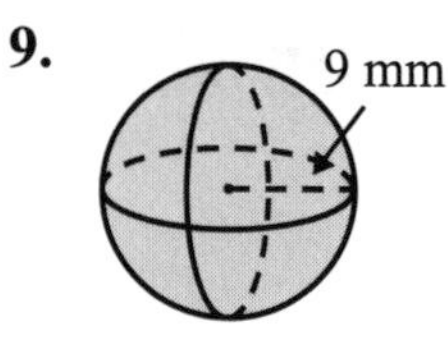

10.

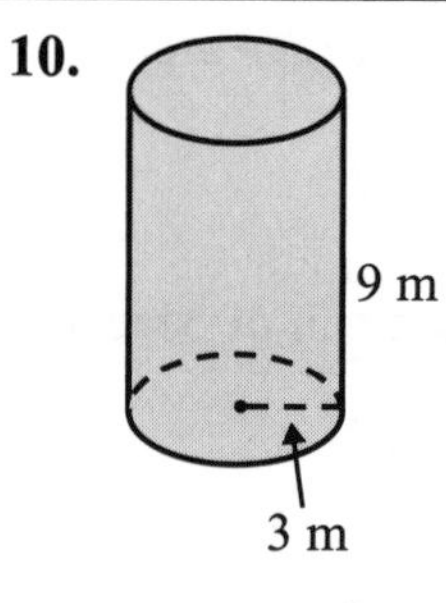

Applications

Solve. Use $\pi \approx 3.14$.

11. ***Recreation:*** A rectangular tent with straight sides has a pyramidal shaped roof. The dimensions of the rectangular portion are 12 ft long, 10 ft wide, and 6 ft high. The peak of the pyramid is 2 ft above the top edge of the walls. What is the volume of the inside of the tent?

2 ft
6 ft
10 ft
12 ft

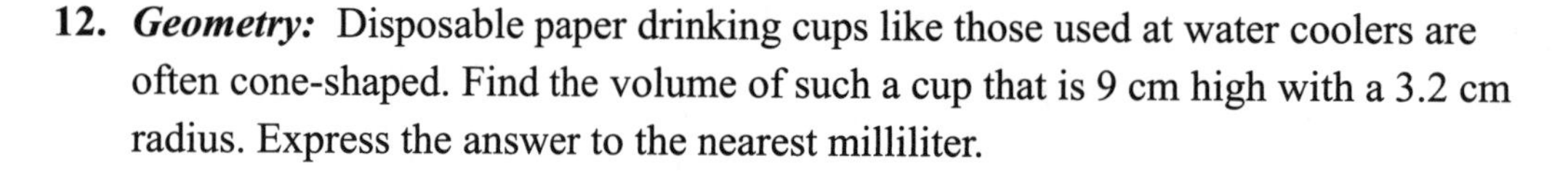

12. ***Geometry:*** Disposable paper drinking cups like those used at water coolers are often cone-shaped. Find the volume of such a cup that is 9 cm high with a 3.2 cm radius. Express the answer to the nearest milliliter.

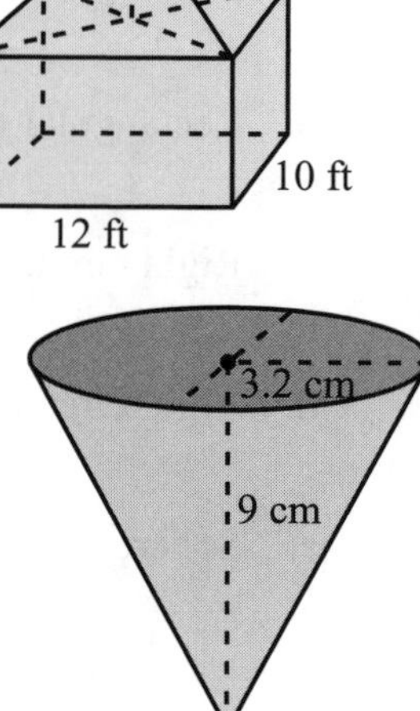

Writing & Thinking

13. Discuss the type of units used for volume and explain why.

14. List the steps and formulas you would use to find the volume of an ice cream cone (assuming the ice cream itself forms a perfect half sphere).

6.9 Similar and Congruent Triangles

Similar Triangles

1. In similar triangles, the corresponding ______________________________

____

2. In similar triangles, the ______________________________

DEFINITION

Properties of Congruent Triangles

Two triangles are congruent if:

1. ______________________________
2. ______________________________

PROPERTIES

Determining Congruent Triangles

1. **Side-Side-Side (SSS)**

If two triangles are such that the ______________________________

____

2. **Side-Angle-Side (SAS)**

If two triangles are such that the ______________________________

____

____

3. **Angle-Side-Angle (ASA)**

If two triangles are such that ______________________________

____

____

DEFINITION

Watch and Work

Watch the video for Example 5 in the software and follow along in the space provided.

Example 5 Determining Whether Triangles are Congruent

Determine whether triangles *PQR* and *MNO* are congruent.

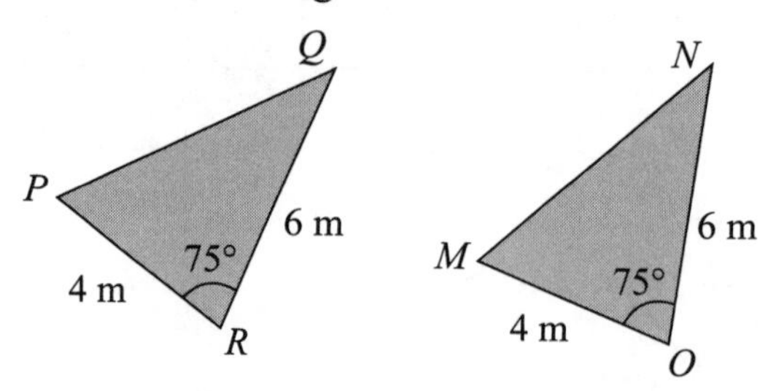

Solution

Now You Try It!

Use the space provided to work out the solution to the next example.

Example A Determining Whether Triangles are Congruent

Determine whether triangles *JKL* and *MNO* are congruent. If they are congruent, state the property that confirms that they are congruent.

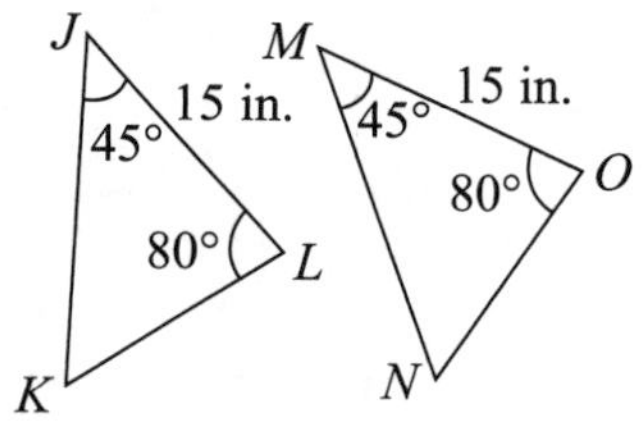

6.9 Exercises

Concept Check

True/False. Determine whether each statement is true or false. If a statement is false, explain how it can be changed so the statement will be true. (**Note:** There may be more than one acceptable change.)

1. Similar triangles have corresponding sides that are equal.

2. If $\triangle ABC \cong \triangle DEF$, then the measure of angle C equals the measure of angle D.

3. If $\triangle ABC \sim \triangle DEF$, then $AC = DF$.

4. Congruent triangles have corresponding angles that are equal.

Practice

Determine whether each pair of triangles is similar. If the pair of triangles is similar, explain why and indicate the similarity by using the ~ symbol.

5.

6.

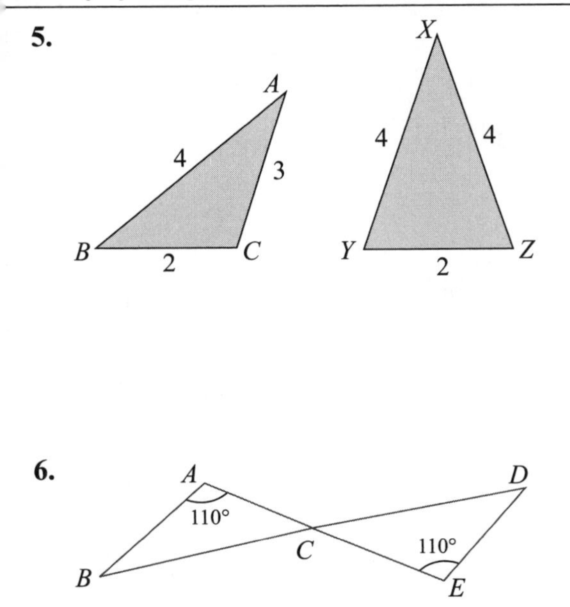

Find the values for x and y.

7. $\triangle ABC \sim \triangle XYZ$

8. $\triangle ABE \sim \triangle CDE$

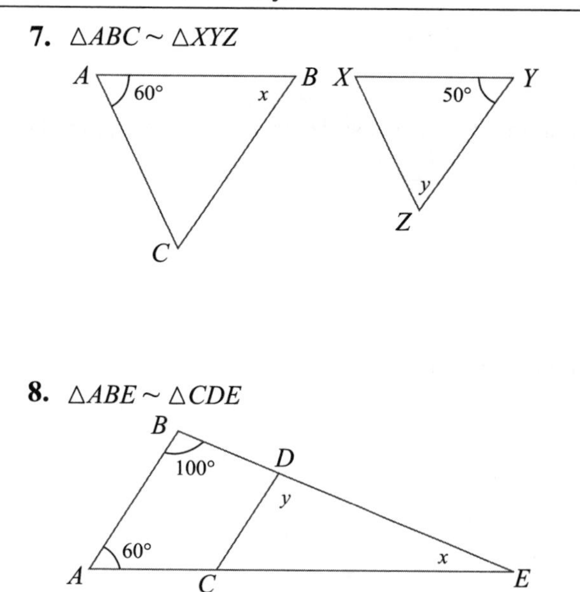

Determine whether each pair of triangles is congruent. If the pair of triangles is congruent, state the property that confirms that they are congruent.

9.

10.

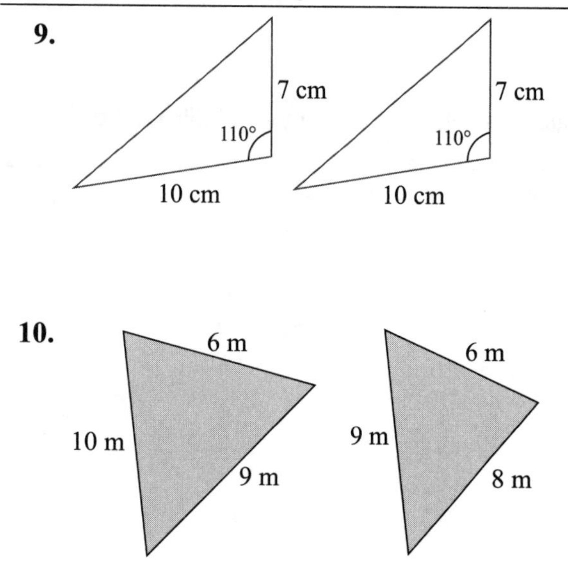

Applications

Solve.

11. ***Construction:*** A child's playhouse is built to look like a smaller version of the family house, where the ends of the roofs have similar proportions. The width of the main house (AB) is 32 feet and the length from the peak to the gutter of the roof for one of the sides is 20 feet. If the width of the playhouse (DF) is 12 feet, what is the length from the peak to the gutter (DE) of the playhouse roof?

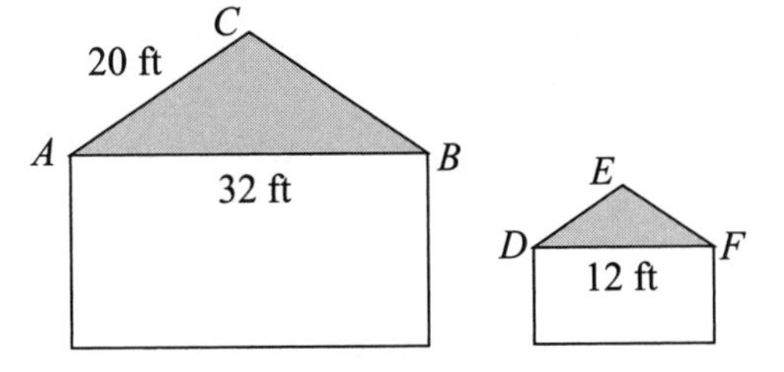

12. ***Holiday Decorating:*** Your neighbors are hanging their holiday lights. The ladder they are currently using is 12 feet long and when leaned up against the house just reaches the top of their 8-foot tall porch. How long of a ladder will they need to reach the top of their chimney which is at a height of 32 feet? (Assume that both ladders are placed such that they make the same angle with the ground.)

Writing & Thinking

13. Determine the errors in the following statement. Assume $\triangle ABC \sim \triangle DEF$.

a. Corresponding angles are congruent. This means $m\angle A = m\angle D$, $m\angle B = m\angle F$, and $m\angle C = m\angle E$.

b. Corresponding sides are the same length.

6.10 Square Roots and the Pythagorean Theorem

Terminology of Radicals

The symbol $\sqrt{\ }$ is called ____________________

The number under the radical sign is called ____________________

The complete expression, such as $\sqrt{49}$, is called ______________________________

DEFINITION

Square Root

For any real number a and any nonnegative real number b, ______________________________

If a is positive, then we ____________________

Thus, __________________

DEFINITION

Terms Related to Right Triangles

Right triangle: A triangle containing ____________________

Hypotenuse: ____

Leg: ____

Leg

Hypotenuse

90°

Leg

DEFINITION

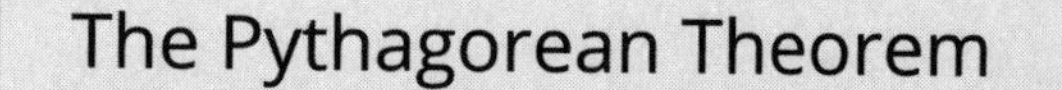

The Pythagorean Theorem

In a right triangle, the square of the length of the hypotenuse is equal to ______________________________

____

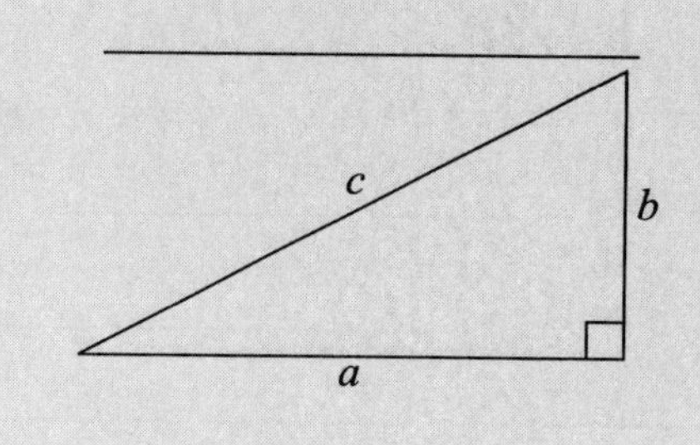

THEOREM

Watch and Work

Watch the video for Example 6 in the software and follow along in the space provided.

Example 6 Finding the Length of the Hypotenuse

Find the length of the hypotenuse of a right triangle with legs of length 12 cm and 5 cm.

Solution

Now You Try It!

Use the space provided to work out the solution to the next example.

Example A Finding the Length of the Hypotenuse

Find the length of the hypotenuse of a right triangle with legs of length 8 cm and 15 cm.

6.10 Exercises

Concept Check

True/False. Determine whether each statement is true or false. If a statement is false, explain how it can be changed so the statement will be true. (**Note:** There may be more than one acceptable change.)

1. 49 is a perfect square.

2. In the expression $\sqrt{81}$, the number 9 is the radicand.

3. The Pythagorean Theorem can be used to find the length of the longest side of a right triangle if the lengths of the two legs are known.

4. The Pythagorean Theorem works for any type of triangle.

Practice

Evaluate each expression.

5. $\sqrt{36}$

6. $\sqrt{169}$

For each problem, **a.** use your understanding of square roots to estimate the value of each square root. Then, **b.** use your calculator to find the value of each square root accurate to the nearest ten-thousandth.

7. a. The nearest whole numbers to $\sqrt{13}$ are ____ and ____.

b. 🖩 Find the value of $\sqrt{13}$.

Use the Pythagorean Theorem to determine whether or not each triangle is a right triangle.

8.

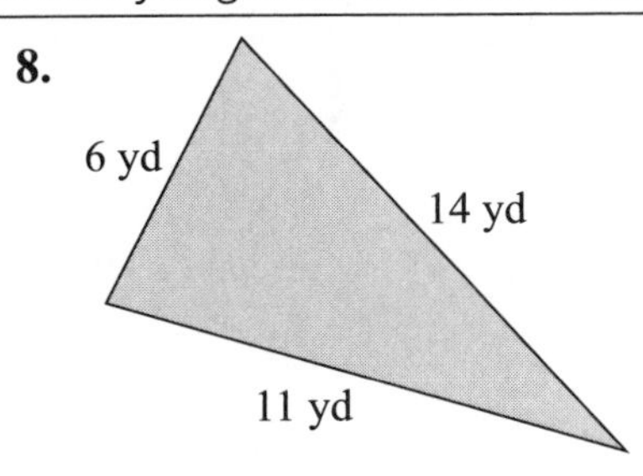

Find the hypotenuse for the right triangle accurate to the nearest hundredth.

9.

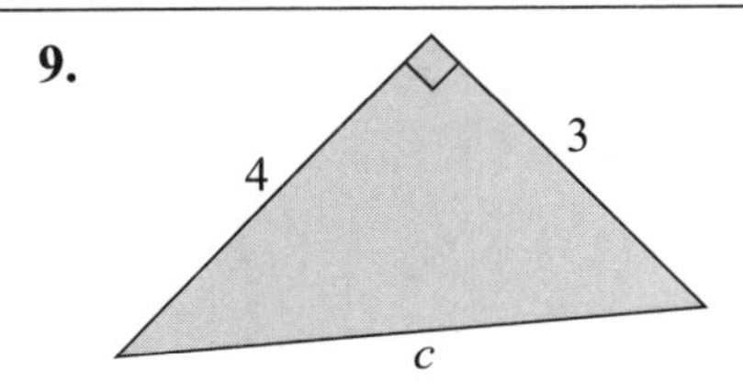

Applications

Solve.

10. ***Safety:*** The base of a fire engine ladder is 30 feet from a building and reaches to a third floor window 50 feet above ground level. Find the length of the ladder to the nearest hundredth of a foot.

11. ***Baseball:*** The shape of home plate in the game of baseball can be created by cutting off two triangular pieces at the corners of a square, as shown in the figure. If each of the triangular pieces has a hypotenuse of 12 inches and legs of equal length, what is the length of one side of the original square, to the nearest tenth of an inch?

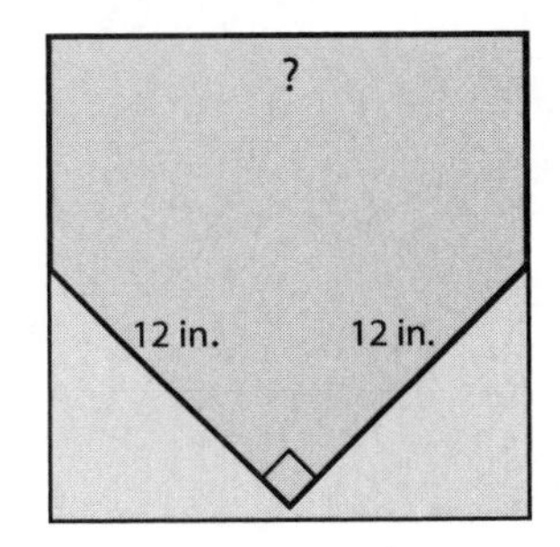

Writing & Thinking

12. Explain the connection between a perfect square and its square root. Give an example.

13. Determine the difference between a radical sign, radicand, and radical expression.

Name: Date:

Chapter 6 Project

Metric Cooking

An activity to demonstrate the use of metric to US conversions in real life.

Your grandma lives outside of the country and has emailed you her famous apple pie recipe to use at an upcoming party. As you start to bake the pie on the day of the party, you realize grandma's recipe is written using only metric units. Looking through the supplies in your kitchen, you find a scale that measures weight in ounces and some measuring spoons that measure volume in $\frac{1}{8}$, $\frac{1}{4}$, and $\frac{1}{2}$ of a teaspoon and 1 whole tablespoon. In order to successfully bake the pie in time for the party, you must quickly convert the metric measurements to US measurements. You start with the pie crust ingredients.

Pie Crust

Ingredient	Metric Measurement	US Measurement
Flour	280 g	______ oz
Vegetable Shortening	90 g	______ oz
Unsalted Butter	50 g	______ oz
Cold Water	90 ml	_____ tbsp
Salt	5 ml	______ tsp

1. Fill in the third column of the pie crust table by converting the measurements of each ingredient using the correct conversion factors. Use 1 ml = 0.068 tbsp and 1 ml = 0.203 tsp. (**Note:** tbsp stands for tablespoon and tsp stands for teaspoon.)

2. The recipe requires the oven to be preheated to 230 °C, but your oven measures degrees in Fahrenheit. What temperature should you preheat your oven to?

   While the oven is preheating, you begin to prepare the ingredients for the apple filling.

Apple Filling

Ingredient	Metric Measurement	US Measurement
Apples	1 kg	______ oz
Sugar	100 g	______ oz
Cornstarch	15 ml	______ tsp
Cinnamon	2.5 ml	______ tsp
Salt	0.5 ml	______ tsp
Nutmeg	0.5 ml	______ tsp
Butter	25 g	______ oz

3. Fill in the third column of the apple filling table by converting the measurements of each ingredient using the correct conversion factors. Use 1 ml = 0.068 tbsp and 1 ml = 0.203 tsp.

4. Since the measuring spoons can only measure $\frac{1}{8}$, $\frac{1}{4}$, and $\frac{1}{2}$ of a teaspoon and 1 whole tablespoon, what is the most reasonable way to round each US volume measurement in each of the tables?

In addition to the usual recipe, your grandma has listed several variations on the recipe that depend on the taste of the apples. For bland apples, you should add 20 ml of lemon juice to the filling. For sour apples, you should increase the sugar to 140 g.

5. You taste the apples and decide you need to increase the sugar to 140 g. You have already added 100 g. How many more ounces of sugar do you need to add to reach 140 g?

6. You notice that the recipe requires a pie pan with a diameter of 23 cm. After measuring the pie pan you discover it is 9 inches in diameter. Is the pie pan the right size for the apple pie? What is its diameter in inches? Round to the nearest whole number.

CHAPTER 7

Solving Linear Equations and Inequalities

Math @ Work

You are required to make decisions every day of your life. Some of these decisions are simple, such as deciding what to eat for breakfast or what to wear during the day. These simple decisions will likely affect your life for only a day. However, certain decisions are difficult to make, such as how much to spend on your first house or which job offer to take. These difficult decisions can affect the rest of your life. The skills you learn while studying algebra will help develop your reasoning and problem-solving skills, which in turn can increase your confidence as you make some of life's big decisions. These skills can even help you realize that small financial decisions can eventually add up to significant costs, such as the cup of coffee you buy every day on the way to school or work.

Suppose you are a traveling sales representative and the company you work for will reimburse travel costs up to \$325 per week for expenses incurred by driving your own car. The company policy states that you will be reimbursed for fuel costs and receive \$0.565 per mile driven. Due to the reimbursement cap, you will want to determine the best routes to take in order to keep your mileage low so all of your mileage is reimbursed by the company, if possible. If you spend \$60 on gas per week on average, how would you determine the maximum number of miles you can drive and stay under the \$325 reimbursement cap?

7.1 Properties of Real Numbers

Properties of Addition and Multiplication

In this table a, b, and c, are real numbers.

Name of Property	For Addition	For Multiplication
Commutative property		
Associative property		
Identity		
Inverse		

PROPERTIES

Properties of Addition and Multiplication (cont.)

In this table a, b, and c, are real numbers.

Zero Factor Law

Distributive Property of Multiplication over Addition

PROPERTIES

Watch and Work

Watch the video for Example 2 in the software and follow along in the space provided.

Example 2 Identifying Properties of Addition and Multiplication

For each of the following equations, state the property illustrated, and show that the statement is true for the value given for the variable by substituting the value in the equation and evaluating.

a. $x+14=14+x$ given that $x=-4$

b. $(3\cdot 6)x=3(6x)$ given that $x=5$

c. $12(y+3)=12y+36$ given that $y=-2$

Solution

Now You Try It!

Use the space provided to work out the solution to the next example.

Example A Identifying Properties of Addition and Multiplication

State the property illustrated and show that the statement is true for the value given for the variable.

a. $x+21=21+x$ given that $x=-7$

b. $(5\cdot 4)x=5(4x)$ given that $x=2$

c. $11(y+3)=11y+33$ given that $y=-4$

7.1 Exercises

True/False. Determine whether each statement is true or false. If a statement is false, explain how it can be changed so the statement will be true. (**Note:** There may be more than one acceptable change.)

1. Changing the order of the numbers in an addition problem is allowed because of the associative property of addition.

2. The equation $(8 \cdot 2) \cdot 5 = 8 \cdot (2 \cdot 5)$ is an example of the associative property of multiplication.

3. The additive identity of all numbers is 1.

4. The commutative property works for division and subtraction.

Practice

Complete the expressions using the given property. Do not simplify.

5. $7 + 3 =$ ______ commutative property of addition

6. $(6 \cdot 9) \cdot 3 =$ ______ associative property of multiplication

7. $19 \cdot 4 =$ ______ commutative property of multiplication

8. $6(5 + 8) =$ ______ distributive property

9. $16 + (9 + 11) =$ ______ associative property of addition

10. $6 \cdot 0 =$ ______ zero factor law

Applications

Solve

11. ***Income:*** Jessica works part-time at a retail store and makes \$11 an hour. During one week, she worked $6\frac{1}{2}$ hours on Monday and $4\frac{1}{4}$ hours on Thursday.

a. Determine the amount of money she earned during the week by evaluating the expression $\$11 \cdot \left(6\frac{1}{2} + 4\frac{1}{4}\right)$.

b. Rewrite this expression to remove the parentheses using one of the properties talked about in this section.

c. What property did you use in Part **b.** to rewrite the expression?

12. ***Budgeting:*** Robin went to the grocery store to buy a few items she needed in order to cook dinner. She bought milk for \$3.99, rolls for \$2.25, a package of steaks for \$12.01, and some marinade for \$1.75. Before getting to the checkout line, Robin remembered that she only had \$20 in her purse. Did she have enough money to buy the food items if the store does not charge sales tax on food?

a. Write an expression to find the total of Robin's food purchases. Do not simplify.

b. Robin doesn't have a calculator to determine the total cost of her items. She wants to make sure that she has enough money to buy them. Rearrange the expression from Part **a.** so that she could quickly find the total using mental math.

c. What properties did you use in Part **b.** to rewrite the expression?

d. Did Robin have enough money to purchase all of the items?

Writing and Thinking

13. a. The distributive property illustrated as $a(b+c) = ab + ac$ is said to "distribute multiplication over addition." Explain, in your own words, the meaning of this phrase.

b. What would an expression that "distributes addition over multiplication" look like? Explain why this would or would not make sense.

7.2 Solving Linear Equations: $x + b = c$ and $ax = c$

Linear Equation in x

If a, b, and c are **constants** and $a \neq 0$ then a **linear equation in x** is ______________________________
____

______

DEFINITION

Addition Principle of Equality

If the same algebraic expression is added to ____
____

Symbolically, if A, B, and C are algebraic expressions, then the equations

____

and

____

____

PROPERTIES

Procedure for Solving Linear Equations that Simplify to the Form $x + b = c$

1. Combine ______________________________________
2. Use the addition principle of equality and ____
   ___________ The objective is to ____
   ____
3. Check your answer by ____

PROCEDURE

Watch and Work

Watch the video for Example 2 in the software and follow along in the space provided.

Example 2 Solving Linear Equations of the Form $x + b = c$

Solve the equation: $x - 3 = 7$

Solution

Now You Try It!

Use the space provided to work out the solution to the next example.

Example A Solving Linear Equations of the Form $x + b = c$

Solve the equation: $x - 5 = 12$

Multiplication (or Division) Principle of Equality

If both sides of an equation are multiplied by (or divided by) the ______________________________

____

Symbolically, if A and B are algebraic expressions and C is any nonzero constant, then the equations

____

____

____

____

PROPERTIES

Procedure for Solving Linear Equations that Simplify to the Form $ax = c$

1. Combine ______________________________
2. Use the **multiplication** (or **division**) **principle of equality** and multiply both sides of the equation by the ______________________________ **(or divide both sides by** __________ __________). The coefficient of the variable will become __________
3. Check your answer by ______________________________

PROCEDURE

7.2 Exercises

Concept Check

True/False. Determine whether each statement is true or false. If a statement is false, explain how it can be changed so the statement will be true. (**Note:** There may be more than one acceptable change.)

1. When an algebraic expression is added to both sides of an equation, the new equation has the same solutions as the original equation.

2. The process of finding the solution set to an equation is called simplifying the equation.

3. A linear equation in x is also called a first-degree equation in x.

4. Equations with the same solutions are said to be equivalent equations.

Practice

Determine whether or not the given number is a solution to the given equation by substituting and then evaluating.

5. $x - 2 = -3$ given that $x = 1$

6. $-10 + x = -14$ given that $x = -4$

Solve each equation.

7. $x-6=1$

8. $x-10=9$

9. $51=17y$

10. $\frac{5x}{7}=65$

Applications

Solve.

11. ***World Languages:*** The Japanese writing system consists of three sets of characters, two with 81 characters (which all Japanese students must know), and a third, Kanji, with over 50,000 characters (of which only some are used in everyday writing). If a Japanese student knows 2107 total characters, solve the equation $x+2(81)=2107$ to determine the number of Kanji characters the student knows.

12. ***Nursing:*** A nurse must give a patient 800 milliliters of intravenous solution over 4 hours. This can be represented by the equation $4x = 800$, where x represents the amount of solution the patient receives per hour in milliliters.

a. Why was multiplication chosen in the equation?

b. Solve the equation to determine the value of x.

c. What does the answer to Part **b.** mean? Write a complete sentence.

Writing & Thinking

13. **a.** Is the expression $6 + 3 = 9$ an equation? Explain.

b. Is 4 a solution to the equation $5 + x = 10$? Explain.

7.3 Solving Linear Equations: $ax + b = c$

Procedure for Solving Linear Equations that Simplify to the Form $ax + b = c$

1. Combine ______________________________
2. Use the **addition principle of equality** and ______________________________ ______________
3. Use the ______________________________ and multiply both sides of the equation by the reciprocal of the coefficient of the variable (**or** ______________________________ **itself**). The ______________________________ will become +1.
4. Check your answer by ______________________________

PROCEDURE

Watch and Work

Watch the video for Example 2 in the software and follow along in the space provided.

Example 2 Solving Linear Equations of the Form $ax + b = c$

Solve the equation: $-26 = 2y - 14 - 4y$

Solution

Now You Try It!

Use the space provided to work out the solution to the next example.

Example A Solving Linear Equations of the Form $ax + b = c$

Solve the equation.

$-18 = 2y - 8 - 7y$

7.3 Exercises

Concept Check

True/False. Determine whether each statement is true or false. If a statement is false, explain how it can be changed so the statement will be true. (**Note:** There may be more than one acceptable change.)

1. If an equation of the form $ax + b = c$ uses decimal or fractional coefficients, the addition and multiplication principles of equality cannot be used.

2. The first step in solving $2x + 3 = 9$ is to add 3 to both sides.

3. To solve an equation that has been simplified to $4x = 12$, you need to multiply both sides by $\frac{1}{4}$, or divide both sides by 4.

4. When solving a linear equation with decimal coefficients, one approach is to multiply both sides in such a way to give integer coefficients before solving.

Practice

Solve each equation.

5. $3x+11=2$

6. $-5x+2.9=3.5$

7. $\frac{2}{5}-\frac{1}{2}x=\frac{7}{4}$

8. $\frac{y}{3}-\frac{2}{3}=7$

Applications

Solve.

9. ***Music:*** The tickets for a concert featuring the new hit band, Flying Sailor, sold out in 2.5 hours. If there were 35,000 tickets sold, solve the equation $35{,}000-2.5x=0$ to find the number of tickets sold per hour.

10. ***Movies:*** All snacks (candy, popcorn, and soda) cost $3.50 each at the local movie theater. Admission tickets cost $7.50 each. After a long week, Carlos treats himself to a night at the movies. His movie night budget is $25 and he spends all his movie money. Solve the equation $\$3.50x+\$7.50=\$25.00$ to determine how many snacks Carlos can buy.

Writing & Thinking

11. Find the error(s) made in solving each equation and give the correct solution.

a.
$$\frac{1}{3}x + 4 = 9$$
$$3 \cdot \frac{1}{3}x + 4 = 3 \cdot 9$$
$$x + 4 = 27$$
$$x + 4 - 4 = 27 - 4$$
$$x = 23$$

b.
$$5x + 3 = 11$$
$$(5x - 3) + (3 - 3) = 11 - 3$$
$$2x + 0 = 8$$
$$\frac{2x}{2} = \frac{8}{2}$$
$$x = 4$$

7.4 Solving Linear Equations: $ax + b = cx + d$

Procedure for Solving Linear Equations that Simplify to the Form $ax + b = cx + d$

1. Simplify each side of the equation by ____

   ____

2. Use the **addition principle of equality** and ____

   ____

3. Use the **multiplication** (or **division**) **principle of equality** and ______________________

   ______________________ of the coefficient of the variable (**or divide** ______________________

   ____________ **itself**). The coefficient of the variable ______________________

4. Check your answer by ____

PROCEDURE

Type of Equation	Number of Solutions
conditional	______________________
identity	______________________
contradiction	______________

Table 1

Watch and Work

Watch the video for Example 9 in the software and follow along in the space provided.

Example 9 Determining Types of Equations

Determine whether the equation $3(x-25)+3x=6(x+10)$ is a conditional equation, an identity, or a contradiction.

Solution

Now You Try It!

Use the space provided to work out the solution to the next example.

Example A Determining Types of Equations

Determine whether the equation $-3(x - 4) = 12 - 2x - x$ is a conditional equation, an identity, or a contradiction.

7.4 Exercises

Concept Check

True/False. Determine whether each statement is true or false. If a statement is false, explain how it can be changed so the statement will be true. (**Note:** There may be more than one acceptable change.)

1. Every linear equation has exactly one solution.

2. If a linear equation simplifies to a statement that is always true, then the original equation is called an identity.

3. If an equation has no solution, it is called an identity.

4. The most general form of a linear equation is $ax + b = cx + d$.

Practice

Solve each equation.

5. $3x+2=x-8$

6. $2(z+1)=3z+3$

7. $x-0.1x+0.8=0.2x+0.1$

8. $0.6x-22.9=1.5x-18.4$

Determine whether each equation is a conditional equation, an identity, or a contradiction.

9. $-2x+13=-2(x-7)$

10. $3x+9=-3(x-3)+6x$

Applications

Solve.

11. ***Event Planning:*** Caitlyn and Steve are planning their wedding reception and must decide between two catering halls. The first site, A Wedding Space, rents for \$800 for one day and charges \$50 per person for dinner. The second venue, A Wedding Place, costs \$1000 to rent for one day and charges \$40 per person for the same dinner. Solve the equation $800 + 50x = 1000 + 40x$ to determine how many guests they can invite so that the cost they pay will be the same at both wedding catering halls.

12. ***Personal Finance:*** The value of a new car depreciates at a rate of about \$250 per month. Suppose a car originally costs \$30,000. The car was bought with a \$1000 down payment and a loan with 0% financing for 60 months with payments of \$200 a month. Solve the equation $30{,}000 - 250t = 29{,}000 - 200t$ to determine how many months it will take for the value of the vehicle to equal the amount owed on the loan?

Writing & Thinking

13. Answer each question.

    a. Simplify the expression $3(x+5)+2(x-7)$.

    b. Solve the equation $3(x+5)+2(x-7)=31$.

    c. How are the methods you used to answer questions **a.** and **b.** similar? How are they different?

7.5 Working with Formulas

1. **Formulas** are general rules or ______________ stated ______________.

2. We say that the formula $d = rt$ is ______________ d ______________ r and t. Similarly, the formula $A = \frac{1}{2}bh$ is solved for __________ in terms of ______________, and the formula $P = R - C$ (profit is equal to revenue minus cost) is solved for __________ in terms of __________ ________.

Watch and Work

Watch the video for Example 6 in the software and follow along in the space provided.

Example 6 Solving for Different Variables

Given $V = \frac{k}{P}$, solve for P in terms of V and k.

Solution

Now You Try It!

Use the space provided to work out the solution to the next example.

Example A Solving for Different Variables

Given $P = \frac{I}{rt}$ solve for t in terms of I, r, and P.

7.5 Exercises

Concept Check

True/False. Determine whether each statement is true or false. If a statement is false, explain how it can be changed so the statement will be true. (**Note:** There may be more than one acceptable change.)

1. When using formulas, typically it does not matter if capital or lower case letters are used: $A = a$, $C = c$, etc.

2. If the perimeter and length are known, $P = 2l + 2w$ can be used to find the width of a rectangle.

3. Rate of interest is stated as an annual rate in percent form.

Applications

n the following application problems, read the descriptions carefully and then substitute the values given in the problem or the corresponding variables in the formulas. Evaluate the resulting expression for the unknown variable.

elocity

f an object is shot upward with an initial velocity v_0 in feet per second, the velocity v in feet per second is given y the formula $v = v_0 - 32t$, where t is time in seconds. (v_0 is read "v sub zero." The $_0$ is called a subscript.)

4. An object projected upward with an initial velocity of 106 feet per second has a velocity of 42 feet per second. How many seconds have passed?

nvestments

The total amount of money in an account with P dollars invested in it is given by the formula $A = P + Prt$, where r s the rate expressed as a decimal and t is time (one year or part of a year).

5. If $1000 is invested at 6% interest, find the total amount in the account after 6 months.

Cost

The total cost C of producing x items can be found by the formula $C = ax + k$, where a is the cost per item and k is he fixed costs (rent, utilities, and so on).

6. Find the total cost of producing 30 items if each costs $15 and the fixed costs are $580.

Solve each formula for the indicated variable.

7. $P = a + b + c$; solve for b.

8. $P = 3s$; solve for s.

9. $I = Prt$; solve for t.

10. $A = P(1 + rt)$; solve for r.

7.6 Applications: Number Problems and Consecutive Integers

Consecutive Integers

Integers are **consecutive** if each ______________________________

Three consecutive integers can be represented as

____

where *n* is an integer.

DEFINITION

Consecutive Even Integers

Even integers are **consecutive** if each is ______________________________

Three consecutive even integers can be represented as

____

where *n* is an **even** integer.

DEFINITION

Consecutive Odd Integers

Odd integers are **consecutive** if each is ______________________________

Three consecutive odd integers can be represented as

____

where *n* is an **odd** integer.

DEFINITION

Watch and Work

Watch the video for Example 6 in the software and follow along in the space provided.

Example 6 Application: Calculating Living Expenses

Joe wants to budget $\frac{2}{5}$ of his monthly income for rent. He found an apartment he likes for \$800 a month. What monthly income does he need to be able to afford this apartment?

Solution

Now You Try It!

Use the space provided to work out the solution to the next example.

Example A Application: Calculating Living Expenses

Jim plans to budget $\frac{3}{7}$ of his monthly income to send his son Taylor to private school. If the school he'd like Taylor to attend costs $1200 a month, what monthly income does Jim need to be able to afford the school?

7.6 Exercises

Concept Check

True/False. Determine whether each statement is true or false. If a statement is false, explain how it can be changed so the statement will be true. (**Note:** There may be more than one acceptable change.)

1. If an odd integer is divided by 2, the remainder will be 1.

2. To find 3 consecutive odd integers, you could use n, $n + 1$, and $n + 3$.

3. Odd integers are integers that are divisible by 1.

4. Even integers are consecutive if each is 2 more than the previous even integer.

Practice

Read each problem carefully, translate the various phrases into algebraic expressions, set up an equation, and solve the equation.

5. Five less than a number is equal to 13 decreased by the number. Find the number.

6. Twice a number increased by 3 times the number is equal to 4 times the sum of the number and 3. Find the number.

7. Find three consecutive integers whose sum is 93.

8. Find three consecutive odd integers such that the sum of twice the first and three times the second is 7 more than twice the third.

Applications

Solve.

9. ***Expenses:*** A collect call from a landline in Ohio to another landline in Ohio has a connection fee of \$2.75 and a charge of \$0.36 per minute. Mr. Anderson made a collect call which cost the receiver of the call \$9.95. This situation can be modeled by $\$9.95 = \$2.75 + \$0.36m$.

a. The unknown value is represented by the variable m in the equation. What is the unknown value in this situation?

b. Solve the equation for the variable.

c. What does the answer to Part **b.** mean? Write a complete sentence.

10. ***Event Planning:*** Robin is in charge of purchasing desserts for a dinner party that her nonprofit organization is throwing. She decides to buy a cake and several specialty cupcakes from Barbara's Bombtastic Bakery. She needs to buy one 8-inch round cake which costs \$19.50. She has \$45 to spend and will spend the leftover amount on cupcakes, which are \$8.50 for a box of 4. How many boxes of cupcakes can Robin purchase?

    a. What is the unknown value in this problem? Let the variable c represent this unknown value.

    b. Write an equation to represent this situation.

    c. Solve the equation for the variable.

    d. What does the answer to Part **c.** mean? Write a complete sentence.

Writing & Thinking

11. **a.** How would you represent four consecutive odd integers?

    b. How would you represent four consecutive even integers?

    c. Are these representations the same? Explain.

7.7 Applications: Distance-Rate-Time, Interest, Average, and Cost

Watch and Work

Watch the video for Example 6 in the software and follow along in the space provided.

Example 6 Application: Calculating Cost

A jeweler paid $350 for a ring. He wants to price the ring for sale so that he can give a 30% discount on the marked selling price and still make a profit of 20% on his cost. What should be the marked selling price of the ring?

Solution

Now You Try It!

Use the space provided to work out the solution to the next example.

Example A Application: Calculating Cost

A used car salesman paid \$1500 for a car. He plans to sell the car at a 25% discount on the marked selling price, but still wants to make a profit of 15% on his cost. What should be the marked selling price of the car?

7.7 Exercises

Concept Check

True/False. Determine whether each statement is true or false. If a statement is false, explain how it can be changed so the statement will be true. (**Note:** There may be more than one acceptable change.)

1. When using the formula $I = Pr$, the value of r should be written as a percent.

2. In the distance-rate-time formula $d = r \cdot t$, the value t stands for the time spent traveling.

3. Profit can be determined by subtracting the cost from the selling price.

4. The concept of average can be used to find unknown numbers.

Applications

Solve.

5. ***Traveling by Car:*** Jamie plans to take the scenic route from Los Angeles to San Francisco. Her GPS tells her it is a 420-mile trip. If she figures her average speed will be 48 mph, how long will the trip take her?

6. ***Investing:*** Amanda invests $25,000, part at 5% and the rest at 6%. The annual return on the 5% investment exceeds the annual return on the 6% investment by $40. How much did she invest at each rate?

7. ***Shopping:*** A particular style of shoe costs the dealer $81 per pair. At what price should the dealer mark them so he can sell them at a 10% discount off the selling price and still make a 25% profit?

8. ***Exam Scores:*** Marissa has five exam scores of 75, 82, 90, 85, and 77 in her chemistry class. What score does she need on the final exam to have an average grade of 80 (and thus earn a grade of B)? (All exams have a maximum of 100 points.)

Writing & Thinking

Each of the following problems is given with an incorrect answer. Explain how you can tell that the answer is incorrect without needing to solve the problem or do any algebra; then, solve the problem correctly.

9. The perimeter of an isosceles triangle is 16 cm. Since the triangle is isosceles, two sides have the same length; the third side is 2 cm shorter than one of the two equal sides. Find the length of one of the two equal sides. **Incorrect answer: 9 cm**

10. Leela found a used textbook, which was marked down 50% from the price of the new textbook. If the used textbook cost $60, how much did the new textbook cost? **Incorrect answer: $90**

7.8 Solving Linear Inequalities

1. The set of all real numbers between a and b is called an ________________.

Type of Interval	Algebraic Notation	Interval Notation	Graph
Open Interval	________	(a, b)	
Closed Interval	$a \le x \le b$	________	
________________	$\begin{cases} a \le x < b \\ a < x \le b \end{cases}$	$[a, b)$ $(a, b]$	
Open Interval	$\begin{cases} x > a \\ x < b \end{cases}$	________	
Half-open Interval	____	$[a, \infty)$ $(-\infty, b]$	

Table 1

2. In an **open interval**, ________________.

3. In a **closed interval**, ________________.

4. In a **half-open interval**, ________________.

5. **Linear inequalities** are inequalities that ________________.

6. A **solution** to an inequality is any number that ________________

________.

Addition Principle for Solving Linear Inequalities

If A and B are algebraic expressions and C is a real number, then ______________

and

____

(If a real number is added to both sides of an inequality, the new inequality is ______________ ______________.)

PROPERTIES

Multiplication Principle for Solving Linear Inequalities

If A and B are algebraic expressions and C is a positive real number, then ______________

and

____

If A and B are algebraic expressions and C is a negative real number, then ______________

and

____

(In other words, if both sides are multiplied by a ______________ ______________.)

PROPERTIES

Steps for Solving Linear Inequalities

1. Combine ______________
2. Use the addition principle of inequality to ______________ ______________
3. Use the multiplication (or division) principle of inequality to multiply (or divide) both sides by the coefficient of the variable so that ______________
   If this coefficient is negative, ______________
4. A quick (and generally satisfactory) check is to ______________ ______________
   If the statement is false, ______________

PROCEDURE

Watch and Work

Watch the video for Example 9 in the software and follow along in the space provided.

Example 9 Solving Linear Inequalities

Solve the inequality $6x + 5 \leq -1$ and graph the solution set. Write the solution set using interval notation.

Solution

Now You Try It!

Use the space provided to work out the solution to the next example.

Example A Solving Linear Inequalities

Solve the inequality $4x+8>-16$ and graph the solution set. Write the solution set using interval notation.

7.8 Exercises

Concept Check

True/False. Determine whether each statement is true or false. If a statement is false, explain how it can be changed so the statement will be true. (**Note:** There may be more than one acceptable change.)

1. If only one endpoint is included in an interval, it is called a half-open interval.

2. When both sides of a linear inequality are multiplied by a negative constant, the sense of the inequality should stay the same.

3. To check the solution set of a linear inequality, every solution in the solution set must be checked in the original inequality.

4. The infinity symbol ∞ does not represent a specific number.

Practice

Graph each interval on a real number line and tell what type of interval it is.

5. $x \leq -3$

6. $-1.5 \leq x < 3.2$

Solve each inequality and graph the solution set. Write each solution set using interval notation.

7. $x + 1 > 5$

8. $-2x \geq 6$

9. $4x - 7 \geq 9$

10. $5x + 6 \geq 2x - 2$

Applications

Solve.

11. ***Test Scores:*** A statistics student has grades of 82, 95, 93, and 78 on four hour-long exams. He must average 90 or higher to receive an A for the course. What scores can he receive on the final exam and earn an A if:

a. The final is equivalent to a single hour-long exam (100 points maximum)?

b. The final is equivalent to two hourly exams (200 points maximum)?

12. ***Postage:*** Allison is going to the post office to buy 34¢ stamps and 3¢ adjustment stamps. Since the current postage rate is 49¢, she will need 5 times as many 3¢ adjustment stamps as 34¢ stamps. If she has $12.25 to spend, what is the largest number of 34¢ stamps she can buy?

Writing & Thinking

13. **a.** Write a list of three situations where inequalities might be used in daily life.

    b. Illustrate theses situations with algebraic inequalities and appropriate numbers.

7.9 Compound Inequalities

Union and Intersection

The **union** (symbolized $\cup$, as in $A \cup B$) of two (or more) sets is ______________________________

____

The **intersection** (symbolized $\cap$, as in $A \cap B$) of two (or more) sets is ______________________________

____

The word **or** is used to indicate ______________________ and the word **and** is used to indicate ______________________.

DEFINITION

1. If the elements in a set can be counted, the set is said to be ______________________. If the elements cannot be counted, the set is said to be ______________________.

Watch and Work

Watch the video for Example 10 in the software and follow along in the space provided.

Example 10 Solving Compound Inequalities Containing OR

Solve the compound inequality $3x + 4 \geq 1$ or $2x - 7 \leq -3$ and graph the solution set. Write the solution set using interval notation.

Solution

Now You Try It!

Use the space provided to work out the solution to the next example.

Example A Solving Compound Inequalities Containing OR

Solve the inequality $4x - 3 \geq 1$ or $5x + 2 < 12$ and graph the solution set. Write the solution set using interval notation.

7.9 Exercises

Concept Check

True/False. Determine whether each statement is true or false. If a statement is false, explain how it can be changed so the statement will be true. (**Note:** There may be more than one acceptable change.)

1. The union of two sets is the set of all elements that belong to both sets.

2. The intersection of two sets is the set of elements that belong to just one set or the other, but not both.

3. A null set contains no elements.

4. The solution set of a compound inequality containing "and" is the union of the solution sets of the two inequalities.

Practice

5. Find the union and intersection of $A = \{2, 4, 6, 8\}$ and $B = \{1, 2, 3, 4\}$

6. Graph the set $\{x \mid x > 3 \text{ or } x \leq 2\}$ on a real number line.

Solve each compound inequality and graph its solution set. Write each solution set using interval notation.

7. $\{x \mid x+3>2 \text{ and } x-1<5\}$

8. $\{x \mid x<-1 \text{ or } 2x+1 \geq 3\}$

9. $-4<x+5<6$

7.10 Absolute Value Equations

Absolute Value

The **absolute value** of a number is ______________________________

DEFINITION

Solving Absolute Value Equations

For $c > 0$:

a. If $|x| = c$, then ____________________

b. If $|ax + b| = c$, then __________________________

DEFINITION

Solving Equations with Two Absolute Value Expressions

If $|a| = |b|$, then ________________________

More generally,

if $|ax + b| = |cx + d|$, then ____________________________________

DEFINITION

7.10 Exercises

Concept Check

True/False. Determine whether each statement is true or false. If a statement is false, explain how it can be changed so the statement will be true. (**Note:** There may be more than one acceptable change.)

1. Equations involving absolute value can only have one solution.

2. If two numbers have the same absolute value, they must be equal to each other.

3. There is no number that has a negative absolute value.

4. If $|a| = |b|$, we can only rewrite it as $a = b$.

Practice

Solve each absolute value equation.

5. $|z| = -\frac{1}{5}$

6. $|y+5| = -7$

7. $|-2x+1| = -3$

8. $3\left|\frac{x}{3}+1\right| - 5 = -2$

9. $\left|\frac{x}{3}-4\right| = \left|\frac{5x}{6}+1\right|$

Name: Date:

7.11 Absolute Value Inequalities

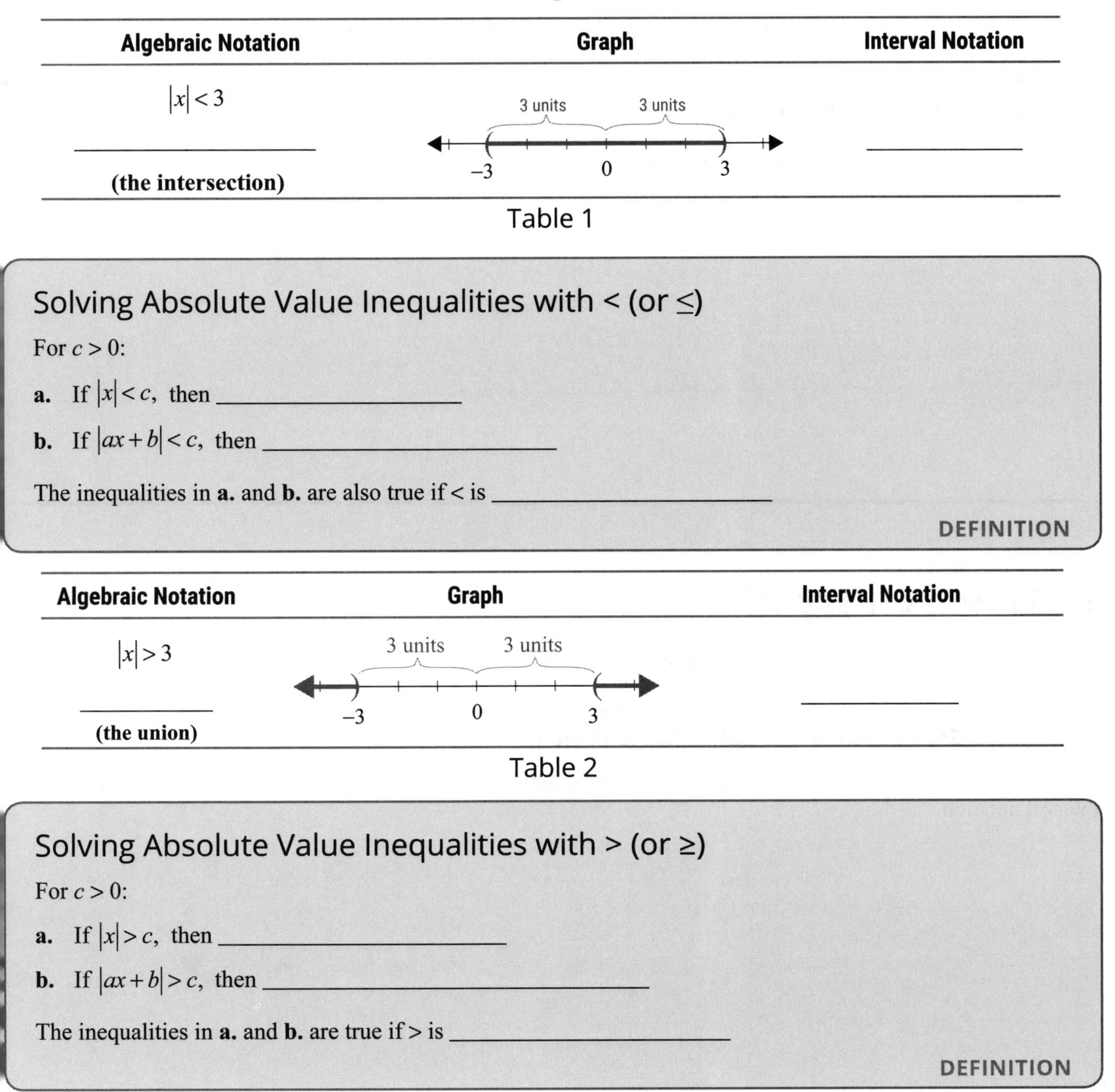

Algebraic Notation	Graph	Interval Notation
$\lvert x\rvert < 3$ ________________ **(the intersection)**	3 units 3 units −3 0 3	____________

Table 1

Solving Absolute Value Inequalities with < (or ≤)

For $c > 0$:

a. If $\lvert x\rvert < c$, then __________________

b. If $\lvert ax+b\rvert < c$, then ______________________

The inequalities in **a.** and **b.** are also true if < is ______________________

DEFINITION

Algebraic Notation	Graph	Interval Notation
$\lvert x\rvert > 3$ __________ **(the union)**	3 units 3 units −3 0 3	____________

Table 2

Solving Absolute Value Inequalities with > (or ≥)

For $c > 0$:

a. If $\lvert x\rvert > c$, then ______________________

b. If $\lvert ax+b\rvert > c$, then ____________________________

The inequalities in **a.** and **b.** are true if > is ______________________

DEFINITION

Watch and Work

Watch the video for Example 9 in the software and follow along in the space provided.

Example 9 Solving Absolute Value Inequalities

Solve the absolute value inequality and graph the solution set: $|2x-5|-5 \geq 4$

Solution

Now You Try It!

Use the space provided to work out the solution to the next example.

Example A Solving Absolute Value Inequalities

Solve the absolute value inequality $|3x+2|+1 \geq 8$ and graph the solution set. Write the solution set using interval notation.

7.11 Exercises

Concept Check

True/False. Determine whether each statement is true or false. If a statement is false, explain how it can be changed so the statement will be true. (**Note:** There may be more than one acceptable change.)

1. If the solution is a union, there are two statements or inequalities, both of which must be true.

2. If the solution to a compound inequality is $-4 < x < 6$, then the solution is a union.

3. For a number to have absolute value greater than 2, its distance from 0 must be less than 2.

4. The inequality $|2x + 9| < -2$ has no solution.

Practice

Solve each of the absolute value inequalities and graph the solution sets. Write each solution using interval notation.

5. $|x| \geq -2$

6. $|x - 3| > 2$

7. $|x + 2| \leq -4$

8. $|3x + 4| - 1 < 0$

9. $4 \leq |3x + 1| - 6$

10. $3|4x+5|-5>10$

Writing & Thinking

A set of real numbers is described. **a**. Sketch a graph of the set on a real number line. **b**. Represent each set using absolute value notation. **c**. Represent each set using interval notation. If the set is one interval, state what type of interval it is.

11. The set of real numbers between -10 and 10, inclusive

12. The set of real numbers within 7 units of 4

Chapter 7 Project

A Linear Vacation

An activity to demonstrate the importance of solving linear equations in real life.

*he process of finding ways to use math to solve real-life problems is called **mathematical modeling**. In the following activity ou will be using linear equations to model some real-life scenarios that arise during a family vacation.*

or each question, be sure to write a linear equation in one variable and then solve.

1. Penny and her family went on vacation to Florida and decided to rent a car to do some sightseeing. The cost of the rental car was a fixed price per day plus $0.29 per mile. When she returned the car, the bill was $209.80 for three days and they had driven 320 miles. What was the fixed price per day to rent the car?

2. Penny's son Chase wanted to go to the driving range to hit some golf balls. Penny gave the pro-shop clerk $60 for three buckets of golf balls and received $7.50 in change. What was the cost of each bucket?

3. Penny's family decided to go to the Splash Park. They purchased two adult tickets and two child tickets. The adult tickets were $1\frac{1}{2}$ times the price of the child tickets and the total cost for all four tickets was $85. What was the cost of each type of ticket?

4. Penny's family went shopping at a nearby souvenir shop where they decided to buy matching T-shirts. If they bought four T-shirts and a $2.99 bottle of sunscreen for a total cost of $54.95, before tax, how much did each T-shirt cost?

5. Penny and her family went out to eat at a local restaurant. Three of them ordered a fried shrimp basket, but her daughter Meghan ordered a basket of chicken tenders, which was $4.95 less than the shrimp basket. If the total order before tax was $46.85, what was the price of a shrimp basket?

6. While on the beach, Penny and her family decided to play a game of volleyball. Penny and her son beat her husband and daughter by two points. If the combined score of both teams was 40, what was the score of the winning team?

CHAPTER 8

Graphing Linear Equations and Inequalities

Math @ Work

Linear equations (or linear functions) can be used to represent many real-world situations. They play an especially important role in the business world, where they can be used to model supply, demand, cost, or profit. Consider the relationship between supply and demand. When a company raises the price of a product, fewer people may be inclined to buy that product. However, the profit earned from the sale of each item will be higher so they might have a higher profit overall. Similarly, if they decrease the price of a product, more people may buy it, also resulting in a higher profit. When a company creates models based on industry research, the information provided by these models can be used to determine the best price for a product to ensure the highest profit is made.

Suppose that a pharmaceutical company is examining the demand of a certain drug. When the price was set at \$55 per unit, 1.2 million units sold per year. When the price was set at \$40 per unit, sales increased to 1.8 million units per year. Using this information, the company can create a linear function to model the relationship between the price of the drug and the demand. If the company has a goal of selling 2.0 million units per year, at what price would they need to set the cost of the drug?

8.1 The Cartesian Coordinate System

1. Descartes based his system on a relationship between ______________ in a plane and ____________________ of real numbers.

2. In the ordered pair (x, y), x is called the ______________________ and y is called the ______________ ______________.

3. In an ordered pair of the form (x, y), the ____________________________ is called the **independent variable** and the ______________________________ is called the **dependent variable**.

4. The Cartesian coordinate system relates algebraic equations and ordered pairs to geometry. In this system, two number lines intersect at right angles and separate the plane into four ________________. The **origin**, designated by the ordered pair $(0, 0)$, is ____. The horizontal number line is called the ____________________ or _____________. The vertical number line is called the ____________________ or _____________.

One-to-One Correspondence

____

____

DEFINITION

Watch and Work

Watch he video for Example 4 in the software and follow along in the space provided.

Example 4 Finding Ordered Pairs

Complete the table so that each ordered pair will satisfy the equation $y = -3x + 1$.

x	y	(x, y)
0		
	4	
$\frac{1}{3}$		
3		

Solution

Now You Try It!

Use the space provided to work out the solution to the next example.

Example A Finding Ordered Pairs

Complete the table so that each ordered pair will satisfy the equation $y = -3x + 2$.

x	y	(x, y)
0		
	1	
−2		
	0	

Solution

8.1 Exercises

True/False. Determine whether each statement is true or false. If a statement is false, explain how it can be changed so the statement will be true. (**Note:** There may be more than one acceptable change.)

1. The graph of every ordered pair that has a positive x-coordinate and a negative y-coordinate can be found in Quadrant IV.

2. To find the y-value that corresponds with $x = 2$, substitute 2 for x into the given equation and solve for y.

3. If $(-7, 3)$ is a solution of $y = 3x + 24$, then $(-7, 3)$ satisfies $y = 3x + 24$.

4. If point $A = (0, 4)$, then point A lies on the x-axis.

Practice

List the set of ordered pairs corresponding to the points on the graph.

5.

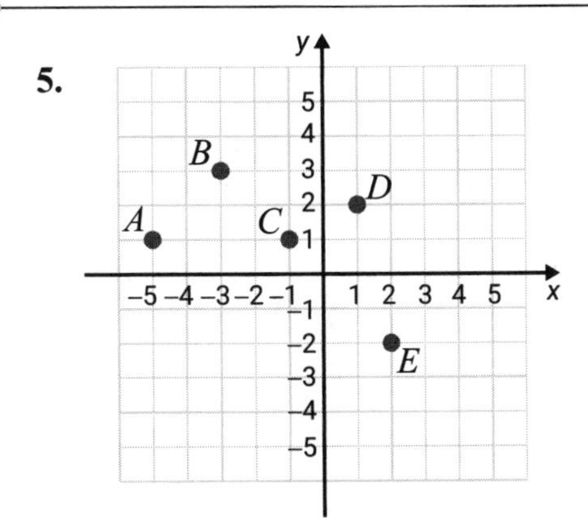

Plot each set of ordered pairs and label the points.

6. $\{A(4,-1), B(3,2), C(0,5), D(1,-1), E(1,4)\}$

Determine the missing coordinate in each of the ordered pairs so that the point will satisfy the equation given.

7. $x - 2y = 2$

a. $(0, \_\_)$

b. $(4, \_\_)$

c. $(\_\_, 0)$

d. $(\_\_, 3)$

Complete the tables so that each ordered pair will satisfy the given equation. Plot the resulting sets of ordered pairs.

8. $y = 2x - 3$

x	y
0	
	–1
–2	
	3

Determine which, if any, of the ordered pairs satisfy the given equation.

9. $2x - 3y = 7$

a. $(1, 3)$

b. $\left(\frac{1}{2}, -2\right)$

c. $\left(\frac{7}{2}, 0\right)$

d. $(2, 1)$

The graph of a line is shown. List any three points on the line. (There is more than one correct answer.)

10.

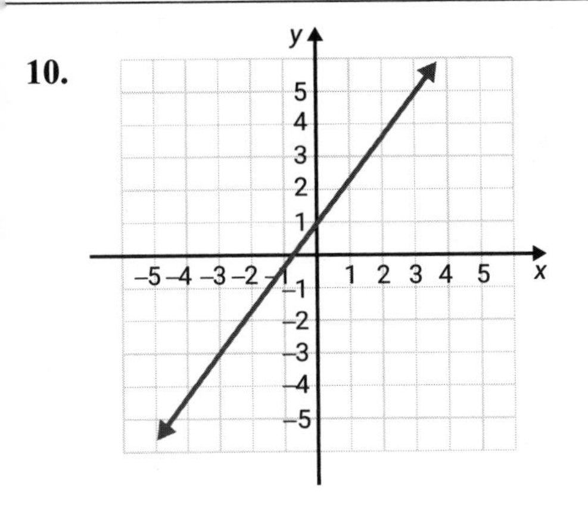

Applications

Solve.

11. ***Exchange Rates:*** At one point in 2017, the exchange rate from US dollars to Euros was $E = 0.85D$ where E is Euros and D is dollars.

a. Make a table of ordered pairs for the values of D and E if D has the values \$100, \$200, \$300, \$400, and \$500.

b. Plot the points corresponding to the ordered pairs.

12. ***Temperature:*** Given the equation $F = \frac{9}{5}C + 32$ where C is temperature in degrees Celsius and F is the corresponding temperature in degrees Fahrenheit:

a. Make a table of ordered pairs for the values of C and F if C has the values $-20°$, $-10°$, $-5°$, $0°$, $5°$, $10°$, and $15°$.

b. Plot the points corresponding to the ordered pairs.

8.2 Graphing Linear Equations in Two Variables

Standard Form of a Linear Equation

Any equation of the form

____

where A, B, and C are real numbers and A and B are not both equal to 0, is called the standard form of a linear equation.

DEFINITION

To Graph a Linear Equation in Two Variables

1. Locate any two points that ____________________
2. ____
3. ______________________________
4. To check: Locate a third point that ______________________________________ __________

PROCEDURE

Watch and Work

Watch the video for Example 2 in the software and follow along in the space provided.

Example 2 Graphing a Linear Equation in Two Variables

Graph: $2x + 3y = 6$

Solution

Now You Try It!

Use the space provided to work out the solution to the next example.

Example A Graphing a Linear Equation in Two Variables

Graph: $3x + 2y = 6$

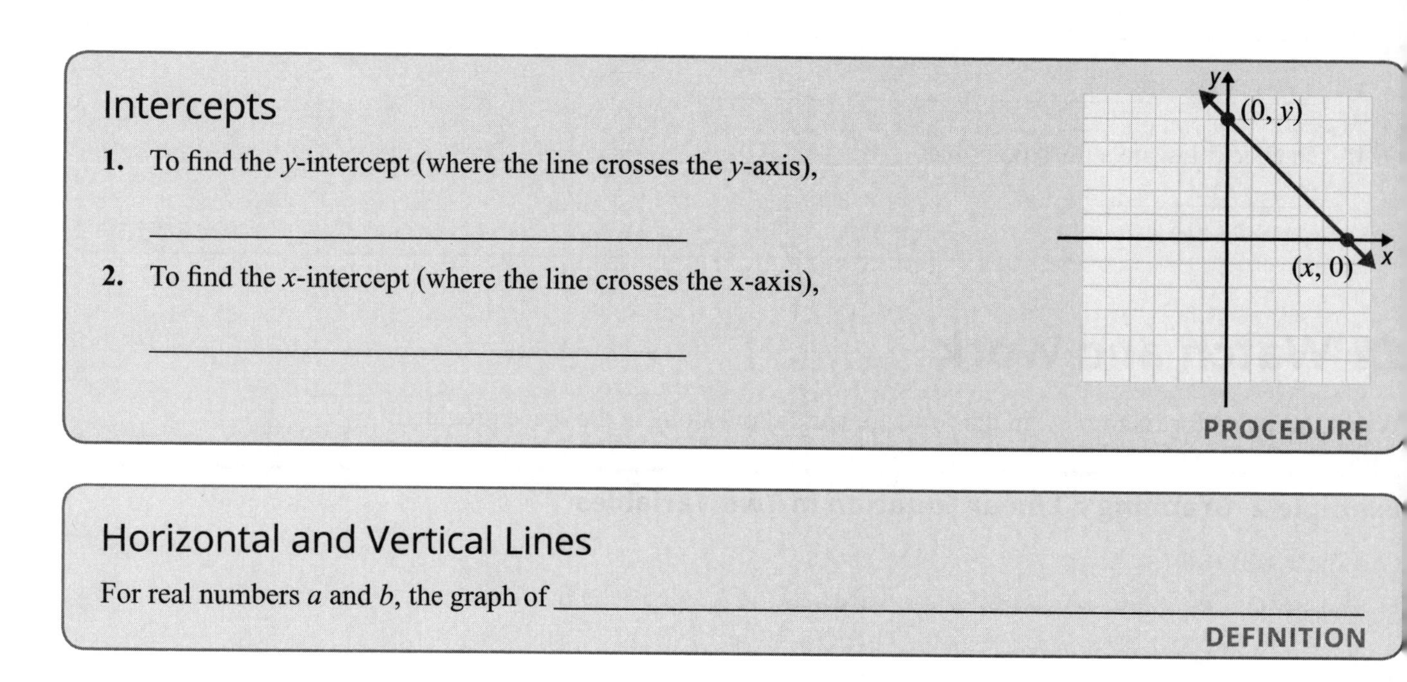

Intercepts

1. To find the y-intercept (where the line crosses the y-axis), ____
2. To find the x-intercept (where the line crosses the x-axis), ____

PROCEDURE

Horizontal and Vertical Lines

For real numbers a and b, the graph of ____

DEFINITION

8.2 Exercises

Concept Check

True/False. Determine whether each statement is true or false. If a statement is false, explain how it can be changed so the statement will be true. (**Note:** There may be more than one acceptable change.)

1. The y-intercept is the point where a line crosses the y-axis.

2. The terms ordered pair and point are used interchangeably.

3. A horizontal line does not have a y-intercept.

4. All x-intercepts correspond to an ordered pair of the form $(0, y)$.

Practice

Graph each linear equation by locating at least two ordered pairs that satisfy the given equation.

5. $x + y = 3$

6. $x = 1$

7. $y = -3$

Graph each linear equation by locating the x-intercept and the y-intercept.

8. $y = 4x - 10$

9. $3x - 7y = -21$

Applications

Solve.

10. ***Chemistry:*** The amount of potassium in a clear bottle of a popular sports drink declines over time when exposed to the UV lights found in most grocery stores. The amount of potassium in a container of this sports drink is given by the equation $y = -30x + 360$, where y represents the mg of potassium remaining after x days on the shelf. Find both the x-intercept and y-intercept, and interpret the meaning of each in the context of this problem.

11. ***Education:*** Mr. Adler has found that the grade each student gets in his Introductory Algebra course directly correlates with the amount of time spent doing homework, and is represented by the equation $y = 7x + 30$, where y represents the numerical score the student receives on an exam (out of 100 points) after spending x hours per week doing homework. Find the y-intercept and interpret its meaning in this context.

Writing & Thinking

12. Explain, in your own words, why it is sufficient to find the x-intercept and y-intercept to graph a line (assuming that they are not the same point).

13. Explain, in your own words, how you can determine if an ordered pair is a solution to an equation.

8.3 Slope-Intercept Form

1. For a line, the ______________________ is called the **slope of the line**.

Slope

Let $P_1(x_1, y_1)$ and $P_2(x_2, y_2)$ be two points on a line. The slope can be calculated as follows.

slope =

Note: ______________________ is standard notation for representing the slope of a line.

FORMULA

Watch and Work

Watch the video for Example 2 in the software and follow along in the space provided.

Example 2 Finding the Slope of a Line

Find the slope of the line that contains the points $(1, 3)$ and $(5, 1)$, and then graph the line.

Solution

Now You Try It!

Use the space provided to work out the solution to the next example.

Example A Finding the Slope of a Line

Find the slope of the line that contains the points $(0, 5)$ and $(4, 2)$, and then graph the line.

Positive and Negative Slope

Lines with positive slope go ______________________________

Lines with negative slope go ______________________________

DEFINITION

Horizontal and Vertical Lines

The following two general statements are true for horizontal and vertical lines.

1. For horizontal lines (of the form ____________________

2. For vertical lines (of the form ____________________

DEFINITION

Slope-Intercept Form

____________ is called the slope-intercept form for the equation of a line, where m is the slope and $(0, b)$ is the y-intercept.

DEFINITION

8.3 Exercises

Practice

Find the slope of the line determined by each pair of points.

1. $(1, -2); (1, 4)$

2. $(-3, 7); (4, -1)$

Determine whether the equation $x = -3$ represents a horizontal line or a vertical line and give its slope.

3. $x = -3$

Write each equation in slope-intercept form. Find the slope and y-intercept, and then use them to draw the graph.

4. $y = 2x - 1$

5. $3y - 9 = 0$

Find an equation in slope-intercept form for the line passing through (0,3) with the slope $m = -\frac{1}{2}$.

6. $(0, 3); m = -\frac{1}{2}$

Applications

Solve.

7. ***Purchases:*** John bought his new car for $35,000 in the year 2014. He knows that the value of his car has depreciated linearly. If the value of the car in 2017 was $23,000, what was the annual rate of depreciation of his car? Show this information on a graph. (When graphing, use years as the x-coordinates and the corresponding values of the car as the y-coordinates.)

8. ***Cell Phones:*** The number of people in the United States with mobile cellular phones was about 198 million in 2011 and about 232 million in 2016. If the growth in the usage of mobile cellular phones was linear, what was the approximate rate of growth per year from 2011 to 2016. Show this information on a graph. (When graphing, use years as the x-coordinates and the corresponding numbers of users as the y-coordinates.)[1]

Writing & Thinking

9. **a.** Explain in your own words why the slope of a horizontal line must be 0.

   b. Explain in your own words why the slope of a vertical line must be undefined.

1 Source: https://www.statista.com/statistics/231612/number-of-cell-phone-users-usa/

8.4 Point-Slope Form

Point-Slope Form

An equation of the form

____

is called the point-slope form for the equation of a line that ____________________

____

FORMULA

Finding the Equation of a Line Given Two Points

To find the equation of a line given two points on the line:

1. Use the formula

____

2. Use this slope, m, and ____________________

____

PROCEDURE

Watch and Work

Watch the video for Example 4 in the software and follow along in the space provided.

Example 4 Finding Equations of Lines Using a Graph

Write an equation in standard form for the following line.

Solution

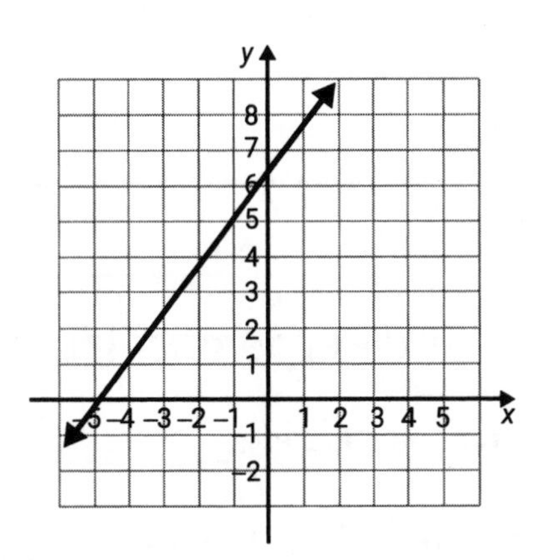

Now You Try It!

Use the space provided to work out the solution to the next example.

Example A Finding Equations of Lines Using a Graph

Write an equation in standard form for the following line.

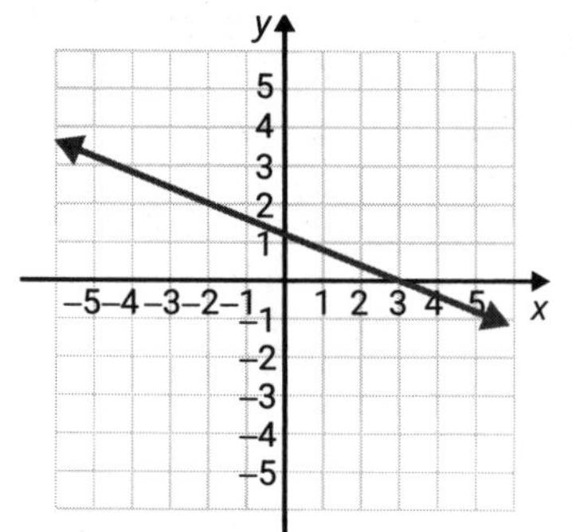

Parallel and Perpendicular Lines

Parallel lines are lines that ______________________ and who have the ____________________.

Perpendicular lines are lines that ________________________________ and whose slopes are

________________________________. Horizontal lines are perpendicular to ________________.

DEFINITION

8.4 Exercises

Practice

Find a. the slope, b. a point on the line, and c. the graph of the line for the following equations in point-slope form.

1. $y-1=2(x-3)$

Find an equation in standard form for the line passing through the given point with the given slope. Graph the line.

2. $(3,4);\ m=3$

Find an equation in slope-intercept form for the line passing through the two given points.

3. $(-5,1);\ (2,0)$

Find an equation in standard form for the line shown.

4.

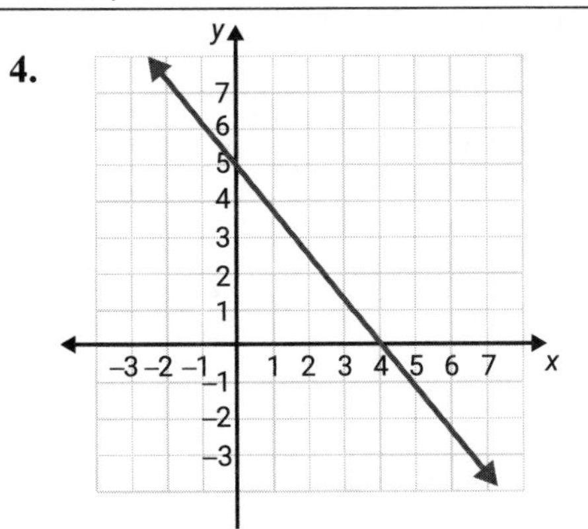

Find an equation in slope-intercept form that satisfies each set of conditions.

5. Find an equation in slope-intercept form for the horizontal line through the point $(-2, 6)$.

Determine whether the pair of lines is **a.** parallel, **b.** perpendicular, or **c.** neither. Graph both lines. (**Hint:** Write the equations in slope-intercept form and then compare slopes.)

6. $\begin{cases} y = -2x + 3 \\ y = -2x - 1 \end{cases}$

Applications

Solve.

7. ***Transportation:*** The cost for an airline to fly from Raleigh, NC, to Nashville, TN, is \$5000. The airline charges \$100 for the one-way ticket from Raleigh to Nashville.

a. Find an equation for the profit P made by the airline on this one-way flight if they sell t tickets.

b. Use the equation found in part **a.** to determine the number of tickets that must be sold for the airline to "break even;" that is, for the profit to be equal to 0?

8. ***Interest:*** Natalie invested some money in a simple interest savings fund. After 2 years, she earned \$120 in interest. After 5 years, she earned \$300 in interest.

a. Write two ordered pairs from the information given where x represents the time in years and y represents the amount of interest earned.

b. Find the slope of the line which contains the two ordered pairs from Part **a.**

c. Write the point-slope equation that models the situation.

d. Rewrite this equation in $y = mx + b$ form.

8.5 Introduction to Functions and Function Notation

Relation, Domain, and Range

A **relation** is a set of ______________________________

The **domain**, ***D***, of a relation is the set of ______________________________

The **range**, ***R***, of a relation is the set of ______________________________

DEFINITION

Functions

A **function** is a relation in which ______________________________

____

DEFINITION

Vertical Line Test

If any vertical line intersects the graph of a relation at more than one point, then the relation is

____

PROCEDURE

Linear Function

A linear function is a function represented by an equation of the form

____

The domain of a linear function is ______________________________

____

DEFINITION

1. In function notation, instead of writing y, write _____, read "f **of** x."

Watch and Work

Watch the video for Example 6 in the software and follow along in the space provided.

Example 6 Evaluating Functions

For the function $g(x) = 4x + 5$, find:

a. $g(2)$

b. $g(-1)$

c. $g(0)$

Solution

Now You Try It!

Use the space provided to work out the solution to the next example.

Example A Evaluating Functions

For the function $g(x) = 3x - 2$, find:

a. $g(3)$

b. $g(-2)$

c. $g(0)$

8.5 Exercises

Concept Check

True/False. Determine whether each statement is true or false. If a statement is false, explain how it can be changed so the statement will be true. (**Note:** There may be more than one acceptable change.)

1. If the domain of a linear function is not explicitly stated, the implied domain is the set of all values of x that produce real values for y.

2. A relation is a function in which each domain element has exactly one corresponding range element.

3. In a function, the range elements can have more than one corresponding domain element.

4. If $s = \{(1, -6), (3, 5), (4, 0), (1, 2)\}$, then s is a function.

Practice

List the sets of ordered pairs that correspond to the points. State the domain and range and indicate if the relation is a function.

5.

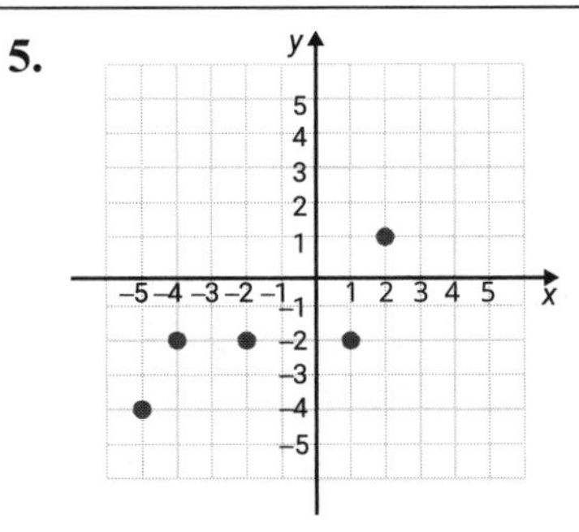

Graph the relation. State the domain and range and indicate which of the relation is a function.

6. $h = \{(1, -5), (2, -3), (-1, -3), (0, 2), (4, 3)\}$

Use the vertical line test to determine whether the graph represents a function. State the domain and range using interval notation.

7.

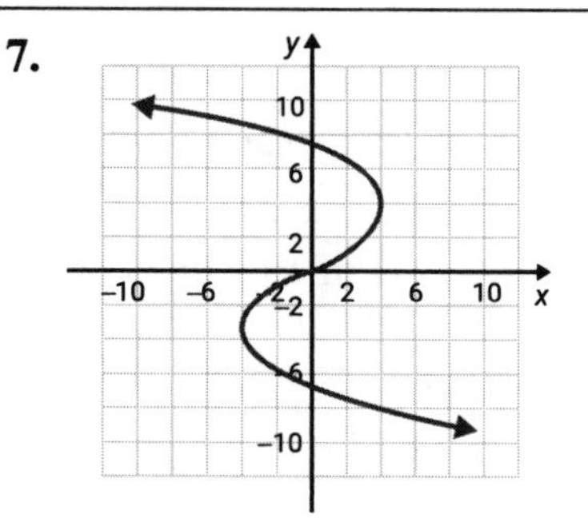

State the domain of the function.

8. $h(x)=\dfrac{7}{3x}$

Find the values of the function as indicated.

9. $F(x)=6x^2-10$

a. $F(0)$

b. $F(-4)$

c. $F(4)$

Applications

Solve.

10. ***Nursing:*** A nurse hangs a 1000-milliliter IV bag which is set to drip at 120 milliliters per hour. Create a model of this situation to represent the amount of IV solution left in the bag after x hours.

a. The y-intercept is the amount of IV solution in the bag initially (time = 0). What is the y-intercept?

b. The slope is equal to the rate that the IV solution is dispensed per hour. What is the slope? (**Hint:** Consider whether the amount of IV solution in the bag is increasing or decreasing and how this would affect the slope.)

c. Write an equation in slope-intercept form to model this situation.

d. Write the equation from Part **c.** using function notation.

e. State the domain and range of the function.

f. State any additional restrictions that should be made on the domain for it to make sense in the context of this problem.

g. How much IV solution is left in the bag after 5 hours?

8.6 Graphing Linear Inequalities in Two Variables

Half-plane A straight line separates ____________________

The points on one side of the line are in ____________________

____

Boundary line The line ____________________

Closed half-plane If the boundary line is ____________________

____

Open half-plane If the boundary line is ____________________

____

Graphing Linear Inequalities

1. First, graph ____________ (__________ if the inequality is __________ if the inequality is __________).

2. Next, determine which ____________________

Method 1

a. Test any one ____________________

b. If the test-point satisfies the inequality, shade the ____________________

____

Note: The point $(0, 0)$, if it is not on the boundary line, is usually the easiest point to test.

Method 2

a. Solve the inequality for y ____________________

b. If the solution shows y ____________________

c. If the solution shows y ____________________

Note: If the boundary line is vertical, then ____________

If the solution shows ____________________

If the solution shows ____________________

3. The shaded half-plane (and the line if it is solid) ____________________

PROCEDURE

Watch and Work

Watch the video for Example 2 in the software and follow along in the space provided.

Example 2 Graphing Linear Inequalities

Graph the solution set to the inequality $y > 2x$.

Solution

Now You Try It!

Use the space provided to work out the solution to the next example.

Example A Graphing Linear Inequalities

Graph the solution set to the inequality $x > -2$.

8.6 Exercises

Concept Check

True/False. Determine whether each statement is true or false. If a statement is false, explain how it can be changed so the statement will be true. (**Note:** There may be more than one acceptable change.)

1. A solid boundary line indicates that the points on that line are included in the solution.

2. If the solution set is an open half-plane, then the boundary line is included in the solution.

3. The boundary line is solid when the inequality uses a < or > symbol.

4. The slope of an inequality is used to determine whether the boundary line is included in the solution.

Practice

Graph the solution set of each of the linear inequalities.

5. $x + y \leq 7$

6. $5x - y < 4$

7. $x + 4 \geq 0$

8. $\frac{1}{2}x - y > 1$

9. $2x - \frac{4}{3}y > 8$

Applications

Solve.

10. ***Grades:*** The grade for a 1-credit-hour survey class is based on an exam and a project, which are worth a maximum of 50 points each. The sum of the two scores must be at least 75 points for a student to earn a passing grade.

a. Let the amount of points earned on the exam be represented by the variable x and the amount of points earned on the project be represented by the variable y. Create a linear inequality to describe the solution set for a passing grade.

b. Graph the linear inequality from part **a.**

c. A student earns 45 points on their final exam and 22 points on their project. Plot this point on the graph. Did this student earn a passing grade?

d. Are there any points in the solution set which do not make sense for this situation?

Writing & Thinking

11. Explain in your own words how to test to determine which side of the graph of an inequality should be shaded.

12. Describe the difference between a closed and an open half-plane.

Chapter 8 Project

What's Your Car Worth?

An activity to demonstrate the use of linear models in real life.

When buying a new car, there are a number of things to keep in mind: your monthly budget, length of the warranty, routine maintenance costs, potential repair costs, cost of insurance, etc.

One thing you may not have considered is the depreciation, or reduction in value, of the car over time. If you like to purchase a new car every 3 to 5 years, then the retention value of a car, or the portion of the original price remaining, is an important factor to keep in mind. If your new car depreciates in value quickly, you may have to settle for less money if you choose to resell it later or trade it in for a new one.

Below is a table of original Manufacturer's Suggested Retail Price (MSRP) values and the anticipated retention value after 3 years for three 2017 mid-price car models.

Car Model	2017 MSRP	Expected Value in 2020	Rate of Depreciation (slope)	Linear Equation
Mini Cooper	$21,800	$ 9,590		
Toyota Camry	$23,955	$11,211		
Ford Taurus	$28,220	$11,234		

1. The x-axis of the graph is labeled "Years after Purchase." Recall that the MSRP value for each car is for the year 2017 when the car was purchased.

   a. What value on the x-axis will correspond to the year 2017?

   b. Using the value from Part **a.** as the x-coordinate and the MSRP values in column two as the y-coordinates, plot three points on the graph corresponding to the value of the three cars at time of purchase.

   c. What value on the x-axis will correspond to the year 2020?

   d. Using the value from Part **c.** as the x-coordinate and the expected car values in column three as the y-coordinates, plot three points on the graph corresponding to the value of the three cars in 2020.

2. Draw a line segment on the graph connecting the pair of points for each car model. Label each line segment after the car model it represents and label each point with a coordinate pair, (x, y). Consider using a different color when plotting

each line segment to help you identify the three models.

3. Use the slope formula, $m = \frac{y_2 - y_1}{x_2 - x_1}$, to answer the following questions.

 a. Calculate the rate of depreciation for each model by calculating the slope (or rate of change) between each pair of corresponding points using the slope formula and enter it into the appropriate row of column 4 of the table.

 b. Are the slopes calculated above positive or negative? Explain why.

 c. Interpret the meaning of the slope for the Toyota Camry making sure to include the units for the variables.

 d. Which car model depreciates in value the fastest? Explain how you determined this.

4. Use the slope-intercept form of an equation, $y = mx + b$, for the following problems.

 a. Write an equation to model the depreciation in value over time of each car (in years). Place these in column five of the table.

 b. What does the y-intercept represent for each car?

5. Use the equations from problem 4 for the following problems.

 a. Predict the value of the Mini Cooper 4 years after purchase.

 b. Predict the value of the Ford Taurus 2½ years after purchase.

6. Determine from the graph how long it takes from the time of purchase until the Ford Taurus and the Mini Cooper have the same value? (It may be difficult to read the coordinates for the point of intersection, but you can get a rough idea of the value from the graph. You can find the exact point of intersection by setting the two equations equal to one another and solving for x.)

 a. After how many years are the car values for the Ford Taurus and the Mini Cooper the same? (Round to the nearest tenth.)

 b. What is the approximate value of both cars at this point in time? (Round to the nearest 100 dollars.)

7. How long will it take for the Toyota Camry to fully depreciate (reach a value of zero)?

 a. For the first method, extend the line segment between the two points plotted for the Toyota Camry until it intersects the horizontal axis. The x-intercept is the time at which the value of the car is zero.

 b. Substitute 0 for y in the equation you developed for the Toyota Camry and solve for x. (Round to the nearest year.)

 c. Compare the results from Parts **a.** and **b.** Are the results similar? Why or why not?

8. How long will it take for the Ford Taurus to fully depreciate? (Repeat Problem 7 for the Ford Taurus. Round to the nearest year.)

9. Why is there such a difference in depreciation for the Camry and the Taurus? Do some research on a reliable Internet site and list two reasons why cars depreciate at different rates.

10. Based on what you have learned from this activity, do you think retention value will be a significant factor when you purchase your next car? Why or why not?

CHAPTER 9

Systems of Linear Equations

Math @ Work

Sometimes a single equation does not provide enough information to reach a solution. When this occurs, we must construct more than one equation and use more than one variable, and use these equations to find an appropriate solution, if one exists. Such sets of equations are called systems.

Chinese mathematicians recorded methods for solving systems of two equations sometime between the 10th and 2nd century BCE, as described in the book *The Nine Chapters on Mathematical Art*. However, it wasn't until the 18th and 19th centuries that methods for solving systems of equations were introduced to the Western world by the European mathematicians Gabriel Cramer, Carl Frederick Gauss, and Wilhelm Jordan. Thanks in large part to their work, algorithms have been derived for calculating solutions to systems with large numbers of equations and variables. With the invention of computers, the time it takes to solve these systems has been drastically reduced, and solving equations now plays a prominent role in the fields of engineering, physics, chemistry, computer science, and economics.

Suppose a farmer is building a fenced-in pasture for his cows and has 180 meters of fencing to use. For aesthetic reasons, the farmer wants the pasture to be in the shape of a rectangle where the length is two times the width. The farmer can use this information, along with the fact that the perimeter of a rectangle is $P = 2l + 2w$, to determine the dimensions of the pasture. What are the dimensions of the desired pasture?

9.1 Systems of Linear Equations: Solutions by Graphing

1. Form a system of linear equations or ______________________________

2. The **solution of a system** of linear equations is the set of ordered pairs (or points) that satisfy ______________________________

To Solve a System of Linear Equations by Graphing

1. Graph both linear equations on ______________________________
2. Observe the point of ______________________________
   a. If the slopes of the two lines are different, then ______________________________
   The system has ______________________________
   b. If the lines have the same slope and different y-intercepts, then ______________________________
   The system has ______________________________
   c. If the lines are the same line, then ______________________________
   ____
3. Check the solution (if there is one) in ______________________________

PROCEDURE

Consistent and Inconsistent Systems of Linear Equations

1. A system is **consistent** if ______________________________
2. A system is **inconsistent** ______________________________

DEFINITION

Watch and Work

Watch the video for Example 3 in the software and follow along in the space provided.

Example 3 Solving Systems (One Solution/A Consistent System)

Solve the system of equations by graphing.

$$\begin{cases} y = -x + 4 \\ y = 2x + 1 \end{cases}$$

Solution

Now You Try It!

Use the space provided to work out the solution to the next example.

Example A Solving Systems (One Solution/A Consistent System)

Solve the system of equations by graphing.

$$\begin{cases} y = 2x + 6 \\ y = -x - 3 \end{cases}$$

Dependent and Independent Systems of Linear Equations

If the graphs of two linear equations are

a. the same line, then the ______________________________

b. different lines, then the ______________________________

DEFINITION

9.1 Exercises

Concept Check

True/False. Determine whether each statement is true or false. If a statement is false, explain how it can be changed so the statement will be true. (**Note:** There may be more than one acceptable change.)

1. To check a solution, substitute it into one of the equations. If the solution satisfies one equation it will satisfy all of the equations.

2. A system of equations with graphs that are parallel lines has exactly one solution.

3. A system of equations with graphs that intersect at one point has exactly one solution.

4. A system of equations with graphs that are the same line has infinitely many solutions.

Practice

Determine which of the given points, if any, lie on both of the lines in the systems of equations by substituting each point into both equations.

5. $\begin{cases} 2x+4y-6=0 \\ 3x+6y-9=0 \end{cases}$

a. $(1, 1)$

b. $(2, 0)$

c. $\left(0, \frac{3}{2}\right)$

d. $(-1, 3)$

The graphs of the lines represented by system of equations are given. Determine the solution of the system by looking at the graph. Check your solution by substituting into both equations.

6. $\begin{cases} x+2y=4 \\ x-y=-2 \end{cases}$

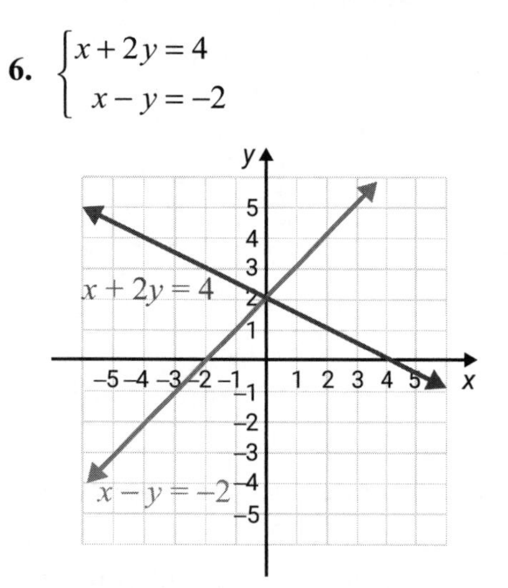

Solve each system of equations by graphing.

7. $\begin{cases} x-2y=4 \\ x=4 \end{cases}$

8. $\begin{cases} 2x+y=0 \\ 4x+2y=-8 \end{cases}$

Applications

ach of the following applications has been modeled using a system of equations. Solve the system graphically.

9. ***Swimming Pools:*** OSHA recommends that swimming pool owners clean their pool decks with a solvent composed of a 12% chlorine solution and a 3% chlorine solution. Fifteen gallons of the solvent consists of 6% chlorine. How much of each of the mixing solutions were used?
Let $x =$ the number of gallons of the 12% solution
and $y =$ the number of gallons of the 3% solution.

The corresponding modeling system is $\begin{cases} x + y = 15 \\ 0.12x + 0.03y = 0.06(15) \end{cases}$

10. ***School Supplies:*** A student bought a calculator and a textbook for a course in algebra. He told his friend that the total cost was \$170 (without tax) and that the calculator cost \$20 more than twice the cost of the textbook. What was the cost of each item?
Let $x =$ the cost of the calculator
and $y =$ the cost of the textbook.

The corresponding modeling system is $\begin{cases} x + y = 170 \\ x = 2y + 20 \end{cases}$

Writing & Thinking

11. Explain, in your own words, why the answer to a consistent system of linear equations can be written as an ordered pair.

Name: Date:

9.2 Systems of Linear Equations: Solutions by Substitution

To Solve a System of Linear Equations by Substitution

1. Solve one of the equations for ____________________
2. Substitute the resulting ____________________
3. Solve this new equation, if possible, and then ____________________
____
4. Check the solution in ____________________

PROCEDURE

Watch and Work

Watch the video for Example 3 in the software and follow along in the space provided.

Example 3 Solving Systems by Substitution (No Solution)

Use the method of substitution to solve the following system of linear equations.

$$\begin{cases} 3x + y = 1 \\ 6x + 2y = 3 \end{cases}$$

Solution

Now You Try It!

Use the space provided to work out the solution to the next example.

Example A Solving Systems by Substitution (No Solution)

Solve the system:

$$\begin{cases} 2x+y=1 \\ 10x+5y=4 \end{cases}$$

Solution

9.2 Exercises

Concept Check

True/False. Determine whether each statement is true or false. If a statement is false, explain how it can be changed so the statement will be true. (**Note:** There may be more than one acceptable change.)

1. The method of substitution reduces the problem from one of solving two equations in two variables to solving one equation in one variable.

2. The method of substitution is most often used when one of the equations is impossible to graph.

3. The method of substitution is more accurate than the graphing method.

4. When using the method of substitution, you should always solve the first equation for x.

Practice

Use the method of substitution to solve each system.

5. $\begin{cases} x + y = 6 \\ y = 2x \end{cases}$

6. $\begin{cases} 3x - 7 = y \\ 2y = 6x - 14 \end{cases}$

7. $\begin{cases} 4x = y \\ 4x - y = 7 \end{cases}$

8. $\begin{cases} 3y + 5x = 5 \\ y = 3 - 2x \end{cases}$

Applications

Each of the following applications has been modeled using a system of equations. Use the method of substitution to solve each system.

9. ***Rectangles:*** The perimeter of a rectangle is 50 meters and the length is 5 meters longer than the width. Find the dimensions of the rectangle.

Let x = the length and y = the width.

The corresponding modeling system is $\begin{cases} 2x+2y=50 \\ x-y=5 \end{cases}$

10. ***Health & Fitness:*** A fitness center manager is trying to decide whether to charge an enrollment fee of $25 with a monthly rate of $50 or an enrollment fee of $100 with a monthly rate of $25. After how many months would it be more profitable for the manager to choose the lower enrollment fee and the higher monthly rate? Round up to the nearest month.

The corresponding modeling system is $\begin{cases} y=50x+25 \\ y=25x+100 \end{cases}$

Writing & Thinking

11. Explain the advantages of solving a system of linear equations

a. by graphing.

b. by substitution.

9.3 Systems of Linear Equations: Solutions by Addition

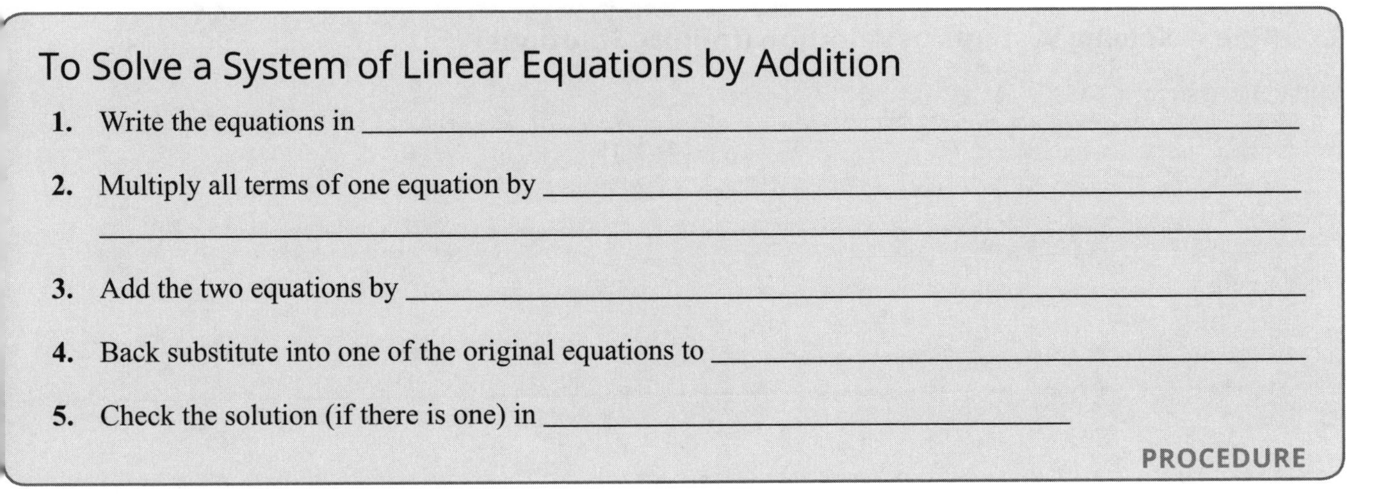

To Solve a System of Linear Equations by Addition

1. Write the equations in ______
2. Multiply all terms of one equation by ______ ______
3. Add the two equations by ______
4. Back substitute into one of the original equations to ______
5. Check the solution (if there is one) in ______

PROCEDURE

Watch and Work

Watch the video for Example 2 in the software and follow along in the space provided.

Example 2 Solving Systems by Addition (Infinite Solutions)

Use the method of addition to solve the following system of linear equations.

$$\begin{cases} 3x - \frac{1}{2}y = 6 \\ 6x - y = 12 \end{cases}$$

Solution

Now You Try It!

Use the space provided to work out the solution to the next example.

Example A Solving Systems by Addition (Infinite Solutions)

Solve the system.

$$\begin{cases} 6x + 3y = 15 \\ 2x + y = 5 \end{cases}$$

Solution

Guidelines for Deciding which Method to Use when Solving a System of Linear Equations

1. The graphing method is helpful in "seeing" the ______________________________
____
2. Both the substitution method and the ______________________________
3. The substitution method may be reasonable and efficient if ______________________________
____
4. The addition method is particularly efficient if ______________________________
____

PROCEDURE

9.3 Exercises

Concept Check

True/False. Determine whether each statement is true or false. If a statement is false, explain how it can be changed so the statement will be true. (**Note:** There may be more than one acceptable change.)

1. When using the method of addition, the solution only needs to be checked in one of the original equations.

2. It's possible for a system of equations to have no solutions.

3. Both the addition method and the substitution method gives approximate solutions.

4. The graphing method is helpful in "seeing" the geometric relationship between the lines and finding approximate solutions.

Practice

Solve each system of linear equations.

5. $\begin{cases} 2x+y=3 \\ 4x+2y=7 \end{cases}$

6. $\begin{cases} y=2x+14 \\ x=14-3y \end{cases}$

7. $\begin{cases} 4x-2y=8 \\ 2x-y=4 \end{cases}$

Write an equation for the line determined by the two given points by using the formula $y = mx + b$ to set up a system of equations with m and b as the unknowns.

8. $(2,3),(1,-2)$

Applications

Each of the following applications has been modeled using a system of equations. Use the method of substitution or the method of addition to solve each system.

9. ***Baseball:*** A minor league baseball team has a game attendance of 4500 people. Tickets cost \$5 for children and \$8 for adults. The total revenue made at this game was \$26,100. How many adults and how many children attended the game?
Let x = number of adults
and y = number of children.

The system that models the problem is $\begin{cases} x+y=4500 \\ 8x+5y=26{,}100 \end{cases}$

10. ***Acid Solutions:*** How many liters each of a 30% acid solution and a 40% acid solution must be used to produce 100 liters of a 36% acid solution?
Let x = amount of 30% solution
and y = amount of 40% solution.

The system that models the problem is $\begin{cases} x+y=100 \\ 0.30x+0.40y=0.36(100) \end{cases}$

Writing & Thinking

11. Explain, in your own words, why the answer to a system with infinite solutions is written as an ordered pair with variables.

9.4 Applications: Distance-Rate-Time, Number Problems, Amounts, and Costs

Watch and Work

Watch the video for Example 2 in the software and follow along in the space provided.

Example 2 Application: Distance-Rate-Time (Times Unknown)

Two buses leave a bus station traveling in opposite directions. One leaves at noon and the second leaves at 1 p.m. The first one travels at an average speed of 55 mph and the second one at an average speed of 59 mph. At what time will the buses be 226 miles apart?

Solution

Now You Try It!

Use the space provided to work out the solution to the next example.

Example A Application: Distance-Rate-Time (Times Unknown)

One man takes the eastbound line at 9 a.m., and his wife takes the westbound line at 10:00 a.m. The husband's train averages 25 mph, while his wife's train averages 30 mph. At what time will the husband and wife be 135 miles apart?

Solution

9.4 Exercises

Practice

Solve each problem by setting up a system of two equations in two unknowns and solve.

1. The sum of two numbers is 40. The sum of twice the larger and 4 times the smaller is 108. Find the numbers.

2. The difference between two numbers is 17. Four times the smaller is equal to 7 more than the larger. What are the numbers?

Applications

Solve

3. ***Boating:*** Jessica drove her speedboat upriver this morning. It took her 1 hour going upriver and 54 minutes going down river. If she traveled 36 miles each way, what would have been the rate of the boat in still water and what was the rate of the current (in miles per hour)?

4. ***Coin Collecting:*** A bag contains pennies and nickels only. If there are 182 coins in all and their value is \$3.90, how many pennies and how many nickels are in the bag?

5. ***Basketball Admission:*** Tickets for the local high school basketball game were priced at $3.50 for adults and $2.50 for students. If the income for one game was $9550 and the attendance was 3500, how many adults and how many students attended that game?

6. ***Age:*** When they got married, Elvis Presley was 11 years older than his wife Priscilla. One year later, Priscilla was two-thirds of Elvis' age. How old was each of them when they got married?

9.5 Applications: Interest and Mixture

Watch and Work

Watch the video for Example 3 in the software and follow along in the space provided.

Example 3 Application: Solving a Mixture Problem

How many ounces each of a 10% salt solution and a 15% salt solution must be used to produce 50 ounces of a 12% salt solution?

Solution

Now You Try It!

Use the space provided to work out the solution to the next example.

Example A Application: Solving a Mixture Problem

How many ounces each of a 12% chlorine solution and a 18% chlorine solution must be used to produce 150 ounces of a 14% chlorine solution?

9.5 Exercises

True/False. Determine whether each statement is true or false. If a statement is false, explain how it can be changed so that the statement will be true. (**Note:** There may be more than one acceptable change.)

1. When interest is calculated on an annual basis, we have $t = 0$ and the formula becomes $I = Pr$.

2. Problems involving mixture occur in the sciences such as physics and chemistry.

3. In an interest problem, time can be given in parts of a year.

4. When two or more items are mixed, the final mixture should satisfy certain conditions of percentage of concentration.

Practice

Solve each problem by setting up a system of two equations in two unknowns and solving the system.

5. ***Investing:*** Carmen invested $9000, part in a 6% passbook account and the rest in a 10% certificate account. If her annual interest was $680, how much did she invest at each rate?

6. ***Investing:*** On two investments totaling $9500, Darius lost 3% on one and earned 6% on the other. If his net annual receipts were $282, how much was each investment?

7. ***Chemistry:*** How many liters each of a 12% iodine solution and a 30% iodine solution must be used to produce a total mixture of 90 liters of a 22% iodine solution?

8. ***Food Science:*** A candymaker is making truffles using a mixture of a melted dark chocolate that is 72% cocoa and milk chocolate that is 42% cocoa. If she wants 6 pounds of melted chocolate that is 52% cocoa, how much of each type of chocolate does she need?

Writing & Thinking

9. Your friend has $20,000 to invest and decided to invest part at 4% interest and the rest at 10% interest. Why might you advise him (or her) to invest all of it

   a. at 4%? **b.** at 10%?

9.6 Systems of Linear Equations: Three Variables

A Solving Systems in Three Variables

1. The general form is ____________________ where A, B, and C are not all equal to 0.

2. The solutions to such equations are called ____________________ and are of the form __________

To Solve a System of Three Linear Equations in Three Variables

1. Select two equations and eliminate ____________________

2. Select a different pair of equations and eliminate ____________________

3. Steps 1 and 2 give two linear equations in two variables. Solve these equations by __________

____

4. Back substitute the values found in Step 3 into ____________________

____

5. Check the solution (if one exists) in ____________________

PROCEDURE

Solutions to Systems of Equations

1. There will be ____________________ (Graphically, the three planes intersect ____________________.)

2. There will be ____________________ (Graphically, the three planes intersect ____________________.)

3. There will be __________ (Graphically, there are ____________________.)

PROPERTIES

Watch and Work

Watch the video for Example 2 in the software and follow along in the space provided.

Example 2 Solving a System with Three Variables (No Solution)

Solve the system of equations.

$$\begin{cases} 3x - 5y + z = 6 & \text{(I)} \\ x - y + 3z = -1 & \text{(II)} \\ 2x - 2y + 6z = 5 & \text{(III)} \end{cases}$$

Solution

Now You Try It!

Use the space provided to work out the solution to the next example.

Example A Solving a System with Three Variables (No Solution)

Solve.

$$\begin{cases} 7x + 2y - z = 7 \\ 3x - y - 2z = 1 \\ 9x - 3y - 6z = -3 \end{cases}$$

9.6 Exercises

Concept Check

True/False. Determine whether each statement is true or false. If a statement is false, explain how it can be changed so the statement will be true. (**Note:** There may be more than one acceptable change.)

1. To find the solution for a system of three linear equations in three variables, start by choosing two variables and eliminating one equation.

2. An ordered triple believed to be a solution to a system of three linear equations in three variables should be checked in all three original equations.

3. If two distinct planes intersect, that intersection forms a straight line.

4. Two distinct planes will always intersect.

Practice

Solve each system of equations. State which systems, if any, have no solution or an infinite number of solutions.

5. $\begin{cases} x+y-z=0 \\ 3x+2y+z=4 \\ x-3y+4z=5 \end{cases}$

6. $\begin{cases} x+y-2z=4 \\ 2x+y=1 \\ 5x+3y-2z=6 \end{cases}$

7. $\begin{cases} 3x+y+4z=-6 \\ 2x+3y-z=2 \\ 5x+4y+3z=2 \end{cases}$

8. $\begin{cases} 2x+y-z=-3 \\ -x+2y+z=5 \\ 2x+3y-2z=-3 \end{cases}$

Applications

Solve.

9. The sum of three integers is 189. The first integer is 28 less than the second. The second integer is 21 less than the sum of the first and third integers. Find the three integers.

10. ***Money:*** A wallet contains \$218 in \$10, \$5, and \$1 bills. There are forty-six bills in all and four more fives than tens. How many bills of each kind are there?

Writing & Thinking

11. Is it possible for three linear equations in three unknowns to have exactly two solutions? Explain your reasoning in detail.

9.7 Matrices and Gaussian Elimination

1. ________________________________ is called a **matrix** (plural **matrices**).
2. Matrices are usually named with capital letters, and each number in the matrix is called an ____________
3. Entries written ____________________ are said to form a **row**, and entries written ________________ are said to form a **column**.
4. We say that the **dimension** of the matrix is ____
5. A matrix with ______________________________________ is called a **square matrix**.

Elementary Row Operations

1. __________________________
2. Multiply a row by ________________________
3. Multiply a row by ____

If any elementary row operation is applied to a matrix, the new matrix is said to be _______________ to the original matrix.

PROPERTIES

Watch and Work

Watch the video for Example 1 in the software and follow along in the space provided.

Example 1 Creating Coefficient and Augmented Matrices

For the system $\begin{cases} y + z = 6 \\ x + 5y - 4z = 4 \\ 2x - 6y + 10z = 14 \end{cases}$

a. Write the corresponding coefficient matrix and the corresponding augmented matrix.

b. In the augmented matrix in Example **1a.**, interchange rows 1 and 2 and multiply row 3 by $\frac{1}{2}$.

Solution

Now You Try It!

Use the space provided to work out the solution to the next example.

Example A Creating Coefficient and Augmented Matrices

a. Write the coefficient matrix and the augmented matrix for the following system.

$$\begin{cases} 2x+y+z=4 \\ x+2y+z=1 \\ 3x+y-z=-3 \end{cases}$$

b. In the augmented matrix you found in Part **a.**, interchange rows 1 and 3 and multiply row 2 by 3.

Solution

Strategy for Gaussian Elimination

1. Write the augmented ______________________________
2. Use elementary row operations to transform ____
3. Solve the corresponding system of equations by ________________________________

PROCEDURE

6. If the final matrix, in triangular form, has a row with all entries 0, then the system has ________________

____

7. If the triangular form of the augmented matrix shows the coefficient entries in one or more rows to be all 0s and the constant not 0, then the system has ______________________

9.7 Exercises

Concept Check

True/False. Determine whether each statement is true or false. If a statement is false, explain how it can be changed so that the statement will be true. (**Note:** There may be more than one acceptable change.)

1. A matrix that has 3 rows and 5 columns is a 5×3 matrix.

2. A matrix with the same number of rows as columns, such as a 3×3 matrix, is called a square matrix.

3. Interchanging two equations in a system of linear equations will change the solution of the system.

4. In a system of linear equations, adding like terms of one equation to another equation will not change the solution of the system.

Practice

Write the coefficient matrix and the augmented matrix for the given system of linear equations.

5. $\begin{cases} 7x - 2y + 7z = 2 \\ -5x + 3y = 2 \\ 4y + 11z = 8 \end{cases}$

Write the system of linear equations represented by the augmented matrix. Use x, y, and z as the variables.

6. $\left[\begin{array}{cc|c} -3 & 5 & 1 \\ -1 & 3 & 2 \end{array}\right]$

Use the Gaussian elimination method to solve the given system of linear equations.

7. $\begin{cases} x + 2y = 3 \\ 2x - y = -4 \end{cases}$

8. $\begin{cases} x + y + 3z = 2 \\ 2x - y + z = 1 \\ 4x + y + 7z = 5 \end{cases}$

Applications

Set up a system of linear equations that represents the information and solve the system using Gaussian elimination.

9. The sum of three integers is 169. The first integer is twelve more than the second integer. The third integer is fifteen less than the sum of the first and second integers. What are the integers?

10. ***Pizza:*** A pizzeria sells three sizes of pizzas: small, medium, and large. The pizzas sell for \$6.00, \$8.00, and \$9.50, respectively. One evening they sold 68 pizzas for a total of \$528.00. If they sold twice as many medium-sized pizzas as large-sized pizzas, how many of each size did they sell?

Writing & Thinking

11. Suppose that Gaussian elimination with a system of three linear equations in three unknowns results in the following triangular matrix. Discuss how you can use back substitution to find that the system has an infinite number of solutions and these solutions satisfy the equation $x + 5y = 6$. (**Hint:** Solve the second equation for z.)

$$\left[\begin{array}{ccc|c} 1 & 2 & -1 & 4 \\ 0 & 3 & 1 & 2 \\ 0 & 0 & 0 & 0 \end{array}\right]$$

9.8 Systems of Linear Inequalities

To Solve a System of Two Linear Inequalities

1. For each inequality, graph the boundary line and ______________________

2. Determine the region of the graph that is ______________________

   ____

   This region is called the ______________________

   ____

3. To check, pick one test-point in the ______________________

Note: If there is no intersection, then the system has no solution.

PROCEDURE

Watch and Work

Watch the video for Example 1 in the software and follow along in the space provided.

Example 1 Solving Systems of Linear Inequalities

Solve the system of linear inequalities graphically. $\begin{cases} x \leq 2 \\ y \geq -x+1 \end{cases}$

Solution

Now You Try It!

Use the space provided to work out the solution to the next example.

Example A Solving Systems of Linear Inequalities

Solve the system of linear inequalities graphically.

$$\begin{cases} y \geq 2 \\ x - y < 4 \end{cases}$$

Solution

Possible Solutions to Systems of Linear Inequalities

When the boundary lines are parallel there are three possibilities:

1. The common region will be in the form of ____________________
2. The common region will be a ____________________ ____________________
3. There will be ____________________

9.8 Exercises

Concept Check

True/False. Determine whether each statement is true or false. If a statement is false, explain how it can be changed so the statement will be true. (**Note:** There may be more than one acceptable change.)

1. When boundary lines are parallel, the system of linear inequalities has no solution.

2. If two half-planes overlap, that region is the union of the graphs.

3. Half-planes are the graphs of linear inequalities.

4. If the graphs of two linear inequalities have no intersection, then the system has no solution.

Practice

Solve the systems of two linear inequalities graphically.

5. $\begin{cases} y > 2 \\ x \geq -3 \end{cases}$

6. $\begin{cases} y > 3x + 1 \\ -3x + y < -1 \end{cases}$

7. $\begin{cases} 2x - 3y \geq 0 \\ 8x - 3y < 36 \end{cases}$

8. $\begin{cases} y > x - 4 \\ y < x + 2 \end{cases}$

Applications

Solve.

9. ***Fundraising:*** Robin is planning a charity ball to raise money for her favorite charity. There are two different ticket options. The VIP option includes dinner, dancing, and cocktails for \$150 per ticket. The regular option includes dancing and cocktails for \$75 per ticket. Robin wants to make at least \$14,000 in ticket sales. The ballroom that is being used for the charity event has a maximum capacity of 150 people.

   a. Write two linear inequalities to describe the situation. Let the variable x represent the number of VIP tickets sold and let the variable y represent the number of regular tickets sold.

   b. Graph the two linear inequalities on the same coordinate plane.

   c. Describe the solution set for the situation.

   d. Can Robin reach her sales goal if she only sells tickets for the regular option? Explain why or why not.

Writing & Thinking

10. Graph the inequalities and explain how you can tell that there is no solution.

$$\begin{cases} y \le 2x - 5 \\ y \ge 2x + 3 \end{cases}$$

Chapter 9 Project

Don't Put All Your Eggs in One Basket!

An activity to demonstrate the use of linear systems in real life.

Have you ever heard the phrase "Don't put all your eggs in one basket"? This is a common saying that is often quoted in the investment world—and it's true. In an ever-changing economy it is important to diversify your investments. Splitting your money up into two or more funds may keep you from losing it all if one of the funds performs poorly. You may be thinking that you are too young to consider investments and saving money for retirement, but it is never too soon—especially in today's economy where interest rates are extremely low. Low rates means it takes even longer to build up your nest egg. So start saving now and be sure to have more than one basket to put your eggs in! For this activity, if you need help understanding some of the investment terms, use the following link as a resource: http://www.investopedia.com/

Let's suppose that you received a total of $5000 in cash as a graduation present from your relatives. You also have an additional $2500 that you saved from your summer job. You are thinking about investing the $7500 in two investment funds that have been recommended to you. One is currently earning 4% interest annually (conservative fund) and the other is earning 8% annually (aggressive fund). Keep in mind that interest rates fluctuate as the economy changes and there are few guarantees on the amount you will actually earn from any investment. Also, note that higher rates of interest typically indicate a higher risk on your investment.

1. If you want to earn $400 total in interest on your investments this year, how much money would you have to invest in each fund? Let the variable x be the amount invested in Fund 1 and the variable y be the amount invested in Fund 2. Recall that to calculate the interest on an investment, use the formula $I = Prt$, where P is the principal or amount invested, r is the annual interest rate, and t is the amount of time invested, which for our problem will be 1 year ($t = 1$). Use the table below to help you organize the information. Note that interest rates have to be converted to decimals before using them in an equation.

	Principal	Interest Rate	Interest
Fund 1	x	0.04	$0.04x$
Fund 2	y	0.08	$0.08y$
Total	**a.**		**b.**

a. Fill in the total amount available for investment in the bottom row of the table.

b. Fill in the total amount of interest desired in the bottom row of the table.

c. What does $0.04x$ represent in the context of this problem?

d. What does $0.08y$ represent in the context of this problem?

e. Using the principal column of the table, write an equation in standard form involving the variables x and y to represent the total amount available for investment.

f. Using the interest column of the table, write an equation in standard form involving the variables x and y to represent the total amount of interest desired.

g. Solve the linear system of two equations derived in Parts **e.** and **f.** to determine the amount to invest in each fund to earn $400 in interest. (You may use any method you choose: substitution, addition/elimination, or graphing)

h. Check to make sure that your solution to the system is correct by substituting the values from Part **g.** for x and y into both equations and verify that the equations are true statements.

2. Suppose you decide you want to earn more interest on your investment. You now want to earn $500 in interest next year instead of $400. Using a table similar to the one in Problem 1, organize the information and follow a similar format to determine the amounts to invest in each of the funds that will earn $500 in interest in a year.

3. Compare the results you obtained from Problems 1 and 2. How did the amounts in each investment change when your desired interest increased by $100?

4. Suppose you decide that $500 is not enough interest and you want to earn an additional $100 on your investments for a total of $600 in interest. Using a table similar to the one in Problem 1, organize the information and follow a similar format to determine the amounts to invest in each of the funds that will earn $600 in interest in a year.

5. Compare the results from Problem 4 to the results from Problem 1 and 2.

a. How much are you investing in Fund 1 to earn $600 in interest?

b. How much are you investing in Fund 2 to earn $600 in interest?

c. How do your results contradict the advice provided to you at the start of this activity?

d. Is it possible to make more than $600 in interest on your $7500 investment using these two funds? Explain why or why not?

e. What is the smallest amount of interest you can earn on your investment using these two funds? How did you determine this?

6. How much interest would you earn if you split the initial principal of $7500 equally between th[e] two funds?

7. If you actually had $7500 to invest in these two funds earning 4% and 8% respectively, how would you invest the money? Explain your reasoning.

CHAPTER 10

Exponents and Polynomials

Math @ Work

Physics is a natural science that deals with matter and its motion through space along with the concepts of energy and force. Within physics, scientists often use polynomials to define the rules that govern nature. Consider the formula for force, $F = ma$ (which means force equals mass times acceleration). This formula defines why a basketball bounces higher when slammed into the ground than when given a simple dribble. It also explains why a baseball may dent a car, but a ball of paper thrown at the same speed will not. Knowing how to perform operations with polynomials and how to factor polynomials is necessary to understand the information provided by these rules of nature.

Suppose you are bowling with friends and using a 5 kg bowling ball (which is approximately 11 pounds). After you roll the ball down the lane, the ball's acceleration can be represented by the polynomial $a = 4t^2 - 12t + 7$ m/s^2, where t represents time in seconds. (We can assume that this is the ball's acceleration only until it strikes the pins.) How would you use the formula $F = ma$ to find a polynomial that represents the force of the ball at time t? (Note that the force of the bowling ball is measured in Newtons N which is kg · m/s^2.)

10.1 Rules for Exponents

The Product Rule for Exponents

If a is a nonzero real number and m and n are integers, then

____

In words, to multiply powers with the same base, ____

PROPERTIES

The Exponent 0

If a is a nonzero real number, then ____________________

The expression _______ is _______________________

DEFINITION

Quotient Rule for Exponents

If a is a nonzero real number and m and n are integers, then

____

In words, to divide two powers with the same base, ____

PROPERTIES

Rule for Negative Exponents

If a is a nonzero real number and n is an integer, then

____

PROPERTIES

Watch and Work

Watch the video for Example 6 in the software and follow along in the space provided.

Example 6 Negative Exponents

Use the rule for negative exponents to simplify each expression so that it contains only positive exponents.

a. 5^{-1}

b. x^{-3}

c. $x^{-9} \cdot x^{7}$

Solution

Now You Try It!

Use the space provided to work out the solution to the next example.

Example A Negative Exponents

Use the rule for negative exponents to simplify each expression so that it contains only positive exponents.

a. 7^{-1}

b. x^{-7}

c. $x^{-11} \cdot x^{6}$

Summary of the Rules for Exponents

For any nonzero real number a and integers m and n:

1. The exponent 1: ____________________
2. The exponent 0: ____________________
3. The product rule: ____________________
4. The quotient rule:
 ____
5. Negative exponents:
 ____

PROPERTIES

10.1 Exercises

Concept Check

True/False. Determine whether each statement is true or false. If a statement is false, explain how it can be changed so the statement will be true. (**Note:** There may be more than one acceptable change.)

1. If a constant does not have an exponent written, it is assumed that the exponent is 0.

2. If a is a nonzero real number and n is an integer, then $a^{-n} = -a^{n}$.

3. Since the product rule is stated for integer exponents, the rule is also valid for 0 and negative exponents.

4. When using the quotient rule, you should subtract the smaller exponent from the larger exponent.

Practice

Simplify each expression. The final form of the expressions with variables should contain only positive exponents. Assume that all variables represent nonzero numbers.

5. $y^3 \cdot y^8$

7. $x^{-3} \cdot x^0 \cdot x^2$

6. $\dfrac{y^7}{y^2}$

8. $\dfrac{10^4 \cdot 10^{-3}}{10^{-2}}$

Applications

Solve.

9. ***Computers:*** Rylee wants to move all her files to a new hard drive that has 2^{12} GB of storage on it. She wants to designate the same amount of storage for each of 2^4 projects. How much storage should be assigned to each project? Write your answer as a power of two.

10. ***Bacteria:*** Trey is studying patterns in bacteria. For a positive test result in his experiment, bacteria must grow in population at a minimum rate of 3^2 in 24 hours. If the initial population of the bacteria is 3^5 and his final measurement after 24 hours is 3^8, should he mark the test as positive or negative?

Name: Date:

10.2 Power Rules for Exponents

Power Rule for Exponents

If a is a nonzero real number and m and n are integers, then

____

In other words, the value of a power raised to a power can be found by ____

____

PROPERTIES

Rule for Power of a Product

If a and b are nonzero real numbers and n is an integer then

____

In words, a power of a product is found by ____

PROPERTIES

Rule for Power of a Quotient

If a and b are nonzero real numbers and n is an integer, then

____

In words, a power of a quotient (in fraction form) is found by ____

____

PROPERTIES

Watch and Work

Watch the video for Example 5 in the software and follow along in the space provided.

Example 5 Using Two Approaches with Fractional Expressions and Negative Exponents

Simplify: $\left(\frac{x^3}{y^5}\right)^{-4}$

Solution

Now You Try It!

Example A **Using Two Approaches with Fractional Expressions and Negative Exponents**

Simplify:

$$\left(\frac{x^6}{y^3}\right)^{-5}$$

Summary of the Rules for Exponents

For any nonzero real numbers a and b and integers m and n:

1. The exponent 1: ______________
2. The exponent 0: ______________
3. The product rule: ______________
4. The quotient rule: ______________
5. Negative exponents: ______________
6. Power rule: ______________
7. Power of a product: ______________
8. Power of a quotient: ______________

PROPERTIES

10.2 Exercises

Concept Check

True/False. Determine whether each statement is true or false. If a statement is false, explain how it can be changed so the statement will be true. (**Note:** There may be more than one acceptable change.)

1. Taking the reciprocal of a fraction changes the sign of any exponent in the fraction.

2. For an exponent to refer to -7 as the base, -7 must be in parentheses.

3. When simplifying an expression with exponents, the rules for exponents must be used in a specific order or the answer will vary.

4. The expression -8^2 simplifies to -64.

Practice

Use the rules for exponents to simplify each of the expressions. Assume that all variables represent nonzero real numbers.

5. $\left(2^{-3}\right)^{-2}$

6. $-3\left(7xy^2\right)^0$

7. $-2\left(3x^{5}y^{-2}\right)^{-3}$

8. $\left(\frac{x}{2}\right)^{3}$

9. $\left(\frac{2x^{2}y}{y^{3}}\right)^{-4}$

10. $\left(\frac{5a^{4}b^{-2}}{6a^{-4}b^{3}}\right)^{-2}\left(\frac{5a^{3}b^{4}}{2^{-2}a^{-2}b^{-2}}\right)^{3}$

10.3 Applications: Scientific Notation

Scientific Notation

If N is a decimal number, then in scientific notation

DEFINITION

Watch and Work

Watch the video for Example 3 in the software and follow along in the space provided.

Example 3 Application: Scientific Notation

Light travels approximately 3×10^8 meters per second. How many meters per minute does light travel?

Solution

Now You Try It!

Use the space provided to work out the solution to the next example.

Example A Application: Scientific Notation

One mole, a value often used in physics and chemistry, equals 6.02×10^{23} particles for all substances. How many particles would be in 8 moles of carbon?

10.3 Exercises

Concept Check

True/False. Determine whether each statement is true or false. If a statement is false, explain how it can be changed so the statement will be true. (**Note:** There may be more than one acceptable change.)

1. The exponent in the number 1.4×10^4 indicates that the decimal point should be moved 4 places to the right.

2. The exponent in the number 2.5×10^{-3} indicates that the decimal point should be moved 3 places to the right.

3. The number 3.53×10^5 is less than 8.72×10^{-4}.

4. The number 4000 written in scientific notation is 0.4×10^4.

Practice

Write the following numbers in scientific notation.

5. 86,000

6. 0.0000000002368

Write the following numbers in decimal form.

7. 4.2×10^{-2}

8. 3.067×10^{10}

First write each of the numbers in scientific notation. Then perform the indicated operations and leave your answer in scientific notation.

9. $0.0003 \cdot 0.0000025$

10. $23{,}400{,}000{,}000 \cdot 5{,}500{,}000{,}000$

Applications

Solve.

11. ***Speed of Light:*** One light-year is approximately 9.46×10^{15} meters. The distance to a certain star is 4.3 light-years. How many meters is this?

12. ***Atomic Weight:*** The mass of an atom of gold is approximately 3.25×10^{-22} grams. What would be the mass of 2000 atoms of gold? Express your answer in scientific notation.

10.4 Introduction to Polynomials

1. A ____________ is an expression that involves only ____________________________________ with constants and/or variables.

Monomial

A monomial in x is a term of the form

____

where k is a real number and n is __________________________

n is called the ____

DEFINITION

Polynomial

A polynomial is a monomial or the ____

The degree of a polynomial is the ____

____

The coefficient of the term of largest degree is called ______________________________

DEFINITION

Special Terminology for Polynomials

Term	Definition	Examples
____________________	polynomial with ________________	$-2x^3$ and ______________________
____________________	polynomial with ________________	$3x+5$ and ______________________
____________________	polynomial with ________________	x^2+6x-7 and ________________

Polynomials with four or more terms are simply referred to as ______________________

DEFINITION

2. To evaluate a polynomial for a given value of the variable:

1. substitute that value for the ____
2. follow the rules for ____________________________

Watch and Work

Watch the video for Example 4 in the software and follow along in the space provided.

Example 4 Evaluating Polynomials

Given the polynomial $p(y) = 5y^3 + y^2 - 3y + 8$, find $p(-2)$.

Solution

Now You Try It!

Use the space provided to work out the solution to the next example.

Example A Evaluating Polynomials

Given the polynomial $p(y) = 4y^3 + 2y^2 - 8y$ find $p(-2)$

10.4 Exercises

Concept Check

True/False. Determine whether each statement is true or false. If a statement is false, explain how it can be changed so the statement will be true. (**Note:** There may be more than one acceptable change.)

1. A nonzero constant is a monomial with no degree.

2. A polynomial with four terms is a quadrinomial.

3. A monomial is a polynomial with one term.

4. A polynomial must have at least two terms.

Practice

Simplify the polynomials. Write the polynomials in descending order and state the degree and type of each polynomial. Then, state the leading coefficient.

5. $4x^2 - x + x^2$

6. $x^4 + 3x^4 - 2x + 5x - 10 - x^2 + x$

Find the values of the functions as indicated.

7. Given $f(x) = 3x - 10$, find

a. $f(2)$

b. $f(-2)$

c. $f(0)$

A polynomial function is given. Rewrite the polynomial function by substituting for the variable as indicated in the given function notation.

8. Given $p(x)=3x^4+5x^3-8x^2-9x$, find $p(a)$.

Applications

Solve.

9. ***Projectile Motion:*** A ball is thrown from the top of a building towards the ground with an initial velocity (or initial speed) of 12 feet per second. The force of gravity causes the ball to accelerate at a rate of 32 feet per second squared. The distance of the ball from the top of the building can be modeled by the polynomial $p(x) = 12x + 32x^2$ where x is the time, in seconds, since the ball was thrown.

a. Identify the degree of the polynomial.

b. Determine how far the ball is from the top of the building at 1 second.

c. Determine how far the ball is from the top of the building at 2 seconds.

d. Determine how far the ball is from the top of the building at 3 seconds.

10. ***Sales:*** PDQ Tennis Shoe Co. follows a profit model of $P=3x^2-15x+2$, where P is the profit in hundreds of dollars after selling x hundred pairs of tennis shoes. Find each of the profits for PDQ after selling 100 pairs, 500 pairs and 1000 pairs of tennis shoes. What does a negative value for profit represent?

Writing & Thinking

11. Tony was classifying expressions for a homework assignment. He said that $7y^2 + 12y - 3$ was a polynomial. Was he correct or not? Justify your answer.

12. Jeanne thought that $10a - 9 + 6a^2$ was in descending order. Explain Jeanne's error and what the correct descending order should be.

10.5 Addition and Subtraction with Polynomials

1. The **sum** of two or more polynomials is found by combining ________________.

2. A negative sign written in front of a polynomial ____________________ indicates the ____________ ______________________.

Watch and Work

Watch the video for Example 4 in the software and follow along in the space provided.

Example 4 Subtracting Polynomials

Subtract: $(9x^4 - 22x^3 + 3x^2 + 10) - (5x^4 - 2x^3 - 5x^2 + x)$

Solution

Now You Try It!

Use the space provided to work out the solution to the next example.

Example A Subtracting Polynomials

Subtract: $(7x^4 - 18x^3 + 8x^2 - 11) - (-8x^4 - 15x^3 + 3x^2 - 12)$

10.5 Exercises

Concept Check

True/False. Determine whether each statement is true or false. If a statement is false, explain how it can be changed so the statement will be true. (**Note:** There may be more than one acceptable change.)

1. When subtracting one polynomial from another polynomial, only the first term of the polynomial is subtracted.

2. To simplify polynomials that are being added and subtracted, combine like terms.

3. The terms $6a^2$ and $7a^2$ are not like terms because they don't have the same coefficient.

4. Absolute value bars and radical signs are not considered grouping symbols.

Practice

Find the indicated sums.

5. $(2x^2+5x-1)+(x^2+2x+3)$

6. $(-2x^2-3x+9)+(3x^2-2x+8)$

Find the indicated differences.

7. $\left(2x^2+4x+8\right)-\left(x^2+3x+2\right)$

8. $\left(x^4+8x^3-2x^2-5\right)-\left(2x^4+10x^3-2x^2+11\right)$

Simplify

9. $-4(x-6)-(8x+2)-3x$

Complete the following word problems.

10. Add $5x^3-8x+1$ to the difference between $2x^3+14x-3$ and x^2+6x+5.

Applications

Solve.

11. ***Business:*** A manufacturer estimates that it costs $2x^3 + 4x^2 - 35$ dollars to create the amount of items it would take to fill a box which has a side length of x feet. The warehouse manager determines it will cost $1.50x^3 + 5$ dollars to store each box for one month.

    a. Add the two polynomials to determine the manufacturing and storage costs for each box of items for one month.

    b. The warehouse manager knows that each box will be stored for an average of 3 months. Determine the cost to produce a box of items and store it for 3 months.

    c. If the box has a side length of 4 feet, use the expression from Part **b.** to determine how much will it cost to create and store a box of items for 3 months.

Writing & Thinking

12. Explain, in your own words, how to subtract one polynomial from another.

13. Give two examples that show how the sum of two binomials might not be a binomial.

10.6 Multiplication with Polynomials

1. Using the ________________ property $a(b+c)=ab+ac$ with multiplication indicated on the left, we can find the product of a monomial with a polynomial of two or more terms as follows.

$$5x(2x+3)=\_\_\cdot 2x+\_\_\cdot 3=\_\_\_\_\_\_\_\_$$

We can apply the distributive property in the following way to multiply two polynomials.

$$\begin{aligned}(2x-1)(x^2+x-5)&=\_\_(x^2+x-5)\_\_(x^2+x-5)\\&=2x\cdot\_\_+2x\cdot\_+2x\cdot(\_\_)-1\cdot\_\_-1\cdot\_-1(\_\_)\\&=\_\_\_\_\_\_\_\_\_\_\_\_\_\_\_\_\\&=\_\_\_\_\_\_\_\_\_\_\end{aligned}$$

Watch and Work

Watch the video for Example 8 in the software and follow along in the space provided.

Example 8 Multiplying Polynomials

Multiply: $(x-5)(x+2)(x-1)$

Solution

Now You Try It!

Use the space provided to work out the solution to the next example.

Example A Multiplying Polynomials

Multiply: $(x-2)(x+3)(x-6)$

10.6 Exercises

Concept Check

True/False. Determine whether each statement is true or false. If a statement is false, explain how it can be changed so the statement will be true. (**Note:** There may be more than one acceptable change.)

1. The distributive property can only be used to multiply a monomial and a polynomial.

2. The product of $(a+b)$ and $(c+d)$ is $ac+bd$.

3. The FOIL method is a way to remember one specific order that the distributive property can be applied.

Practice

Multiply and simplify, if necessary.

4. $-3x^2(2x^3+5x)$

5. $-4x^3(x^5-2x^4+3x)$

6. $(x+4)(x-3)$

7. $(y+3)(y^2-y+4)$

Applications

Solve.

8. ***Advertising:*** A graphic artist is designing a poster to advertise an upcoming event. The only restrictions regarding the poster size is that it must have a length of $3x$ inches and a width of $2x + 5$ inches. Find a simplified expression for the area of the poster.

9. ***Shipping:*** Armon works for a company that ships artwork worldwide. The size of each item varies, but all of the art is on square canvases. Armon's job is to make the wooden shipping crates for each piece of art. In order to protect the artwork, each crate must be 10 inches deep. The crate must also be 10 inches wider and 12 inches taller than the artwork. Letting x represent the length of one side of the artwork, find the volume of the rectangular shipping crate.

Writing & Thinking

10. We have seen how the distributive property is used to multiply polynomials.

Show how the distributive property can be used to find the product

$$\begin{array}{r} 75 \\ \times 93 \\ \hline \end{array}$$

(**Hint:** $75 = 70 + 5$ and $93 = 90 + 3$)

10.7 Special Products of Binomials

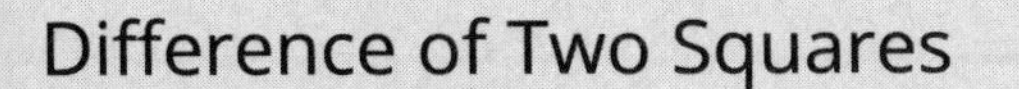

Difference of Two Squares

____

DEFINITION

Squares of Binomials (Perfect Square Trinomials)

$(x+a)^2 = x^2 + 2ax + a^2$ Square of a ______________

$(x-a)^2 = x^2 - 2ax + a^2$ Square of a ______________

DEFINITION

Watch and Work

Watch the video for Example 3 in the software and follow along in the space provided.

Example 3 Squares of Binomials

Find the following products.

a. $(2x+3)^2$ **b.** $(5x-1)^2$ **c.** $(9-x)^2$ **d.** $(y^3+1)^2$

Solution

Now You Try It!

Use the space provided to work out the solution to the next example.

Example A Squares of Binomials

Find the following products.

a. $(3x+5)^2$

b. $(4x-2)^2$

c. $(8-2x)^2$

d. $(2y^3-2)^2$

10.7 Exercises

Concept Check

True/False. Determine whether each statement is true or false. If a statement is false, explain how it can be changed so the statement will be true. (**Note:** There may be more than one acceptable change.)

1. When two binomials are in the form of the sum and difference of the same term, the product will be a trinomial.

2. When the two binomials being multiplied together are the same, the product will be a trinomial.

3. Perfect square trinomials result from squaring a binomial sum or a binomial difference.

4. When finding the product of two binomials that are in the form of the sum and difference of the same two terms, the FOIL method and the difference of two squares formula will produce different results.

Practice

Find each product and identify any that are either the difference of two squares or a perfect square trinomial.

5. $(x-7)^2$

6. $(x+12)(x-12)$

7. $(3x-2)(3x-2)$

8. $(5x-9)(5x+9)$

Applications

Solve.

9. ***Geometry:*** A square is 20 inches on each side. A square x inches on each side is cut from each corner of the square.

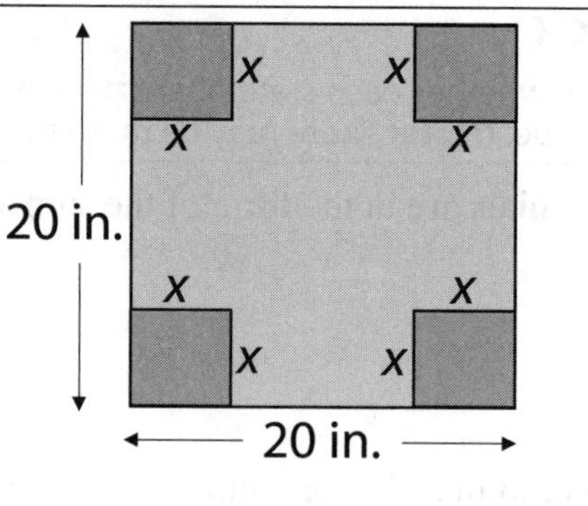

a. Represent the area of the remaining portion of the square in the form of a polynomial function $A(x)$.

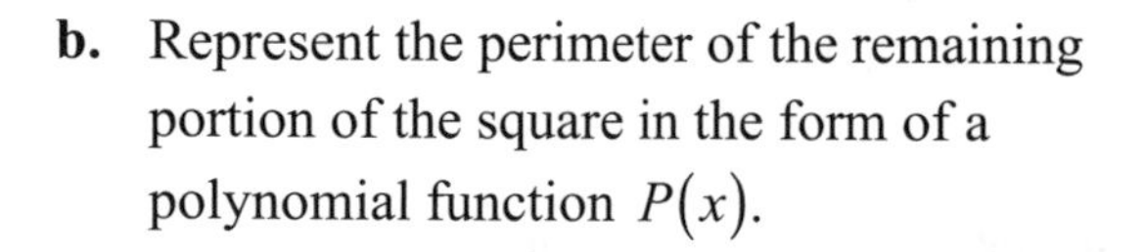

b. Represent the perimeter of the remaining portion of the square in the form of a polynomial function $P(x)$.

10. ***Probability:*** In the case of binomial probabilities, if x is the probability of success in one trial of an event, then the expression $f(x)=15x^4(1-x)^2$ is the probability of 4 successes in 6 trials where $0 \leq x \leq 1$.

a. Represent the expression $f(x)$ as a single polynomial by multiplying the polynomials.

b. If a fair coin is tossed, the probability of heads occurring is $\frac{1}{2}$. That is, $x = \frac{1}{2}$. Find the probability of 4 heads occurring in 6 tosses.

Writing & Thinking

11. A square with sides of length $(x+5)$ can be broken up as shown in the diagram. The sums of the areas of the interior rectangles and squares is equal to the total area of the square: $(x+5)^2$. Show how this fits with the formula for the square of a sum.

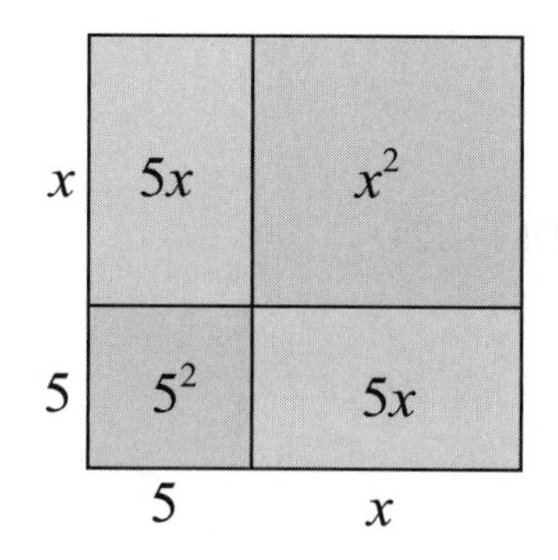

10.8 Division with Polynomials

1. To divide the numerator by the denominator (with a monomial in the denominator), we ______________

____ and simplify each fraction.

The Division Algorithm

For polynomials P and D, the division algorithm gives

____

where Q and R are polynomials and the degree of ________________________

THEOREM

Watch and Work

Watch the video for Example 4 in the software and follow along in the space provided.

Example 4 Using Long Division (Terms Missing)

Simplify $\frac{x^4+9x^2-3x+5}{x^2-x+2}$ by using long division.

Solution

Now You Try It!

Use the space provided to work out the solution to the next example.

Example A Using Long Division (Terms Missing)

Simplify $\frac{7x^3+4x-9}{x-2}$ by using long division.

10.8 Exercises

Concept Check

True/False. Determine whether each statement is true or false. If a statement is false, explain how it can be changed so the statement will be true. (**Note:** There may be more than one acceptable change.)

1. When dividing polynomials, any remainder must be of smaller degree than the divisor.

2. The first step in the division algorithm is to align the polynomials in ascending order.

3. To aid in organization and clarity when dividing polynomials, it is best to fill in any missing powers with ones.

4. The process followed when dividing two polynomials is called the division algorithm with polynomials.

Practice

Express each quotient as a sum (or difference) of fractions and simplify, if possible.

5. $\dfrac{8y^3 - 16y^2 + 24y}{8y}$

6. $\dfrac{20y^5 - 14y^4 + 21y^3 + 42y^2}{4y^2}$

Divide by using the division algorithm. Write the answers in the form $\boldsymbol{Q} + \dfrac{\boldsymbol{R}}{\boldsymbol{D}}$, where the degree of $\boldsymbol{R}$ < the degree of $\boldsymbol{D}$.

7. $\dfrac{x^2 - 2x - 20}{x + 4}$

8. $\dfrac{21x^3 + 41x^2 + 13x + 5}{3x + 5}$

9. $\dfrac{x^4 - 3x^3 + 2x^2 - x + 2}{x - 3}$

10. $\dfrac{x^3 - 27}{x - 3}$

Applications

Solve.

11. ***Geometry:*** A moving company uses a box that has a volume of $x^3 - 2x^2 - 13x - 10$ cubic inches.

    a. If the height of the box is $x + 2$, what is the area of the base of the box?

    b. If the height of the box is $x + 1$, what is the area of the base of the box?

Writing & Thinking

12. Suppose that a polynomial is divided by $(3x - 2)$ and the answer is given as $x^2 + 2x + 4 + \frac{20}{3x - 2}$. What was the original polynomial? Explain how you arrived at this conclusion.

Name: Date:

10.9 Synthetic Division and the Remainder Theorem

Synthetic Division

In summary, synthetic division can be accomplished as follows:

Steps

1. Write only the coefficients of the ____________________

____

Example

$$\begin{array}{r|rrrr} -3 & 5 & 11 & -3 & 1 \\ & & & & \\ \hline \end{array}$$

2. Rewrite the first coefficient (5) as ____________________

____

$$\begin{array}{r|rrrr} -3 & 5 & 11 & -3 & 1 \\ & \downarrow & & & \\ \hline & 5 & & & \end{array}$$

3. Multiply the coefficient (5) by the ____________________

____

$$\begin{array}{r|rrrr} -3 & 5 & 11 & -3 & 1 \\ & \downarrow & -15 & & \\ \hline & 5 \nearrow & -4 & & \end{array}$$

4. Continue to multiply each new coefficient by the constant divisor and ____________________

____

$$\begin{array}{r|rrrr} -3 & 5 & 11 & -3 & 1 \\ & \downarrow & -15 & 12 & -27 \\ \hline & 5 \nearrow & -4 \nearrow & 9 \nearrow & -26 \end{array}$$

5. The constants on the bottom line are the ____________________

____

$$\frac{5x^3+11x^2-3x+1}{x+3} = 5x^2-4x+9+\frac{-26}{x+3}$$
$$= 5x^2-4x+9-\frac{26}{x+3}$$

The Remainder Theorem

If a polynomial $P(x)$ is divided by ____________________

THEOREM

Watch and Work

Watch the video for Example 4 in the software and follow along in the space provided.

Example 4 Using the Remainder Theorem and Synthetic Division

Use synthetic division to show that $(x-6)$ is a factor of $P(x)=x^3-14x^2+53x-30$.

Solution

Now You Try It!

Use the space provided to work out the solution to the next example.

Example A Using the Remainder Theorem and Synthetic Division

Use synthetic division to show that $(x-2)$ is a factor of $P(x)=x^3+x^2+2x-16$.

10.9 Exercises

Concept Check

True/False. Determine whether each statement is true or false. If a statement is false, explain how it can be changed so the statement will be true. (**Note:** There may be more than one acceptable change.)

1. Synthetic division can be used to divide a polynomial by $2x + 3$.

2. At the end of the synthetic division process, the constants on the bottom line are the coefficients of the quotient and the remainder.

3. Synthetic division can be used to find the value of a polynomial for a particular value of x.

4. Synthetic division is only used when the divisor is a first-degree polynomial of the form $(x + c)$ or $(x - c)$

Practice

Divide the following expressions using synthetic division. **a.** Write the answer in the form $Q + \frac{R}{D}$ where R is a constant. **b.** In each exercise, $D = (x - c)$. State the value of c and the value of $P(c)$. (Assume $P(x)$ is the numerator of the fraction.)

5. $\dfrac{x^2 - 12x + 27}{x - 3}$

6. $\dfrac{x^3 - 6x^2 + 8x - 5}{x - 2}$

7. $\dfrac{4x^3 - x^2 + 13}{x - 1}$

8. $\dfrac{3x^4 + 2x^3 + 2x^2 + x - 1}{x + 1}$

Applications

Solve.

9. ***Geometry:*** A moving company uses a box that has a volume of $x^3 + 7x^2 - 6x - 72$ cubic inches.

a. If the height of the box is $x + 4$, what is the area of the base of the box?

b. If the height of the box is $x - 3$, what is the area of the base of the box?

Chapter 10 Project

Math in a Box

An activity to demonstrate the use of polynomials in real life.

Suppose you have a piece of cardboard with length 32 inches and width 20 inches and you want to use it to create a box. You would need to cut a square out of each corner of the cardboard so that you can fold the edges up. But what size square should you cut? Cutting a small square will make a shorter box. Cutting a large square will make a taller box. Look at the diagram below.

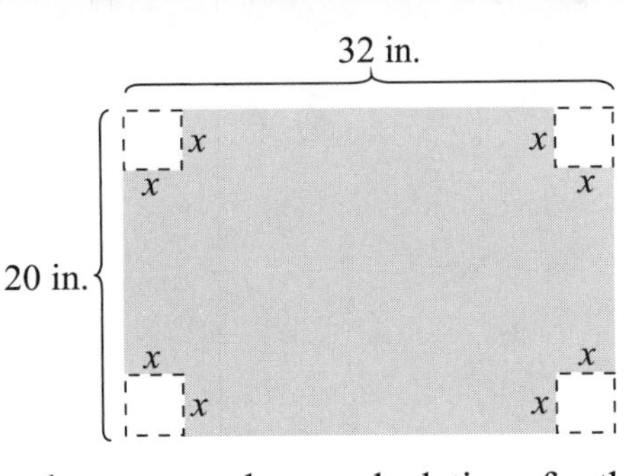

1. Since we haven't determined the size of the square to cut from each corner, let the side length of the square be represented by the variable x. Write a simplified polynomial expression in x and note the degree of the polynomial for each of the following geometric concepts:

   a. The length of the base of the box once the corners are cut out.

   b. The width of the base of the box once the corners are cut out.

   c. The height of the box.

   d. The perimeter of the base of the box.

   e. The area of the base of the box.

   f. The volume of the box.

2. Evaluate the volume expression for the following values of x. (Be sure to include the units of measurement.)

   a. $x = 1$ in.

   b. $x = 2$ in.

   c. $x = 3$ in.

   d. $x = 3.5$ in.

   e. $x = 6$ in.

   f. $x = 7$ in.

3. Based on your volume calculations for the different values of x in Problem 2, if you were trying to maximize the volume of the box, between what two values of x do you think the maximum will be?

4. Using trial and error, see if you can determine the side length x of the square that maximizes the volume of the box. (**Hint:** It will be a value in the interval from problem 3.)

5. Using the value you found for x in Problem 4, determine the dimensions of the box that maximize its volume.

6. Calculate the volume of the box in Problem 5.

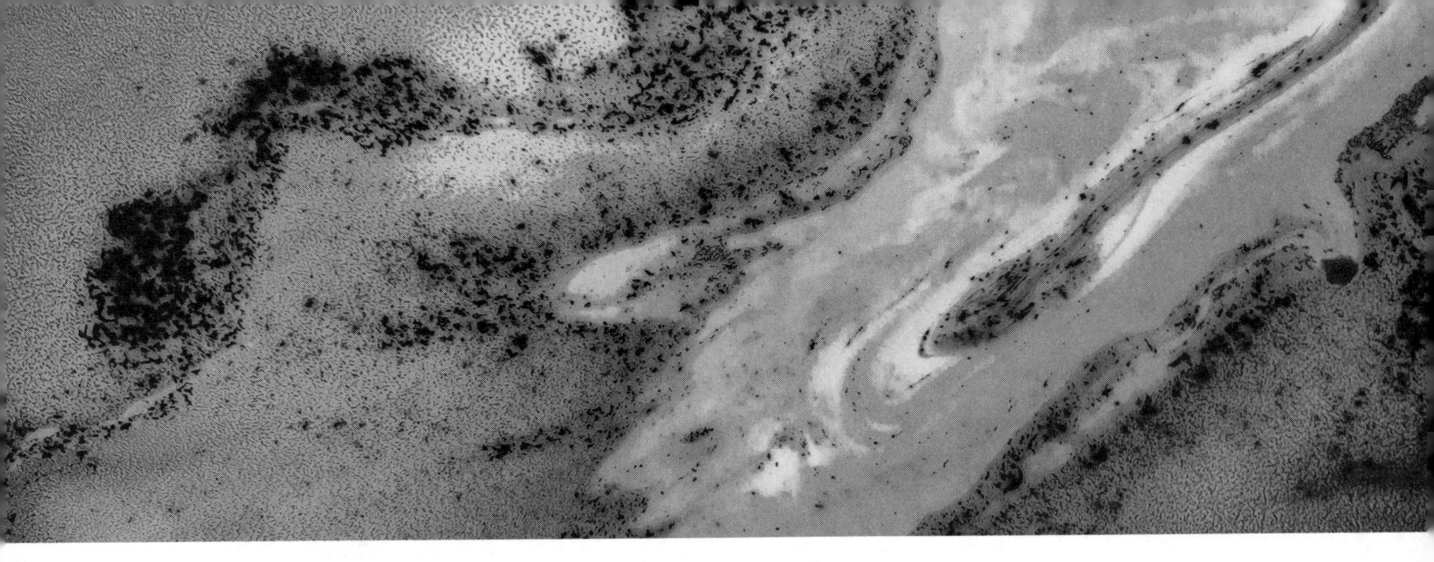

CHAPTER 11

Factoring Polynomials

Math @ Work

Quadratic polynomials can be used to describe a variety of different situations. Many situations in physics can be represented by quadratic polynomials, such as the path of an object that is affected by gravity after being thrown. Quadratic polynomials also show up in economics in situations related to maximizing profit or minimizing cost. Geometry is another area where quadratic polynomials can be useful.

We often know how two measurements relate to each other, but we don't know the exact values of these measurements. Perhaps we know that the length of an object should be four inches longer than the width, or three times as long as the width, no matter what the width is. In situations like this, we can use a variable to represent the width and an expression related to that variable to represent the length. Both of those representations are linear expressions, but if you multiply them together, the result is a quadratic polynomial that represents the area of the rectangle.

Suppose you are entering an art contest where the only requirement is that the canvas size must be 98 square inches. You decide that your painting would look best if the canvas is twice as tall as it is wide. How would you find the dimensions of the canvas?

11.1 Greatest Common Factor (GCF) and Factoring by Grouping

1. The **greatest common factor** (**GCF**) of two or more integers is the ______________________________

____

Procedure for Finding the GCF of a Set of Terms

1. Find the prime factorization of ______________________________
2. List all the factors that are ______________________________
3. Raise each common factor to the ______________________________
4. Multiply these powers to ____________________

Note: If there is no common prime factor or variable, then ____________________

PROCEDURE

Factoring Out the GCF

1. Find the GCF of the ______________________________
2. Divide this monomial factor into ______________________________

The product of the GCF and this new polynomial factor is ______________________________

PROCEDURE

Watch and Work

Watch the video for Example 10 in the software and follow along in the space provided.

Example 10 Factoring Polynomials by Grouping

Factor $xy + 5x + y + 5$ by grouping.

Solution

Now You Try It!

Use the space provided to work out the solution to the next example.

Example A Factoring Polynomials by Grouping

Factor $x^2 + xy + x + y$ by grouping.

11.1 Exercises

Concept Check

True/False. Determine whether each statement is true or false. If a statement is false, explain how it can be changed so the statement will be true. (**Note:** There may be more than one acceptable change.)

1. When finding the GCF of a polynomial, you need to consider only the coefficients.

2. An expression is factored completely if none of its factors can be factored.

3. One way to find the GCF of a set of numbers is to use the prime factorization of each number.

4. Binomials cannot be factored out of algebraic expressions.

Practice

Find the GCF for each set of terms.

5. $\{25, 30, 75\}$

6. $\{8a^3, 16a^4, 20a^2\}$

Factor each polynomial by finding the GCF (or –1 · GCF).

7. $14x + 21$

8. $10x^2y - 25xy$

Factor each of the polynomials by grouping. If a polynomial cannot be factored, write "not factorable."

9. $3x + 3y + ax + ay$

10. $10xy - 2y^2 + 7yz - 35xz$

Applications

Solve.

11. ***Projectile Motion:*** A circus performer is shot vertically into the air with an initial velocity of 48 feet per second. The height of the performer above the ground in feet can be described by the polynomial $48x - 16x^2$ after x seconds.

a. Find the height of the circus performer after 2 seconds.

b. Factor the polynomial $48x - 16x^2$.

c. Use the factored form of the polynomial from Part **b.** to find the height of the circus performer after 2 seconds.

d. Are the answers from Parts **a.** and **c.** the same? Explain why or why not.

Writing & Thinking

12. Explain why the GCF of $-3x^2 + 3$ is 3 and not -3.

11.2 Factoring Trinomials: $x^2 + bx + c$

Watch and Work

Watch the video for Example 1 in the software and follow along in the space provided.

Example 1 Factoring Trinomials with Leading Coefficients of 1

Factor: $x^2 + 8x + 12$

Solution

Now You Try It!

Use the space provided to work out the solution to the next example.

Example A Factoring Trinomials with Leading Coefficients of 1

Factor: $x^2 + 10x + 21$

To Factor Trinomials of the Form $x^2 + bx + c$

To factor $x^2 + bx + c$, if possible, find ______________________________

1. If c is positive, then ______________________________
   a. Both will be ______________________
      Example: ______________________
   b. Both will be______________________
      Example: ______________________
2. If c is negative, ______________________________
   Examples: $x^2 + 6x - 7 = (x+7)(x-1)$ and______________________

PROCEDURE

11.2 Exercises

Concept Check

True/False. Determine whether each statement is true or false. If a statement is false, explain how it can be changed so the statement will be true. (**Note:** There may be more than one acceptable change.)

1. In a trinomial such as $x^2 - 5x + 4$, one would need to find two factors of 4 whose sum is negative 5.

2. In factoring a trinomial with leading coefficient 1, if the constant term is negative, then both factors must be negative.

3. The first step in factoring a trinomial is to look for a common monomial factor.

4. For a trinomial with leading coefficient 1, if no pair exists whose product is the constant and whose sum is the middle term's coefficient, then the trinomial is not factorable.

ractice

Completely factor each trinomial. If a trinomial cannot be factored, write "not factorable."

5. $x^2 - 6x - 27$

6. $a^2 + a + 2$

7. $y^2 - 14y + 24$

8. $2a^4 + 24a^3 + 54a^2$

Applications

Solve.

9. ***Triangles:*** The area of a triangle is $\frac{1}{2}$ the product of its base and its height. If the area of the triangle shown is given by the function $A(x) = \frac{1}{2}x^2 + 24x$, find representations for the lengths of its base and its height (where the base is longer than the height).

10. ***Rectangles:*** The area of the rectangle shown is given by the polynomial function $A(x) = 4x^2 + 20x$. If the width of the rectangle is $4x$, what is the length?

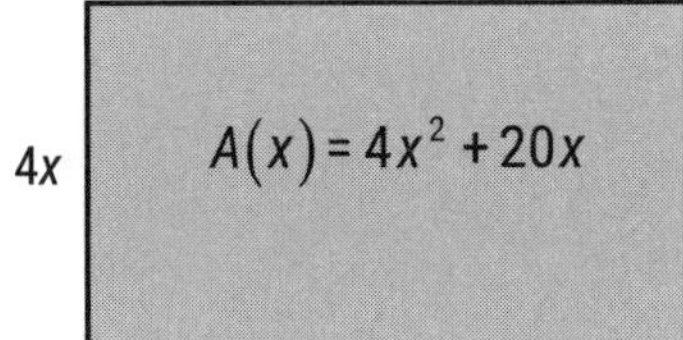

Writing & Thinking

11. Discuss, in your own words, how the sign of the constant term determines what signs will be used in the factors when factoring trinomials.

11.3 Factoring Trinomials: $ax^2 + bx + c$

Guidelines for the Trial-and-Error Method

1. If the sign of the constant term is positive (+), the signs in ______________________________

____

2. If the sign of the constant term is negative (–), the signs in ______________________________

____

Analysis of Factoring by the *ac*-Method

	General Method	Example
	$ax^2 + bx + c$	$2x^2 + 9x + 10$
Step 1:	Multiply ___	Multiply ________
Step 2:	Find two integers whose product is *ac* and ______ ______________________________ ______________________________	Find two integers whose product is 20 and ______________________________ ______________________________
Step 3:	Rewrite the middle term (*bx*) using the ________ ______________________________ ______________________________	______________________________ ______________________________
Step 4:	Factor by grouping the ____________________ ______________________________	Factor by grouping the ________________ ______________________________
Step 5:	Factor out the common binomial factor. This will give ______________________________ ______________________________	Factor out the common binomial factor ______________________________

PROCEDURE

Watch and Work

Watch the video for Example 3 in the software and follow along in the space provided.

Example 3 Using the *ac*-Method

Use the *ac*-method to factor $3x^2 + 19x + 6$.

Solution

Now You Try It!

Use the space provided to work out the solution to the next example.

Example A Using the *ac*-Method

Use the *ac*-method to factor $3a^2 + 14a + 8$.

11.3 Exercises

Concept Check

True/False. Determine whether each statement is true or false. If a statement is false, explain how it can be changed so the statement will be true. (**Note:** There may be more than one acceptable change.)

1. A trinomial is factorable if the middle term is the difference of the inner and outer products of two binomials.

2. The trial-and-error method of factoring a trinomial follows the same steps as the FOIL method of multiplication.

3. The first step in the *ac*-method of factoring is to rewrite the middle term.

4. Factoring can be checked by multiplying the factors and verifying that the product matches the original polynomial.

Practice

Completely factor each polynomial. If a polynomial cannot be factored, write "not factorable."

5. $6x^2 + 11x + 5$

6. $-x^2 + 3x - 2$

7. $x^2 + 8x + 64$

8. $9x^2 - 3x - 20$

9. $12x^2 - 38x + 20$

10. $5a^2 - 7a + 2$

Writing & Thinking

11. It is true that $2x^2 + 10x + 12 = (2x+6)(x+2) = (2x+4)(x+3)$. Explain how the trinomial can be factored in two ways. Is there some kind of error?

12. It is true that $5x^2 - 5x - 30 = (5x-15)(x+2)$. Explain why this is not the completely factored form of the trinomial.

11.4 Special Factoring Techniques

Difference of Two Squares

Consider the polynomial $x^2 - 25$. By recognizing this expression as the **difference of two squares**, we can go directly to the factors:

$$x^2 - 25 = \_\_\_\_\_\_\_\_\_\_\_\_\_\_\_\_\_\_$$

DEFINITION

Sum of Two Squares

The sum of two squares is an expression of the form ____________________________________

DEFINITION

In a perfect square trinomial, both the _______ and _______ terms of the trinomial must be perfect squares. If the first term is of the form x^2 and the last term is of the form a^2, then the middle term must be of the form _______ or ________.

Watch and Work

Watch the video for Example 3 in the software and follow along in the space provided.

Example 3 Factoring Perfect Square Trinomials

Factor completely.

a. $z^2 - 12z + 36$

b. $4y^2 + 12y + 9$

c. $2x^3 - 8x^2y + 8xy^2$

d. $(x^2 + 6x + 9) - y^2$

Solution

Now You Try It!

Use the space provided to work out the solution to the next example.

Example A Factoring Perfect Square Trinomials

Factor completely.

a. $z^2 + 40z + 400$

b. $y^2 - 14y + 49$

c. $3x^2z - 18xyz + 27y^2z$

d. $\left(y^2 + 8y + 16\right) - z^2$

Sum and Difference of Two Cubes

Sum of two cubes: ____________________

Difference of two cubes: ____________________

DEFINITION

11.4 Exercises

Concept Check

True/False. Determine whether each statement is true or false. If a statement is false, explain how it can be changed so the statement will be true. (**Note:** There may be more than one acceptable change.)

1. The expression $x^2 + 20x + 100$ is a perfect square trinomial.

2. When factoring polynomials, always look for a common monomial factor first.

3. The sum of two squares, $(x^2 + a^2)$, is factorable.

4. Sixty-four is a perfect square and a perfect cube.

Practice

Completely factor each of the given polynomials. If a polynomial cannot be factored, write "not factorable."

5. $25 - z^2$

6. $y^2 - 16y + 64$

7. $y^3 + 216$

8. $x^2 + 64y^2$

9. $4x^3 - 32$

Solve.

10. a. Represent the area of the shaded region of the square shown below as the difference of two squares.

b. Use the factors of the expression in Part **a.** to draw (and label the sides of) a rectangle that has the same area as the shaded region.

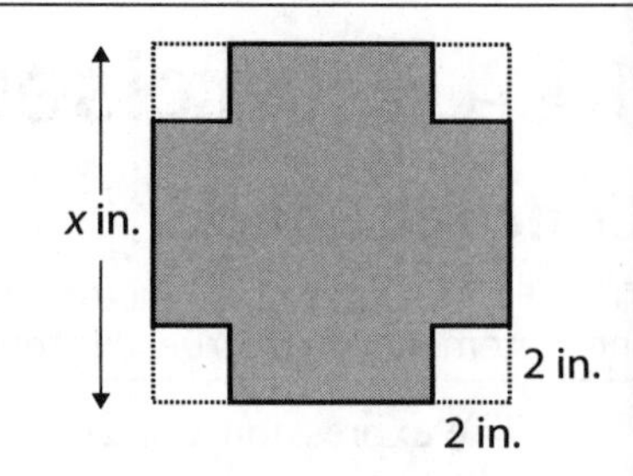

Writing & Thinking

11. a. Show that the sum of the areas of the rectangles and squares in the figure is a perfect square trinomial.

b. Rearrange the rectangles and squares in the form of a square and represent its area as the square of a binomial.

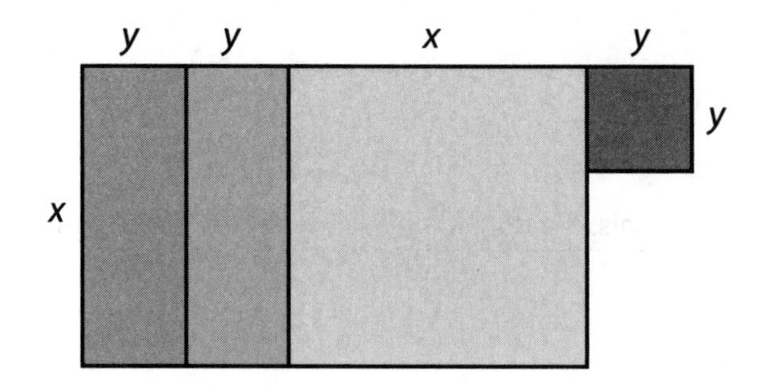

11.5 Review of Factoring Techniques

General Guidelines for Factoring Polynomials

1. **Always look for a common** ______________________

   If the leading coefficient is negative, ______________________

2. **Check the number of terms.**

   a. Two terms:

   1. difference of ______________________
   2. sum of ______________________
   3. difference of ______________________
   4. sum of ______________________

   b. Three terms:

   1. ______________________
   2. ______________________

      Guidelines for the trial-and-error method

      a. If the sign of the constant term is positive, the signs in ______________________

      ____

      b. If the sign of the constant term is negative, the signs in ______________________

      ____

   3. ______________________

      Guidelines for the *ac*-method

      a. Multiply ______________________

      b. Find two integers whose product is *ac* and ______________________

      ____

      c. Rewrite the middle term (bx) using ______________________

      d. ______________________

   c. Four terms:

   Group terms with a common factor and ______________________

3. Check the possibility of ______________________

   Checking: Factoring can be checked by ______________________

   The product should be ______________________

PROCEDURE

11.5 Exercises

Concept Check

True/False. Determine whether each statement is true or false. If a statement is false, explain how it can be changed so the statement will be true. (**Note:** There may be more than one acceptable change.)

1. You should always start by checking the number of terms when factoring a polynomial.

2. If a trinomial is to be factored, the trial-and-error or *ac*-methods can be used.

3. If there are four terms in a polynomial, it cannot be factored.

Practice

Completely factor each of the given polynomials. If a polynomial cannot be factored, write "not factorable."

4. $x^2 - 100$

5. $x^2 + 10x + 25$

6. $x^2 + 16x + 64$

7. $20x^2 - 21x - 54$

8. $2y^2 + 6yz + 5y + 15z$

9. $x^3 + 125$

11.6 Solving Quadratic Equations by Factoring

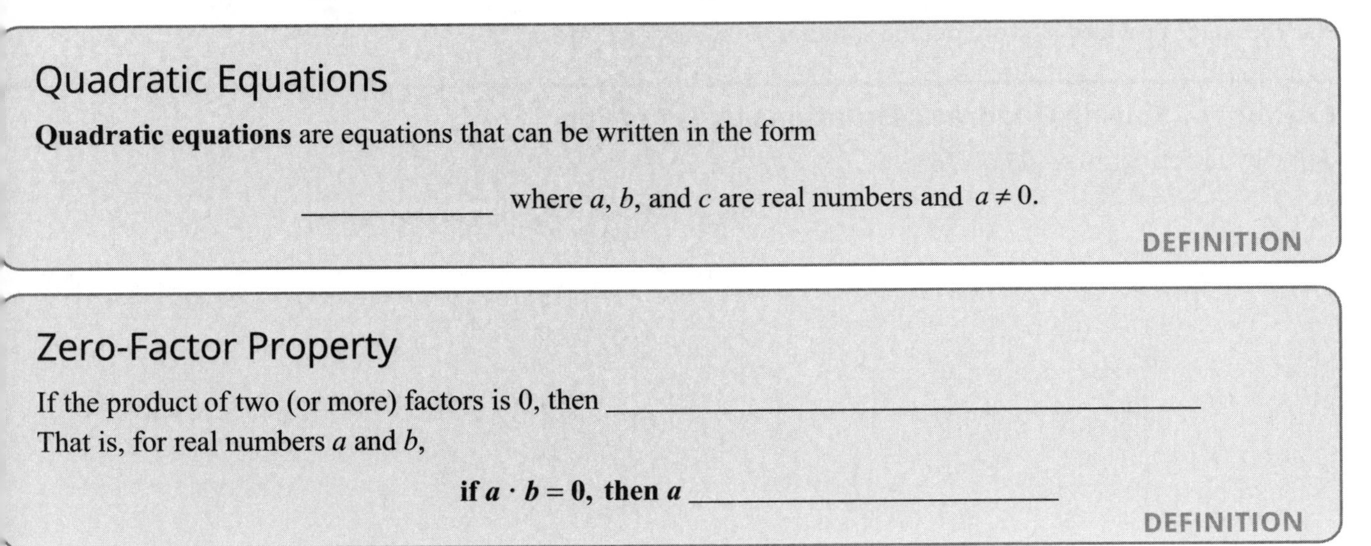

Quadratic Equations

Quadratic equations are equations that can be written in the form

______________ where a, b, and c are real numbers and $a \neq 0$.

DEFINITION

Zero-Factor Property

If the product of two (or more) factors is 0, then ____
That is, for real numbers a and b,

if $a \cdot b = 0$, then a ____________________________

DEFINITION

Watch and Work

Watch the video for Example 4 in the software and follow along in the space provided.

Example 4 Solving Quadratic Equations by Factoring

Solve by factoring: $4x^2 - 4x = 24$

Solution

Now You Try It!

Use the space provided to work out the solution to the next example.

Example A Solving Quadratic Equations by Factoring

Solve by factoring: $9x^2 - 27x = 36$

To Solve a Quadratic Equation by Factoring

1. Add or subtract terms as necessary so that ______________________ and the equation is in the ______________________ where a, b, and c are real numbers and $a \neq 0$.
2. Factor completely. (If there are any fractional coefficients, ______________________ ______________________
3. Set each nonconstant factor equal to ______________________
4. Check each solution, one at a time, in ______________________

PROCEDURE

Factor Theorem

If $x = c$ is a root of a polynomial equation in the form ______________________ ______________________

THEOREM

11.6 Exercises

Concept Check

True/False. Determine whether each statement is true or false. If a statement is false, explain how it can be changed so the statement will be true. (**Note:** There may be more than one acceptable change.)

1. When solving quadratic equations by factoring, it is important that all of the coefficients are integers.

2. The standard form for a quadratic equation is $ax^2 + bx = c$.

3. Not all quadratic equations can be solved by factoring.

4. All quadratic equations have two distinct solutions.

Practice

Solve each equation by factoring.

5. $x^2 - 11x + 18 = 0$

6. $9x^2 + 63x + 90 = 0$

7. $(x-5)(x+3) = 9$

8. Find a polynomial equation with integer coefficients that has $x = 5$ and $x = 7$ as roots.

Applications

Solve.

9. ***Falling Objects:*** A ball is dropped from the top of a building that is 784 feet high. The height of the ball above ground level is given by the polynomial function $h(t) = -16t^2 + 784$ where t is measured in seconds.

a. How high is the ball after 3 seconds? 5 seconds?

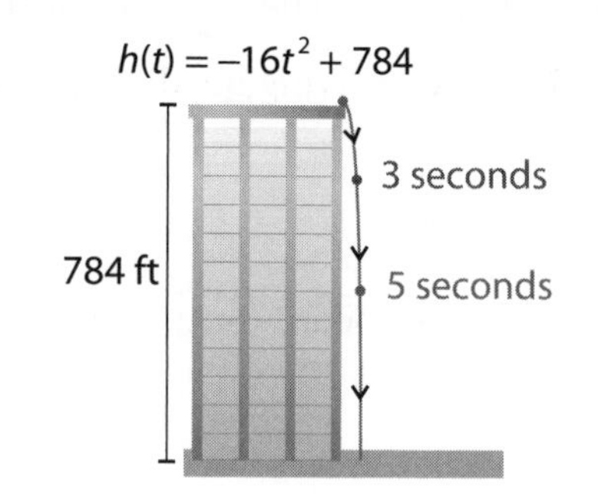

b. How far has the ball traveled in 3 seconds? 5 seconds?

c. When will the ball hit the ground? Explain your reasoning in terms of factors.

10. ***Falling Objects:*** A tennis ball is dropped from a building. The position of the ball after t seconds is given by the polynomial function $s(t) = -4.9t^2 + 490$, where s is the height in meters of the ball.

a. Find $s(0)$. What does this value represent in the context of this problem?

b. How high is the tennis ball 2 seconds after it has been dropped?

c. How long before the tennis ball hits the ground?

Writing & Thinking

11. When solving equations by factoring, one side of the equation must be 0. Explain why this is so.

12. In solving the equation $(x+5)(x-4) = 6$, why can't we just put one factor equal to 3 and the other equal to 2? Certainly $3 \cdot 2 = 6$.

11.7 Applications: Quadratic Equations

Attack Plan for Application Problems

1. Read the ______________________
2. Decide what is asked for and ______________________
   It may help to organize a ______________________
3. Form and then solve an equation that ______________________
4. Check your solution with ______________________

PROCEDURE

Watch and Work

Watch the video for Example 2 in the software and follow along in the space provided.

Example 2 Application: Solving Quadratic Equations

In an orange grove, there are 10 more trees in each row than there are rows. How many rows are there if there are 96 trees in the grove?

Solution

r+10 trees per row

r rows

Now You Try It!

Use the space provided to work out the solution to the next example.

Example A Application: Solving Quadratic Equations

In a theater, there are four less seats in a row than there are rows. How many rows are there if there are 357 seats in the theater?

Consecutive Integers

Integers are **consecutive** if each is ______________________________
Three consecutive integers can be represented as ______________________

For example, ________________

DEFINITION

Consecutive Even Integers

Even integers are consecutive if each is __________________________________
Three consecutive even integers can be represented as __________________________________

For example, ___________________

DEFINITION

Consecutive Odd Integers

Odd integers are consecutive if each is __________________________________
Three consecutive odd integers can be represented as __________________________________

For example, ___________________

DEFINITION

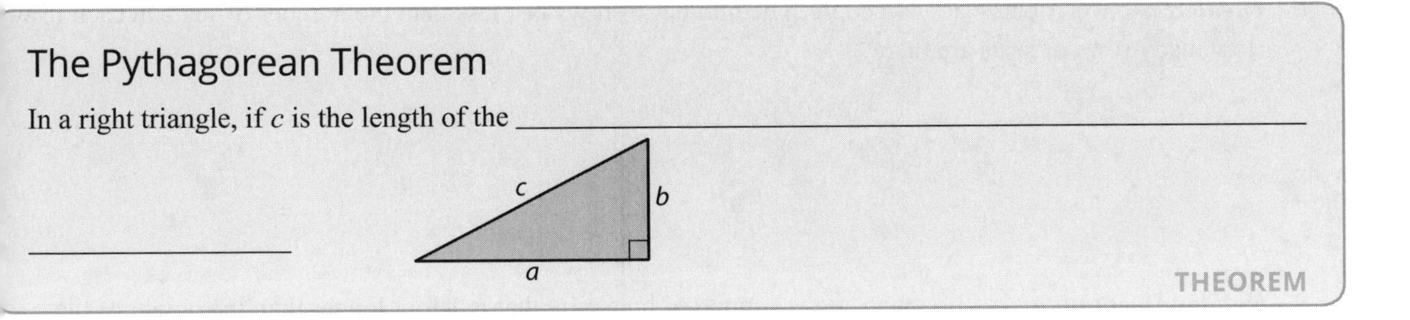

11.7 Exercises

Concept Check

True/False. Determine whether each statement is true or false. If a statement is false, explain how it can be changed so the statement will be true. (**Note:** There may be more than one acceptable change.)

1. The Pythagorean Theorem states that if the two legs of a right triangle are added, the sum will equal the hypotenuse.

2. The expressions n, $n + 1$, and $n + 2$ can represent three consecutive integers.

3. The Pythagorean Theorem can be used with any triangle.

4. The three numbers -10, -8, and -6 are consecutive even integers.

Applications

Write a quadratic equation for each of the following word problems. Then solve the word problem. Remember to check each solution with the wording of the original problem to make sure it is reasonable.

5. One number is 10 more than another. If their product is -25, find the numbers.

6. The square of an integer is equal to seven times the integer. Find the integer.

7. ***Rectangles:*** The length of a rectangular yard is 3 meters greater than the width. If the area of the yard is 54 square meters, find the length and width of the yard.

8. ***Theater:*** A theater can seat 144 people. The number of rows is 7 less than the number of seats in each row. How many rows of seats are there?

9. ***Holiday Decorating:*** A Christmas tree is supported by a wire that is 1 foot longer than the height of the tree. The wire is anchored at a point whose distance from the base of the tree is 49 feet shorter than the height of the tree. What is the height of the tree?

Writing & Thinking

10. The pattern in Kara's linoleum flooring is in the shape of a square 8 inches on a side with right triangles (with legs whose lengths are x inches) placed on each side of the original square so that a new larger square is formed. What is the area of the new square? Explain why you do not need to find the value of x.

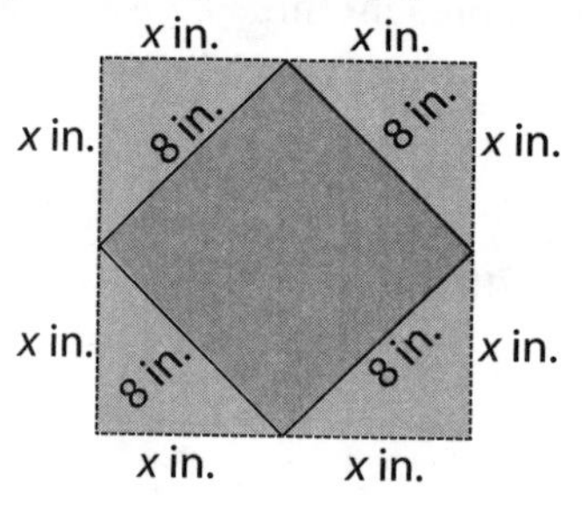

Name: Date:

Chapter 11 Project

Building a Dog Pen

An activity to demonstrate the use of the Pythagorean Theorem and quadratic equations in real life.

Justin's house sits on a lot that is shaped like a trapezoid. He decides to make the back part of the lot usable by building a triangular dog pen for his dog Blackjack. On his lunch break at work he decides to order the materials but realizes that he forgot to write down the actual dimensions of that area of the lot. He wants to get the pen done this weekend because his buddies are coming over to help him, so he has to order the materials today in order for them to arrive on time. He remembers that one of the sides is 5 feet more than the length of the shortest side and the longest side is 10 feet more than the shortest side. Can he figure out the dimensions of the area so that he can order the materials today?

1. Using the diagram of the lot below and the variable x for the length of the shortest side of the dog pen, write an expression for the other two sides of the triangular pen and label them on the diagram.

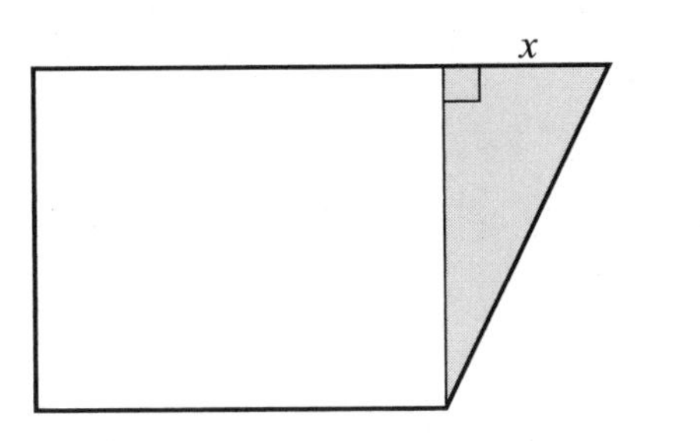

2. Using the Pythagorean Theorem, substitute the three expressions for the sides of the triangle into the formula and simplify the resulting polynomial. Be sure to move all terms to one side of the equation with the other side equal to zero. (Remember that the longest side is the hypotenuse in the formula. Make sure your leading term has a positive coefficient and that your squared binomials result in a trinomial.)

3. Use the equation from Problem 2. Factor the resulting quadratic equation into two linear binomial factors. Set each linear factor equal to zero and solve for x.

a. Factor the resulting quadratic equation into two linear binomial factors.

b. Find the two solutions to the equation from Part **a.** using the zero factor law.

c. Do both of these solutions make sense? Explain your reasoning.

d. Using the solution that makes sense, substitute this value for x and determine the dimensions of the dog pen.

4. To fence in the dog pen, Justin plans to purchase chain link fencing at a cost of \$1.90 per foot.

a. How much fencing will he need?

b. How much will the fencing cost?

5. How much area will the dog pen have?

6. Justin decides to also put a dog house in the pen to protect Blackjack in bad weather. The dog house is rectangular in shape and measures 2.5 feet by 3 feet. Once the dog house is in the pen, how much area will Blackjack have to run in?

7. The sides of the pen form a right triangle and the measurements of the sides of the pen were found using the Pythagorean Theorem. Any three positive integers that satisfy the Pythagorean Theorem are called a **Pythagorean triple**. There are an infinite number of these triples and numerous formulas that can be used to generate them. Do some research on the internet to find one of these formulas and use the formula to generate three more sets of Pythagorean triples. Verify that they are Pythagorean triples by substituting them into the Pythagorean Theorem.

8. Another way to generate a Pythagorean triple is to take an existing triple and multiply each integer by a constant. Take the Pythagorean triple $(3, 4, 5)$ and multiply each integer in the triple by the factors below and verify that the result is also a Pythagorean triple by substituting into the Pythagorean Theorem.

   a. Multiply by 3:

   b. Multiply by 5:

   c. Multiply by 8:

CHAPTER 12

Rational Expressions

Math @ Work

You may be familiar with the phrase "many hands make light work," a quote attributed to John Heywood, an English playwright who lived shortly before Shakespeare. The truth behind this statement has proven itself time and time again. The effects of collaboration are everywhere; collaboration between team members helps companies meet deadlines, the collaboration of multiple organizations after a natural disaster allows more people to receive help, and thousands of collaborating solar cells produce more than enough energy to run the International Space Station.

As company managers look at how best to distribute their resources, they can use rational equations to determine exactly how much time each task will take a team of people to complete. All the managers need to know is the amount of time it takes each individual to complete the entire task and then they can find the time it should take the team to complete the task if they work together. By evaluating these times, managers can make sure each pending deadline has the manpower necessary to complete it in a timely fashion.

Suppose you and your sister are raking leaves. Your sister can rake the yard in 3 hours by herself, and you can rake the yard in $2\frac{5}{8}$ hours working alone. How long will it take the two of you to rake the yard if you work together?

Name: Date:

12.1 Introduction to Rational Expressions

Rational Expressions

A rational expression is an algebraic expression that can be written in the form

DEFINITION

Summary of Arithmetic Rules for Rational Numbers (or Fractions)

A **fraction** (or **rational number**) is a number that can be written in the form

The Fundamental Principle:

The **reciprocal** of $\frac{a}{b}$ is

Multiplication:

Division:

Addition:

Subtraction:

PROPERTIES

The Fundamental Principle of Rational Expressions

If $\frac{P}{Q}$ is a rational expression and P, Q, and K are polynomials where

$Q, K \neq 0$, then

DEFINITION

Watch and Work

Watch the video for Example 3 in the software and follow along in the space provided.

Example 3 Reducing Rational Expressions

Use the fundamental principle to reduce each expression to lowest terms. State any restrictions on the variable by using the fact that no denominator can be 0. This restriction applies to denominators **before and after** a rational expression is reduced.

a. $\dfrac{2x-10}{3x-15}$

b. $\dfrac{x^2-3x-4}{x^2-16}$

c. $\dfrac{y-10}{10-y}$

Solution

Now You Try It!

Use the space provided to work out the solution to the next example.

Example A Reducing Rational Expressions

Reduce each expression to lowest terms. State any restrictions on the variable.

a. $\dfrac{2x-6}{5x-15}$

b. $\dfrac{x^2-x-20}{x^2-25}$

c. $\dfrac{x-5}{5-x}$

Opposites in Rational Expressions

For a polynomial P,

____

In particular,

____

DEFINITION

12.1 Exercises

Concept Check

True/False. Determine whether each statement is true or false. If a statement is false, explain how it can be changed so the statement will be true. (**Note:** There may be more than one acceptable change.)

1. A simplified rational expression cannot have any common factors other than 1 and -1 in both the numerator and denominator.

2. The difference between a rational number and a rational expression is that a rational expression generally has polynomials in the numerator and/or denominator.

3. While a rational number cannot have a zero denominator, a rational expression can have a zero denominator.

4. If a denominator is $x + 5$, it is defined for all values except 5.

Practice

Reduce each expression to lowest terms. State any restrictions on the variable(s).

5. $\dfrac{9x^2y^3}{12xy^4}$

6. $\dfrac{2x-8}{16-4x}$

7. $\dfrac{xy-3y+2x-6}{y^2-4}$

8. Evaluate $\dfrac{3y-4}{y^2+25}$ for $y=3$

Applications

Solve.

9. ***Event Planning:*** The cost of renting a party room with tables, chairs, and simple decorations is \$200 plus \$15 per person attending.

a. Write a rational expression that represents the total price per person for renting the party room, where x is the number of people attending.

b. What is the price per person to rent the party room if 10 people are attending?

c. Determine which values of the variable will make the rational expression from Part **a.** undefined.

d. Considering the context of the given problem, are there any additional restrictions on the variable? If so, explain why these restrictions are in place.

10. ***Rectangles:*** The area of a rectangle (in square feet) is represented by the polynomial function $A(x) = 4x^2 - 4x - 15$. If the length of the rectangle is $(2x + 3)$ feet, find a representation for the width.

$A(x) = 4x^2 - 4x - 15$

$2x + 3$

Writing & Thinking

11. a. Define the term rational expression.

b. Give an example of a rational expression that is undefined for $x = -2$ and $x = 3$ and has a value of 0 for $x = 1$. Explain how you determined this expression.

c. Give an example of a rational expression that is undefined for $x = -5$ and never has a value of 0. Explain how you determined this expression.

12.2 Multiplication and Division with Rational Expressions

To Multiply Rational Expressions

To multiply any two (or more) rational expressions,

1. completely factor each ______________________
2. multiply the numerators and ______________________
____
3. "divide out" any common factors from the ______________________
____

PROCEDURE

Multiplying Rational Expressions

If P, Q, R, and S are polynomials and $Q, S \neq 0$, then

____

DEFINITION

Dividing Rational Expressions

If P, Q, R, and S are polynomials with $Q, R, S \neq 0$, then

____

Note that $\frac{S}{R}$ is the ______________________

DEFINITION

Watch and Work

Watch the video for Example 8 in the software and follow along in the space provided.

Example 8 Dividing with Rational Expressions

Divide and reduce, if possible. Assume that no denominator has a value of 0.

$$\frac{x^2 - 8x + 15}{2x^2 + 11x + 5} \div \frac{2x^2 - 5x - 3}{4x^2 - 1}$$

Solution

Now You Try It!

Use the space provided to work out the solution to the next example.

Example A Dividing with Rational Expressions

Divide and reduce, if possible. Assume that no denominator has a value of 0.

$$\frac{x^2-9x+18}{3x^2+19x+6} \div \frac{3x^2-17x-6}{x^2+x-30}$$

12.2 Exercises

Concept Check

True/False. Determine whether each statement is true or false. If a statement is false, explain how it can be changed so the statement will be true. (**Note:** There may be more than one acceptable change.)

1. The reciprocal of $\frac{x}{x+3}$ is $\frac{-x-3}{x}$.

2. Dividing rational expressions is similar to dividing fractions.

3. There are no restrictions on the denominator $12x^2$.

4. Because $\frac{4x^2}{16x}$ reduces to $\frac{x}{4}$, there are no restrictions on the denominator.

Practice

Perform the indicated operations and reduce to lowest terms. Assume that no denominator has a value of 0.

5. $\frac{x^2-9}{x^2+2x} \cdot \frac{x+2}{x-3}$

6. $\frac{2x^2+x-3}{x^2+4x} \cdot \frac{2x+8}{x-1}$

7. $\frac{x-1}{6x+6} \div \frac{2x-2}{x^2+x}$

8. $\frac{x+3}{x^2+3x-4} \div \frac{x+2}{x^2+x-2}$

Applications

Solve

9. ***Carpentry:*** Erik is building a cubby bookshelf, that is, a bookshelf divided into storage holes (cubbies) instead of shelves. He wants the height of the bookshelf to be $x^2 - 3x - 10$ and the width to be $x^2 + 5x + 6$. Each cubby hole in the bookshelf will have a height of $x + 3$ and a width of $x - 5$.

   a. Write a rational expression to determine how many cubbies high the bookshelf will be.

   b. Write a rational expression to determine how many cubbies wide the bookshelf will be.

   c. Multiply the rational expressions from Parts **a.** and **b.** (and reduce to lowest terms) to obtain a rational expression that gives the total number of cubbies in the entire bookshelf.

12.3 Least Common Multiple of Polynomials

To Find the LCM of a Set of Counting Numbers

1. Find the ______________________________
2. List the prime factors that appear in ______________________________
3. Find the product of these primes using each prime the ______________________________ ______________________________

PROCEDURE

To Find the LCM of a Set of Polynomials

1. Completely factor ____________________ (including prime factors for ____________________).
2. Form the product of all factors that appear, using each factor the ______________________________ ______________________________

PROCEDURE

Watch and Work

Watch the video for Example 5 in the software and follow along in the space provided.

Example 5 Finding the LCM of Polynomials

Find the LCM of the polynomials $y^2 + 10y + 25$, $3y^2 + 15y$, and $5y - 25$.

Solution

Now You Try It!

Use the space provided to work out the solution to the next example.

Example A Finding the LCM of Polynomials

Find the LCM of the polynomials $4y^2 + 4y + 1$, $4y^2 + 2y$, and $8y - 4$.

12.3 Exercises

Concept Check

True/False. Determine whether each statement is true or false. If a statement is false, explain how it can be changed so the statement will be true. (**Note:** There may be more than one acceptable change.)

1. When adding fractions with different denominators, add the denominators.

2. The fraction $\frac{R}{R}$ is equivalent to 1.

3. The least common denominator (LCD) is the least common multiple of the denominators.

4. When finding the LCM of a set of polynomials, you only find the factors of any numerical terms.

Practice

Find the indicated sums and reduce, if possible.

5. $\frac{1}{2}+\frac{1}{10}+\frac{1}{6}$

6. $\frac{3}{4}+\frac{9}{10}+\frac{7}{20}+\frac{1}{2}$

Find the least common multiple (LCM) of each set of polynomials.

7. $x^2-25, \quad 7x+35$

8. $30-3y, \quad y^2-20y+100$

Write a rational expression on the right equivalent to the given rational expression on the left.

9. $\frac{11}{2x+6}=\frac{?}{6(x+3)(x-3)}$

10. $\frac{4}{5x-x^2}=\frac{?}{x(x-5)(x+5)}$

12.4 Addition and Subtraction with Rational Expressions

Addition with Rational Expressions

For polynomials P, Q, and R, with $Q \neq 0$,

____

DEFINITION

Watch and Work

Watch the video for Example 1 in the software and follow along in the space provided.

Example 1 Adding Rational Expressions with a Common Denominator

Find each sum and reduce, if possible. (**Note:** the importance of the factoring techniques we studied earlier will be important here.) State any restrictions on the variable.

a. $\dfrac{x}{x^2-1}+\dfrac{1}{x^2-1}$

b. $\dfrac{1}{x^2+7x+10}+\dfrac{2x+3}{x^2+7x+10}$

Solution

Now You Try It!

Use the space provided to work out the solution to the next example.

Example A Adding Rational Expressions with a Common Denominator

Find each sum and reduce, if possible. State any restrictions on the variable.

a. $\frac{x}{x^2-25}+\frac{5}{x^2-25}$

b. $\frac{2}{x^2+8x+15}+\frac{3x+7}{x^2+8x+15}$

Adding Rational Expressions with Different Denominators

1. Find the ____________
2. Rewrite each fraction in ______________________________
3. Add the numerators and ______________________
4. ______________

PROCEDURE

Placement of Negative Signs

If P and Q are polynomials and $Q \neq 0$, then

____

DEFINITION

Subtraction with Rational Expressions

For polynomials P, Q, and R, with $Q \neq 0$,

____

DEFINITION

12.4 Exercises

Concept Check

True/False. Determine whether each statement is true or false. If a statement is false, explain how it can be changed so the statement will be true. (**Note:** There may be more than one acceptable change.)

1. The LCM of a set of denominators is called the least common denominator.

2. With polynomials, it is most common to place negative signs in the denominator.

3. As with addition, when subtracting rational expressions with different denominators, the first step is to find the LCM of the denominators.

4. You should not use parentheses when subtracting rational expressions.

Practice

Perform the indicated operations and reduce, if possible. Assume that no denominator has a value of 0.

5. $\dfrac{3x-1}{2x-6}+\dfrac{x-11}{2x-6}$

6. $\dfrac{3x}{x-4}+\dfrac{16-x}{4-x}$

7. $\dfrac{x^2+2}{x^2+x-12}+\dfrac{x+1}{12-x-x^2}$

8. $\dfrac{4}{x+5}-\dfrac{2x+3}{x^2+4x-5}$

Applications

Solve.

9. ***Landscaping:*** A landscaper is hired to place large flowering bushes along the borders of a botanical garden. The property is in the shape of a rectangle which measures $7x^2+3$ feet long by $4x^2+5$ feet wide. The bushes are to be placed every $x+2$ feet across the width of the property and every $x-2$ feet along the length of the property.

   a. Write a rational expression to determine how many bushes will go along one length of the property.

   b. Write a rational expression to determine how many bushes will go along one width of the property.

   c. Use the rational expressions from Parts a. and b. to create a rational expression to determine how many bushes will be needed to line the entire property.

10. ***Design:*** Two teams of set designers are jointly creating a set for a scene in a movie. In one hour, the first team can create $\frac{1}{x}$ of the set and the second team can create $\frac{1}{2x-3}$ of the set. If the two teams work together, how much of the set will be completed in one hour?

12.5 Simplifying Complex Fractions

To Simplify Complex Fractions (First Method)

1. Simplify the numerator so that ______________________
2. Simplify the denominator so that ______________________
3. Divide the ______________________

PROCEDURE

To Simplify Complex Fractions (Second Method)

1. Find the LCM of all the denominators in ______________________
2. Multiply both the numerator and denominator of ______________________
3. Simplify both the numerator and ______________________

PROCEDURE

Watch and Work

Watch the video for Example 4 in the software and follow along in the space provided.

Example 4 Second Method for Simplifying Complex Fractions

Simplify the complex fraction. $\dfrac{\dfrac{1}{x+3}-\dfrac{1}{x}}{1+\dfrac{3}{x}}$

Solution

Now You Try It!

Use the space provided to work out the solution to the next example.

Example A Second Method for Simplifying Complex Fractions

Simplify the complex fraction. $\dfrac{\dfrac{1}{x+6}-\dfrac{1}{x}}{1+\dfrac{6}{x}}$

12.5 Exercises

Concept Check

True/False. Determine whether each statement is true or false. If a statement is false, explain how it can be changed so the statement will be true. (**Note:** There may be more than one acceptable change.)

1. When simplifying complex fractions, the answer should always be reduced to lowest terms.

2. Complex fractions are those fractions in which only the denominator consists of one or more fractions itself.

3. Sometimes finding the LCM of all denominators is an important first step for simplifying complex fractions.

4. The LCM of the denominators of $\frac{2}{x-6}$ and $\frac{x}{6}$ is 6.

ractice

implify the following complex fractions.

5. $\dfrac{\frac{2x}{3y^2}}{\frac{5x^2}{6y}}$

7. $\dfrac{\frac{3}{x}+\frac{5}{2x}}{\frac{1}{x}+4}$

6. $\dfrac{\frac{x+3}{2x}}{\frac{2x-1}{4x^2}}$

8. $\dfrac{\frac{7}{x}-\frac{14}{x^2}}{\frac{1}{x}-\frac{4}{x^3}}$

implify the following complex algebraic expressions.

9. $\frac{1}{x+1}-\frac{3}{2x}\cdot\frac{4x}{x+1}$

10. $\frac{x}{x-1}-\frac{3}{x-1}\cdot\frac{x+2}{x}$

Applications

Solve.

11. ***Investing:*** The average percent yield (APY) of an annuity is the annual interest rate earned in a given year that accounts for the effects of compounding. The APY acts as the interest rate for a simple interest account and is larger than the stated interest rate on the compound interest account. The formula to calculate the APY on an annuity after 2 years is

$$\text{APY} = \left(1+\frac{r}{2}\right)^2 - 1,$$

where r is the stated interest rate.

a. Simplify the expression for APY and write as a single rational expression.

b. Using the original formula, calculate the APY for an annuity whose interest rate is 6%. Do not round.

c. Using the expression in Part **a.**, calculate the APY for an annuity whose interest rate is 6%. Do not round.

d. Does the result from Part **c.** match the result from Part **b.**? Explain why or why not.

e. How much larger is the APY than the interest rate?

f. Why do you think the APY is larger than the interest rate? Write a complete sentence.

Writing & Thinking

12. Some complex fractions involve the sum (or difference) of complex fractions. Beginning with the outermost denominator, simplify each of the following expressions.

a. $1+\cfrac{1}{1+\cfrac{1}{1+\cfrac{1}{1+1}}}$

b. $2-\cfrac{1}{2-\cfrac{1}{2-\cfrac{1}{2-1}}}$

c. $x+\cfrac{1}{x+\cfrac{1}{x+\cfrac{1}{x+1}}}$

12.6 Solving Rational Equations

Proportion

A proportion is an ______________________________

In symbols, ______________________

DEFINITION

To Solve an Equation Containing Rational Expressions

1. Find the ______________________
2. List any __________________________ If one of these values appears to be a solution, it ____________________
3. Multiply both sides of the equation by ______________________
4. Solve the resulting equation. (This equation will have ______________________________.)
5. ____________________________________

   (Remember that __________________________.)

PROCEDURE

Watch and Work

Watch the video for Example 3 in the software and follow along in the space provided.

Example 3 Solving Rational Equations

State any restrictions on the variable, and then solve the equation.

$$\frac{1}{x-4}=\frac{3}{x^2-5x}$$

Solution

Now You Try It!

Use the space provided to work out the solution to the next example.

Example A Solving Rational Equations

State any restrictions on the variable, and then solve the equation.

$$\frac{1}{x+2}=\frac{6}{x^2+7x}$$

12.6 Exercises

Concept Check

True/False. Determine whether each statement is true or false. If a statement is false, explain how it can be changed so the statement will be true. (**Note:** There may be more than one acceptable change.)

1. An equation that involves the sum of rational expressions is also a proportion.

2. Multiplying by an LCD can cause extraneous roots.

3. A proportion is properly written if the numerators agree in type and the denominators agree in type.

4. When checking the solutions in the original equation, any solution that gives a 0 denominator cannot be checked.

Practice

5. State any restrictions on x then solve $\frac{8}{x-3}=\frac{12}{2x-3}$

State any restrictions on x, and then solve the equations.

6. $\frac{5x}{4}-\frac{1}{2}=-\frac{3}{16}$

7. $\frac{2}{4x+1}=\frac{4}{x^2+9x}$

8. Solve $S = \dfrac{a}{1-r}$ for r (formula for the sum of an infinite geometric sequence)

Find the lengths of the sides labeled with variables.

9. $\triangle JKL \sim \triangle JTB$

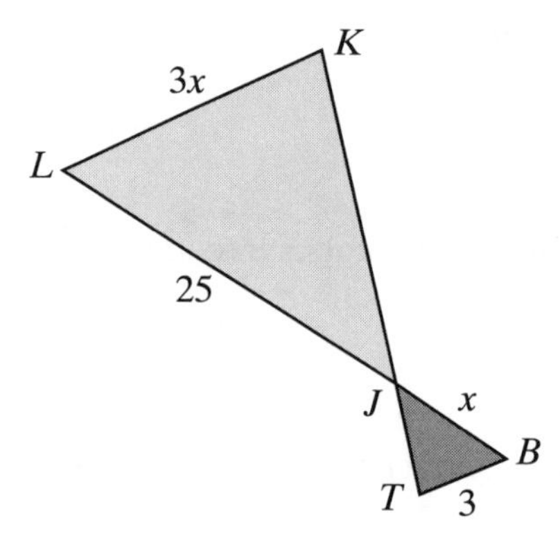

Applications

Solve.

10. ***Computers:*** Making a statistical analysis, Ana found 3 defective computers in a sample of 20 computers. If this ratio is consistent, how many defective computers does she expect to find in a batch of 2400 computers?

11. ***Baking:*** The recipe for Nestle Tollhouse Chocolate Chip Cookies calls for 2 cups of chocolate chips to make 5 dozen cookies. If you want to bake 17 dozen cookies, how many cups of chocolate chips do you need?

12.7 Applications: Rational Expressions

To Solve a Word Problem Containing Rational Expressions

1. Read the problem carefully. Read it ______________________
2. Decide what is asked for and ______________________
   Draw a diagram or ______________________
3. Form and ______________________
4. Check your solution with the ______________________

PROCEDURE

Watch and Work

Watch the video for Example 1 in the software and follow along in the space provided.

Example 1 Fractions

The denominator of a fraction is 8 more than the numerator. If both the numerator and denominator are increased by 3, the new fraction is equal to $\frac{1}{2}$. Find the original fraction.

Solution

Now You Try It!

Use the space provided to work out the solution to the next example.

Example A Fractions

The denominator of a fraction is 3 more than the numerator. If both the numerator and denominator are increased by 4, the new fraction is equal to $\frac{3}{4}$. Find the original fraction.

12.7 Exercises

Concept Check

True/False. Determine whether each statement is true or false. If a statement is false, explain how it can be changed so the statement will be true. (**Note:** There may be more than one acceptable change.)

1. The first step in solving application problems is to assign a variable to the unknown value.

2. If the total amount of work took 5 hours to do, then $\frac{1}{5}$ of the work can be done in one hour.

3. If you know the distance and rate, you can use the formula $t = d - r$ to represent time.

Applications

Solve.

4. What number must be added to both the numerator and denominator of $\frac{16}{21}$ to make the resulting fraction equal to $\frac{5}{6}$?

Solve the following word problems. Remember to check each solution with the wording of the original problem to make sure it is reasonable.

5. ***Travel by Car:*** It takes Rosa, traveling at 30 mph, 30 minutes longer to go a certain distance than it takes Melody traveling at 50 mph. Find the distance traveled.

6. ***Landscaping:*** Toni needs 4 hours to complete the yard work. Her husband, Sonny, needs 6 hours to do the work. How long will the job take if they work together?

7. ***Wind Speed:*** An airplane can fly 650 mph in still air. If it can travel 2800 miles with the wind in the same time it can travel 2400 miles against the wind, find the wind speed.

Writing & Thinking

8. If n is any integer, then $2n$ is an even integer and $2n + 1$ is an odd integer. Use these ideas to solve the following problems.

   a. Find two consecutive odd integers such that the sum of their reciprocals is $\frac{12}{35}$.

   b. Find two consecutive even integers such that the sum of the first and the reciprocal of the second is $\frac{9}{4}$.

12.8 Applications: Variation

Direct Variation

A variable quantity y varies directly as (or is directly proportional to) a variable x if there is a constant k such that

____

The constant k is called the ______________________

DEFINITION

Inverse Variation

A variable quantity y varies inversely as (or is inversely proportional to) a variable x if there is a constant k such that

____

The constant k is called the ______________________

DEFINITION

1. If a variable varies either directly or inversely as more than one other variable, the variation is said to be

____

2. If the combined variation is all direct variation (the variables are multiplied), then it is called ______________

____

Watch and Work

Watch the video for Example 6 in the software and follow along in the space provided.

Example 6 More Variation

The distance an object falls varies directly as the square of the time it falls (until it hits the ground and assuming little or no air resistance). If an object fell 64 feet in 2 seconds, how far would it have fallen by the end of 3 seconds?

Solution

Now You Try It!

Use the space provided to work out the solution to the next example.

Example A More Variation

The distance an object falls varies directly as the square of the time it falls. If an object fell 64 feet in 2 seconds, how many feet will it fall in 5 seconds?

12.8 Exercises

Concept Check

True/False. Determine whether each statement is true or false. If a statement is false, explain how it can be changed so the statement will be true. (**Note:** There may be more than one acceptable change.)

1. The number of hamburgers eaten varies inversely with calories consumed.

2. The equation $y = \frac{k}{x}$ represents direct variation.

3. Distance and time varies directly, which means they are directly proportional.

4. The circumference of a circle varies directly with its radius.

Practice

Use the information given to find the unknown value.

5. If y varies directly as x, and $y = 3$ when $x = 9$, find y if $x = 7$.

6. If y varies inversely as x^2, and $y = -8$ when $x = 2$, find y if $x = 3$.

7. z varies jointly as x and y, and $z = 60$ when $x = 2$ and $y = 3$. Find z if $x = 3$ and $y = 4$.

8. z varies directly as x and inversely as y^2. If $z = 5$ when $x = 1$ and $y = 2$, find z if $x = 2$ and $y = 1$.

Applications

Solve.

9. ***Stretching a Spring:*** The length a hanging spring stretches varies directly with the weight placed on the end. If a spring stretches 5 in. with a weight of 10 lb, how far will the spring stretch if the weight is increased to 12 lb?

10. ***Construction:*** The safe load (L) of a wooden beam supported at both ends varies jointly as the width (w) and the square of the depth (d) and inversely as the length (l). A beam 4 in. wide, 6 in. deep, and 12 ft long supports a load of 4800 lb safely. What is the safe load of a beam of the same material that is 6 in. wide, 10 in. deep, and 15 ft long?

Writing & Thinking

11. Explain, in your own words, the meaning of the following terms.
    - **a.** Direct variation
    - **b.** Joint variation
    - **c.** Inverse variation
    - **d.** Combined variation

    Discuss an example of each type of variation that you have observed in your daily life.

Chapter 12 Project

Let's Be Rational Here!

An activity to demonstrate the use of rational expressions in real life.

'ou may be surprised to learn how many different situations in real life involve working with rational expressions and rational equations. Hopefully after spending the day with Meghan and her friends you'll be convinced of their importance.

t's Saturday and Meghan has a list of things to accomplish today: revise her budget, paint the walls in the spare bedroom, ravel to the lake with her friends and then ski on the lake for the rest of the day, provided the weather stays nice.

1. Meghan decides to tackle the budget first. After reviewing her budget, she decides that she really needs to get a part-time job to earn some extra money. She is remodeling the living room and would like to buy some new furniture. Letting x represent her new monthly combined salary, she estimates that $\frac{1}{4}$ of her new salary will be used for bills and approximately $\frac{1}{5}$ for her car payment. She would like to have \$1100 left over each month, of which \$100 will be saved for the new furniture. What must her new monthly salary be?

a. Let the variable x represent Meghan's new monthly salary. Write an expression to represent the amount of her new salary used for bills. (Remember that the word **of** implies multiplication.)

b. Write an expression to represent the amount of Meghan's new monthly salary used for her car payment.

c. Write an equation that sums the expenditures and leftover balance and set this sum equal to the new monthly salary of x.

d. Find the LCD of the rational expressions from Part **c.** and multiply it by each term in the equation to remove the fractions.

e. Solve the equation from Part **d.** to determine what Meghan's new monthly salary needs to be.

f. Approximately how much of Meghan's new salary will be used to pay her car payment?

2. With the budget done, Meghan prepares to paint the spare bedroom. She just finished painting her bedroom last week and it took her about 4 hours. Her roommate Ashley painted her bedroom a couple of weeks ago and it took her 6 hours. All the bedrooms are similar in size. Meghan realizes that if she gets Ashley to help her with the painting, it will take them less time and they can get to the lake sooner. How long will it take Meghan and Ashley working together to paint the spare bedroom? Use the table below to help you set up the problem.

Person	Time (in hours)	Part of Work Done in 1 Hour
Meghan	4	$\frac{\quad}{\quad}$
Ashley	6	$\frac{\quad}{\quad}$
Together	x	$\frac{1}{x}$

a. Fill in the missing information in column three of the table.

b. Use the entries in the last column of the table to set up an equation to represent the sum of the amount of work done by both Meghan and Ashley in an hour.

c. Solve the equation to determine how long it will take the two girls to paint the spare bedroom when working together. Express the result as a decimal to the nearest tenth. Convert this measurement to hours and minutes.

3. With the spare bedroom painted, the girls call Lucas to let him know they are ready to head to the lake. While he is preparing the boat for the lake, Lucas is trying to decide which route they should travel to get there. If he travels the highway, he can travel 20 mph faster than the scenic route. However, the highway is 30 miles longer than the scenic route, which is 60 miles long. Lucas thinks it should take him the same amount of time to get there using either route. Use the following table to help you organize the information for this problem.

	Distance (miles)	**÷**	**Rate (mph)**	**=**	**Time (hours)**
Highway	90		$x + 20$		___
Scenic Route	60		x		___

a. Fill in the missing information in the table. (**Hint:** Recall that the formula that involves distance, rate, and time is $d = r \cdot t$.)

b. Since Lucas expects the time to be the same for each route, create an equation from the time column and solve for x.

c. How fast must Lucas travel on the highway to get to the lake in the same amount of time as traveling the scenic route?

CHAPTER 13

Roots, Radicals, and Complex Numbers

Math @ Work

While roots and radicals may not be used in everyday situations such as grocery shopping, they are used in a variety of applications that relate to our day to day lives. One such application is the measurement of distances. In 1923, N. Korzenewsky, the leader of an expedition in the deserts of Turkestan, recorded seeing snow-capped mountains 750 km (466 miles) away. A similar but more reliable recorded observation was made in 1933 by Commander C. L. Garner of the Coast and Geodetic Survey. He wrote that instrumental measurements were made in both directions "between Mt. Shasta and Mt. St. Helena in California," which is a distance of 192 miles.

When you watch a sunset over the ocean, do you know how far you are from the horizon? On a clear day, a person looking over the ocean can see an island on the horizon up to 80 miles away (which is approximately the distance from San Diego, California, to San Clemente Island). However, the visible distance varies due to the clarity and temperature of the air. A general rule of thumb for calculating the distance in miles between you and the horizon is the formula $d = 1.32\sqrt{h}$, where h is the height in feet that your eyes are above the ground.

Suppose you and a group of friends are watching the sunset from the roof of a two-story building. One friend wonders how far away the horizon is. You know that your eyes are approximately 28 feet above the ground. How can you use this information to estimate the distance between you and the horizon?

13.1 Evaluating Radicals

Radical Terminology

The symbol $\sqrt{}$ is called ______________________

The number under the ______________________________________

The complete expression, such as $\sqrt{64}$ is called ___________________________________

DEFINITION

Square Root

If a is a nonnegative real number, then

____

____

____

DEFINITION

Cube Root

If a is a real number, then _________________________

DEFINITION

Watch and Work

Watch the video for Example 4 in the software and follow along in the space provided.

Example 4 Evaluating Cube Roots

a. Because $2^3 =$ _, $\sqrt[3]{8} =$ _.

b. Because $(-6)^3 =$ ____, $\sqrt[3]{-216} =$ __.

____

c. Because $\left(\frac{1}{3}\right)^3 =$ __, $\sqrt[3]{\frac{1}{27}} =$ _.

Now You Try It!

Use the space provided to work out the solution to the next example.

Example A Evaluating Cube Roots

Evaluate the following radical expressions.

a. $\sqrt[3]{64}$

b. $\sqrt[3]{-125}$

c. $\sqrt[3]{\frac{1}{1000}}$

13.1 Exercises

Concept Check

True/False. Determine whether each statement is true or false. If a statement is false, explain how it can be changed so the statement will be true. (**Note:** There may be more than one acceptable change.)

1. If a number is squared and the principal square root of the result is found, that square root is always equal to the original number.

2. There is no real number that can be a square root of a negative number.

3. The index is the number underneath the radical sign.

4. The cube root of -27 is a real number.

Practice

Simplify the following square roots and cube roots.

5. $\sqrt{49}$

6. $\sqrt{289}$

7. $\sqrt[3]{1000}$

8. $\sqrt[3]{\dfrac{27}{64}}$

9. $\sqrt{0.04}$

Applications

Solve.

10. ***Area:*** The area of a square tile is 16 square inches.

a. How long are the sides of the square tile?

b. How many tiles would be needed for a four-foot-long and four-inch-high backsplash in a newly designed bathroom?

11. ***Volume:*** The volume of a child's building block is 64 cubic centimeters.

a. Assuming the building block is a perfect cube, find the length of each side of the block.

b. If a child stacks 5 blocks directly on top of each other, find the height of the structure that is created.

Writing & Thinking

12. Discuss, in your own words, why the square root of a negative number is not a real number.

13. Discuss, in your own words, why the cube root of a negative number is a negative number.

13.2 Simplifying Radicals

Properties of Square Roots

If a and b are **positive** real numbers, then

1. ____________________

2. ____________________

PROPERTIES

Simplest Form for Square Roots

A square root is considered to be in **simplest form** when ____________________

____

DEFINITION

Square Root of x^2

If x is a real number, then ____________________

Note: If $x \geq 0$ is given, then ____________________

DEFINITION

Watch and Work

Watch the video for Example 3 in the software and follow along in the space provided.

Example 3 Simplifying Square Roots with Variables

Simplify each of the following square roots. Look for perfect square factors and even powers of the variables. Assume that all variables represent positive real numbers.

a. $\sqrt{81x^4}$

b. $\sqrt{64x^5y}$

c. $\sqrt{18a^4b^6}$

d. $\sqrt{\dfrac{9a^{13}}{b^4}}$

Solution

Now You Try It!

Use the space provided to work out the solution to the next example.

Example A Simplifying Square Roots with Variables

Simplify each of the following square roots. Assume that all variables represent positive real numbers.

a. $\sqrt{16x^8}$

b. $\sqrt{100x^3y^3}$

c. $\sqrt{12x^8y^{12}}$

d. $\sqrt{\dfrac{25z^{18}}{y^8}}$

Simplest Form for Cube Roots

A cube root is considered to be in ____________________

DEFINITION

13.2 Exercises

Concept Check

True/False. Determine whether each statement is true or false. If a statement is false, explain how it can be changed so the statement will be true. (**Note:** There may be more than one acceptable change.)

1. Any variable term with an exponent of 5 has a perfect cube factor within that variable term.

2. The simplest form of a radical expression can be found by using prime factorization.

3. If x is a real number, then $\sqrt{x^2} = x$.

4. The term $7b\sqrt[3]{6c^2}$ is in simplified form.

Practice

Simplify each of the following radical expressions. Assume that all variables represent positive real numbers.

5. $\sqrt{162}$

6. $\sqrt{\dfrac{32}{49}}$

7. $\sqrt{24x^{11}y^2}$

8. $\sqrt[3]{56}$

9. $\sqrt[3]{-8x^8}$

Applications

Use the following two formulas associated with electricity

$I = \sqrt{\dfrac{P}{R}}$ $P =$ power (in watts)

$I =$ current (in amperes)

$E =$ voltage (in volts)

$E = \sqrt{PR}$ $R =$ resistance (in ohms, Ω)

---

10. ***Electricity:*** What is the current in amperes of a light bulb that produces 150 watts of power and has a 25 Ω resistance?

11. ***Electricity:*** If a light bulb has a resistance of 30 Ω and produces 90 watts of power, what is its current in amperes?

Writing & Thinking

12. Under what conditions is the expression $\sqrt{a}$ not a real number?

13. Explain why the expression $\sqrt[3]{y}$ is a real number regardless of whether $y > 0$, $y < 0$, or $y = 0$.

13.3 Rational Exponents

Radical Notation

If n is an integer greater than 1, then ______________________________

The expression $\sqrt[n]{a}$ is called ______________

The symbol $\sqrt[n]{}$ is called ______________

n is called ______________

____

Note: If no index is given, it is ______________________

For example, ______________

DEFINITION

The General Form $a^{\frac{m}{n}}$

If n is an integer greater than 1, m is any integer, and $a^{\frac{1}{n}}$ is a real number, then

____

In radical notation: ______________________

DEFINITION

Watch and Work

Watch the video for Example 4 in the software and follow along in the space provided.

Example 4 Simplifying Radical Notation by Changing to Exponential Notation

Simplify each expression by first changing it into an equivalent expression with rational exponents. Then rewrite the answer in simplified radical form.

a. $\sqrt[4]{\sqrt[3]{x}} =$

b. $\sqrt[3]{a}\sqrt{a} =$

c. $\dfrac{\sqrt{x^3}\sqrt[3]{x^2}}{\sqrt[5]{x^2}} =$

Now You Try It!

Use the space provided to work out the solution to the next example.

Example A Simplifying Radical Notation by Changing to Exponential Notation

Rewrite the expression in simplified radical form.

a. $\sqrt[5]{\sqrt[2]{x}}$

b. $\sqrt[4]{x} \cdot \sqrt{x}$

c. $\dfrac{\sqrt[3]{x^2} \cdot \sqrt{x^3}}{\sqrt[6]{x^5}}$

13.3 Exercises

Concept Check

True/False. Determine whether each statement is true or false. If a statement is false, explain how it can be changed so the statement will be true. (**Note:** There may be more than one acceptable change.)

1. The same rules for exponents apply to both integer exponents and rational exponents.

2. If the cube root of 7 were to be converted into exponential notation it would be $\sqrt[3]{7}$.

3. Any expression to the power 0, such as $\left(\sqrt[4]{x}\right)^0$, is equal to 1.

4. The expression $y^{\frac{1}{2}}$ can be rewritten in radical notation as $\sqrt{y^2}$.

Practice

5. Use radical notation to write an expression that is equivalent to $8^{\frac{1}{3}}$.

6. Use exponential notation to write an expression that is equivalent to $\sqrt{3}$.

Simplify each numerical expression.

7. $100^{-\frac{1}{2}}$

8. $64^{\frac{2}{3}}$

9. Simplify $\dfrac{a^{\frac{1}{2}} \cdot a^{-\frac{3}{4}}}{a^{-\frac{1}{2}}}$. Assume that all variables represent positive real.

10. Simplify $\frac{\sqrt[4]{y^3}}{\sqrt[6]{y}}$ by first changing it into an equivalent expression with rational exponents. Rewrite the answer in simplified radical form. Assume that all variables represent positive real numbers.

Applications

Solve.

11. ***Area:*** The width of a rectangle is $\sqrt[3]{64^2}$ ft and the length is $216^{\frac{2}{3}}$ ft. What is the area of the rectangle?

12. ***Amusement Parks:*** An amusement park is creating signs to indicate the velocity of the roller coaster car on certain hills of the most popular rides. A roller coaster car gains kinetic energy as it goes down a hill. The velocity, or speed, of an object in kilometers per hour (kph) can be determined by $V = \left(\frac{2k}{m}\right)^{\frac{1}{2}}$, where k is the kinetic energy of the object in joules (J) and m is the mass of the object in kilograms (kg).

a. For the most popular roller coaster, the car has a mass of 300 kg and the car has a kinetic energy of 375,000 J on the first hill. What velocity does the car obtain on the first hill?

b. For the second most popular roller coaster, the car has a mass of 350 kg and the car has a kinetic energy of 70,000 on the first hill. What velocity does the car obtain on the first hill?

Writing & Thinking

13. Is $\sqrt[5]{a} \cdot \sqrt{a}$ the same as $\sqrt[5]{a^2}$? Explain why or why not.

14. Assume that x represents a positive real number. Describe what kind of number the exponent n must be for x^n to mean

a. a product.

b. a quotient.

c. 1.

d. a radical state.

13.4 Addition, Subtraction, and Multiplication with Radicals

1. ______________________ have the same index and radicand, or they can be simplified so that they have the same index and radicand.

Watch and Work

Watch the video for Example 3 in the software and follow along in the space provided.

Example 3 Multiplying with Radicals

Multiply and simplify the following expressions. Assume that all variables represent positive real numbers.

a. $\sqrt{5} \cdot \sqrt{15}$

b. $\sqrt{7}\left(\sqrt{7} - \sqrt{14}\right)$

c. $\left(3\sqrt{7} - 2\right)\left(\sqrt{7} + 3\right)$

d. $\left(\sqrt{2x} + 5\right)\left(\sqrt{2x} - 5\right)$

e. $\left(1 - \sqrt{3x + 4}\right)^2$

Solution

Now You Try It!

Use the space provided to work out the solution to the next example.

Example A Multiplying with Radicals

Multiply and simplify the following expressions. Assume that all variables represent positive real numbers.

a. $\sqrt{3} \cdot \sqrt{75}$

b. $\sqrt{13}\left(\sqrt{26} - \sqrt{13}\right)$

c. $\left(2\sqrt{5} - 1\right)\left(\sqrt{5} + 2\right)$

d. $\left(\sqrt{3z} - \sqrt{5}\right)\left(\sqrt{3z} + \sqrt{5}\right)$

e. $\left(2 - \sqrt{5x+1}\right)^2$

13.4 Exercises

Concept Check

True/False. Determine whether each statement is true or false. If a statement is false, explain how it can be changed so the statement will be true. (**Note:** There may be more than one acceptable change.)

1. The radicals $\sqrt{a}$ and $\sqrt[3]{a}$ are like radicals.

2. The radicals $3\sqrt{a}$ and $\sqrt{a}$ are like radicals.

3. The sum $4\sqrt{3}+8\sqrt{5}$ cannot be simplified.

Practice

Simplify the following radical expressions. Assume that all variables represent positive real numbers.

4. $\sqrt{5}+\sqrt{4}-2\sqrt{5}+6$

5. $5\sqrt[3]{x}+9\sqrt[3]{y}-10\sqrt[3]{y}+4\sqrt[3]{x}$

6. $\sqrt{18}-2\sqrt{12}+5\sqrt{2}$

Multiply the following radical expressions and then simplify the results. Assume that all variables represent positive real numbers.

7. $3\sqrt{18}\cdot\sqrt{2}$

8. $(3+\sqrt{2})(5-\sqrt{2})$

9. $\left(\sqrt{5}+2\sqrt{2}\right)^2$

Applications

Solve.

10. ***Radio Circuits:*** For a complete radio circuit, $d=\sqrt{2g}+\sqrt{2h}$, where d equals the visual horizon distance and g and h are the heights of the radio antennas at the respective stations. What is d when $g = 75$ ft and $h = 85$ ft?

11. ***Decorating:*** Mary is making a tile decoration for her wall. Using square tiles of different sizes, Mary created one decoration that is five tiles across, with sides touching. The first tile is 10 in.^{2}, the second is 20 in.^{2}, the third is 30 in.^{2}, the fourth is 20 in.^{2}, and the fifth is 10 in.^{2} What is the length of the decoration?

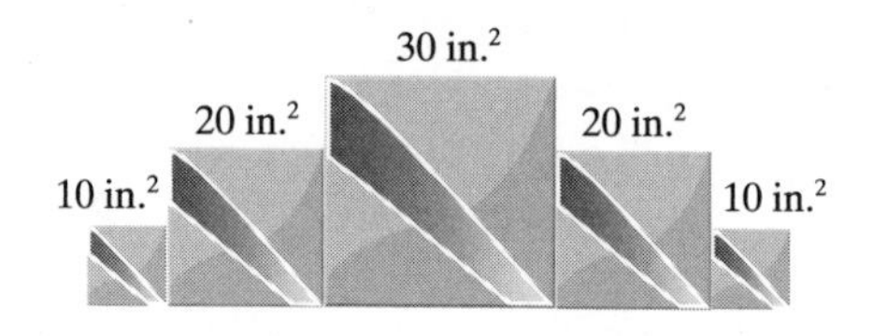

13.5 Rationalizing Denominators

To Rationalize a Denominator Containing a Square Root or a Cube Root

1. If the denominator contains a square root, ______________________________

____

2. If the denominator contains a cube root, ______________________________

____

PROCEDURE

Watch and Work

Watch the video for Example 2 in the software and follow along in the space provided.

Example 2 Rationalizing Denominators Containing Cube Roots

Rationalize each denominator. Assume that all variables represent positive real numbers.

a. $\frac{3}{\sqrt[3]{5}}$

b. $\frac{7}{\sqrt[3]{32y}}$

Solution

Now You Try It!

Use the space provided to work out the solution to the next example.

Example A Rationalizing Denominators Containing Cube Roots

Rationalize each denominator. Assume that all variables represent positive real numbers.

a. $\dfrac{4}{\sqrt[3]{7}}$

b. $\dfrac{5}{\sqrt[3]{9x^5}}$

To Rationalize a Denominator Containing a Sum or Difference Involving Square Roots

If the denominator of a fraction contains a sum or difference involving a square root, rationalize the denominator by multiplying both the numerator and denominator by the ____________________

____

1. If the denominator is of the form $a - b$, ____________________

2. If the denominator is of the form $a + b$, ____________________

The new denominator will be the ____________________

____

PROCEDURE

13.5 Exercises

Concept Check

True/False. Determine whether each statement is true or false. If a statement is false, explain how it can be changed so the statement will be true. (**Note:** There may be more than one acceptable change.)

1. The conjugate of $y-\sqrt{5}$ is $y+\sqrt{5}$.

2. To rationalize the denominator, multiply only the denominator by an expression that will result in a denominator with no radicals.

3. To rationalize a fraction whose denominator is $\sqrt[3]{a}$, you would need to multiply the numerator and the denominator by $\sqrt[3]{a}$.

4. The fraction $\frac{\sqrt{2}}{3}$ is in simplest form.

Practice

Rationalize the denominator and simplify, if possible. Assume that all variables represent positive real numbers.

5. $\frac{-3}{\sqrt{7}}$

6. $\frac{\sqrt{ab}}{\sqrt{9ab}}$

7. $\frac{2}{\sqrt{6}-2}$

8. $\frac{x}{\sqrt{x}+2}$

Applications

Solve.

9. ***Volume of a Cylinder:*** The radius of a cylinder can be expressed in terms of its volume and its height by $r=\sqrt{\frac{V}{\pi h}}$. Rationalize the denominator of this formula.

10. ***Business:*** A company that sells computers learns that their income can be represented by the equation $I=\frac{6500p}{\sqrt{p}-10}$ dollars when they sell their computers for p dollars. Rationalize the denominator of this equation.

Writing & Thinking

11. In your own words, explain how to rationalize the denominator of a fraction containing the sum or difference of square roots in the denominator. Why does this work?

13.6 Solving Radical Equations

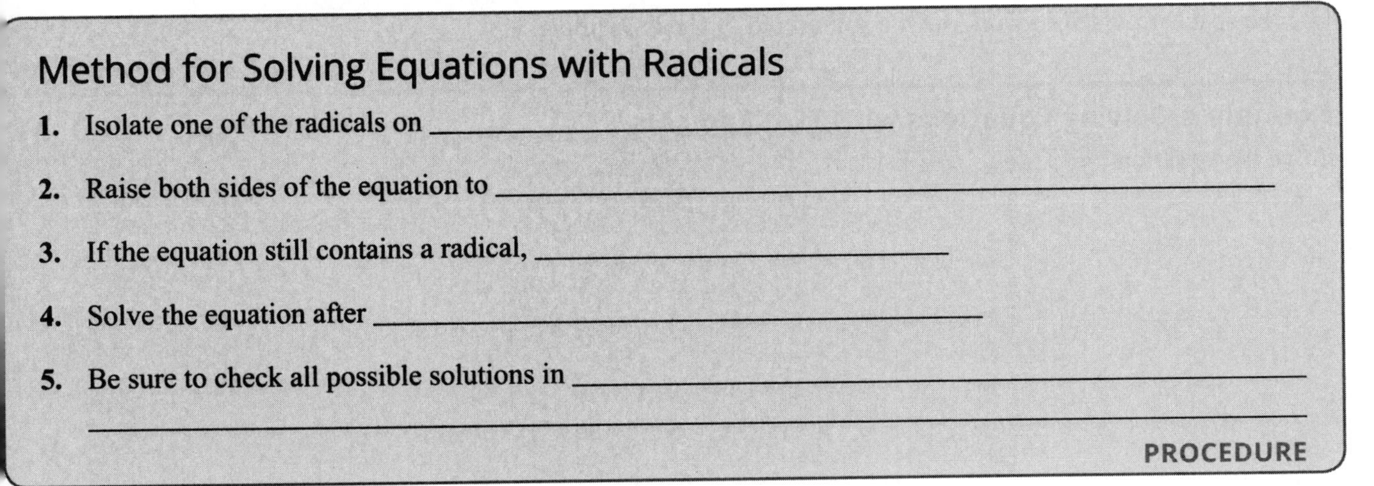

Method for Solving Equations with Radicals

1. Isolate one of the radicals on ______
2. Raise both sides of the equation to ______
3. If the equation still contains a radical, ______
4. Solve the equation after ______
5. Be sure to check all possible solutions in ______

______

PROCEDURE

Watch and Work

Watch the video for Example 6 in the software and follow along in the space provided.

Example 6 Solving Equations with Two Radicals

Solve the equation: $\sqrt{x+4}=\sqrt{3x-2}$

Solution

Now You Try It!

Use the space provided to work out the solution to the next example.

Example A Solving Equations with Two Radicals

Solve the equation: $\sqrt{2x+5}=\sqrt{x+7}$

13.6 Exercises

Concept Check

True/False. Determine whether each statement is true or false. If a statement is false, explain how it can be changed so the statement will be true. (**Note:** There may be more than one acceptable change.)

1. All radical equations will have two solutions.

2. If no true statements result when all possible solutions are checked, then there is no solution.

3. When solving equations with radicals, you should only have to raise both sides of the equation to a power one time.

4. A radical expression set equal to a negative value, such as $\sqrt{x+2} = -4$, has no real solution.

Practice

Solve the following equations. Be sure to check your answers in the original equation.

5. $\sqrt{8x+1} = 5$

6. $5 + \sqrt{x+5} - 2x = 0$

7. $\sqrt{2x-5}=\sqrt{3x-9}$

8. $\sqrt{x}+\sqrt{x-3}=3$

Applications

Solve.

9. ***Landscaping:*** A landscaper is designing a pond in the shape of a right triangle that has a square flower patch along each edge. She knows two of the flower patches will have side lengths of 5 ft and 13 ft and that the remaining flower patch must have a side length a which satisfies the equation $13=\sqrt{a^2+5^2}$. What is the side length of the remaining flower patch?

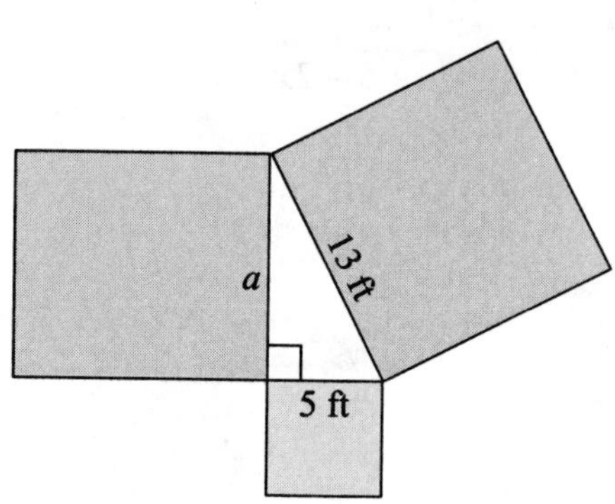

13.7 Functions with Radicals

Relation, Domain, and Range

A **relation** ____________________

The **domain** D of a relation is ____________________

The **range** R of a relation is ____________________

DEFINITION

Function

A **function** is a relation in which ____________________

DEFINITION

The definition of a function can also be stated in the following ways.

1. A function is a relation in which ____________________
2. A function is a relation in which ____________________

Vertical Line Test

If **any** vertical line intersects the graph of a relation at ____________________

____

DEFINITION

Radical Function

A **radical function** is a function of the form ____________________

____

The **domain** of such a function depends on the index n.

1. If n is an even number, the domain is ____________________
2. If n is an odd number, the domain is ____________________

The **range** of the function also depends on the index n.

3. If n is an even number, the range is ____________________
4. If n is an odd number, the range is ____________________

DEFINITION

Watch and Work

Watch the video for Example 3 in the software and follow along in the space provided.

Example 3 Graphing a Radical Function

Graph the function $y = \sqrt{x+5}$.

Solution

Now You Try It!

Jse the space provided to work out the solution to the next example.

Example A Graphing a Radical Function

Graph the function $y = -\sqrt{2x+1}$.

13.7 Exercises

Concept Check

True/False. Determine whether each statement is true or false. If a statement is false, explain how it can be changed so the statement will be true. (**Note:** There may be more than one acceptable change.)

1. If a radical function has an index that is an even number, then the domain is the set of all x such that $g(x) \geq 0$.

2. If a radical function has an odd numbered index, the domain is the set of all positive numbers.

3. Both the domain and the range of a radical function depend on the index.

4. To graph a radical function, you must be aware of its domain and you should plot at least a few points to see the nature of the resulting curve.

Practice

Find each function value as indicated and round decimal values to the nearest ten-thousandth, if necessary.

5. Given $f(x)=\sqrt{2x+1}$, find
   a. $f(2)$
   b. $f(4)$
   c. $f(24.5)$
   d. $f(1.5)$

Use interval notation to indicate the domain of each radical function.

6. $y=\sqrt{x+8}$

7. $f(x)=\sqrt[3]{x+4}$

Determine at least 5 points for the given function and then sketch the graph of the function.

8. $y = \sqrt{2x+6}$

9. $g(x) = \sqrt[3]{x-6}$

Applications

Solve.

10. ***Sports:*** The hang time of an athlete can be represented by

$$t = 2\sqrt{\frac{2h}{g}},$$

where t is the hang time of the athlete in seconds, h is the height of the jump in feet, and g is the acceleration due to gravity. (The gravity constant g can be estimated by using 32 ft/sec^2.)

a. Using a graphing calculator, graph this equation.

b. Identify the domain and range of this function. Does this make sense in the context of the function?

c. Find the hang time of an athlete with a 24-inch vertical leap. Round your answer to the nearest hundredth.

Writing & Thinking

11. Use your graphing calculator to graph the function $g(x)=\sqrt{x}\cdot\sqrt{x}$. Explain why the graph of this function differs from the graph of $f(x)=x$.

13.8 Introduction to Complex Numbers

i and i^2

____

DEFINITION

Using the definition of $\sqrt{-1}$, the following definition for the square root of a negative number can be made.

$\sqrt{-a} = \sqrt{a}\, i$

If a is a positive real number, then

____

Note: The number i is not ______________________________

To avoid confusion, we ______________________________

DEFINITION

Complex Numbers

The **standard form** of a **complex number** is ______________________________

a is called ____________________ and b is called ____________________

If $b = 0$, then $a + bi =$ ______________________________

If $a = 0$, then $a + bi =$ ______________________________

Complex Number: ______________

____

DEFINITION

Equality of Complex Numbers

For complex numbers $a + bi$ and $c + di$,

____

DEFINITION

Watch and Work

Watch the video for Example 3 in the software and follow along in the space provided.

Example 3 Solving Equations

Solve each equation for x and y.

a. $(x+3)+2yi=7-6i$

b. $2y+3-8i=9+4xi$

Solution

Now You Try It!

Use the space provided to work out the solution to the next example.

Example A Solving Equations

Solve each equation for x and y.

a. $(x-7)+4yi=3-8i$

b. $4y+7-5i=11+xi$

Addition and Subtraction with Complex Numbers

For complex numbers $a + bi$ and $c + di$,

____

and

____

DEFINITION

13.8 Exercises

Concept Check

True/False. Determine whether each statement is true or false. If a statement is false, explain how it can be changed so the statement will be true. (**Note:** There may be more than one acceptable change.)

1. An irrational number is a real number.

2. Every real number is a complex number.

3. If $a + bi = c + di$, then $a = d$ and $b = c$.

4. The square root of negative one is one.

Practice

5. Find the real part and imaginary part of $-11+\sqrt{2}\,i$

6. Simplify $\sqrt{-121}$

Solve the equations for *x* and *y*.

7. $x + 3i = 6 - yi$

8. $x + yi + 8 = 2i + 4 - 3yi$

Find each sum or difference as indicated.

9. $(2+3i)+(4-i)$

10. $(4+3i)-(\sqrt{2}+3i)$

Writing & Thinking

11. Answer the following questions and give a brief explanation of your answer.

a. Is every real number a complex number?

b. Is every complex number a real number?

12. List 5 numbers that do and 5 numbers that do not fit each of the following categories (if possible).

a. rational number

b. integer

c. real number

d. pure imaginary number

e. complex number

f. irrational number

13.9 Multiplication and Division with Complex Numbers

Watch and Work

Watch the video for Example 1 in the software and follow along in the space provided.

Example 1 Multiplying with Complex Numbers

Find the following products.

a. $(3i)(2-7i)$

b. $(5+i)(2+6i)$

c. $(\sqrt{2}-i)(\sqrt{2}-i)$

d. $(-1+i)(2-i)$

Solution

Now You Try It!

Use the space provided to work out the solution to the next example.

Example A Multiplying with Complex Numbers

Find the following products.

a. $(4i)(1-7i)$

b. $(4+2i)(3+5i)$

c. $(\sqrt{2}-i)(\sqrt{2}+i)$

1. The two complex numbers $a + bi$ and $a - bi$ are called ____________________ or simply **conjugates** of each other. **the product of two complex conjugates will always be a** ____________________

____

Writing Fractions with Complex Numbers in Standard Form

1. Multiply both the numerator and denominator ____________________

2. Simplify the resulting ____________________

3. Write the ____________________

PROCEDURE

13.9 Exercises

Concept Check

True/False. Determine whether each statement is true or false. If a statement is false, explain how it can be changed so the statement will be true. (**Note:** There may be more than one acceptable change.)

1. Regardless of the value of the exponent, the only possible values for any power of i are i and $-i$.

2. The product $\sqrt{a} \cdot \sqrt{b}$ can be rewritten as $\sqrt{ab}$ as long as a and b are real numbers.

3. When i is squared, the product is 1.

4. The conjugate of $4 - 5i$ is $4 + 5i$.

Practice

Perform the indicated operations and write each result in standard form.

5. $-4i(6-7i)$

6. $(2+7i)(6+i)$

7. $\frac{5}{4i}$

8. $\frac{6+i}{3-4i}$

Simplify the following powers of *i* and write each result in standard form. Assume *k* is a positive integer.

9. i^{13}

10. i^{-3}

Find the indicated products and simplify.

Writing & Thinking

11. Explain why the product of every complex number and its conjugate is a nonnegative real number.

12. What condition is necessary for the conjugate of a complex number, $a + bi$, to be equal to the reciprocal of this number?

Chapter 13 Project

Let's Get Radical!

An activity to demonstrate the use of radical expressions in real life.

There are many different situations in real life that require working with radicals, such as solving right triangle problems, working with the laws of physics, calculating volumes, and even solving investment problems. Let's take a look at a simple investment problem to see how radicals are involved.

The formula for computing compound interest for a principal P that is invested at an annual rate r and compounded annually is given by $A = P(1+r)^n$, *where A is the accumulated amount in the account after n years.*

1. Let's suppose that you have $5000 to invest for a term of 2 years. If you want to be sure and make at least $600 in interest, then at what interest rate should you invest the money?

a. One way to approach this problem would be through trial and error, substituting various rates for r in the formula. This approach might take a while. Using the table below to organize your work, try substituting 3 values for r. Remember that rates are percentages and need to be converted to decimals before using in the formula. Did you get close to $5600 for the accumulated amount in the account after 2 years?

Annual Rate (r)	Principal (P)	Number of Years (n)	Amount, $A = P(1+r)^n$
	$5000	2	
	$5000	2	
	$5000	2	

b. Let's try a different approach. Substitute the value of 2 for n and solve this formula for r. Verify that you get the following result: $r = \sqrt{\frac{A}{P}} - 1$ (**Hint:** First solve for $(1+r)^2$ and then take the square root of both sides of the equation.) Notice that you now have a radical expression to work with. Substitute $5000 for P and $5600 for A (which is the principal plus $600 in interest) to see what your rate must be. Round your answer to the nearest percent.

2. Now, let's suppose that you won't need the money for 3 years.

a. Use $n = 3$ years and solve the compound interest formula for r.

b. What interest rate will you need to invest the principal of $5000 at in order to have at least $5600 at the end of 3 years? (To evaluate a cube root you may have to use the rational exponent of $\frac{1}{3}$ on your calculator.) Round to the nearest percent.

c. Compare the rates needed to earn at least $600 when $n = 2$ years and $n = 3$ years. What did you learn from this comparison? Write a complete sentence.

3. Using the above formulas for compound interest when $n = 2$ years and $n = 3$ years, write the general formula for r for any value of n.

4. Using the formula from Problem 3, compute the interest rate needed to earn at least $3000 in interest on a $5000 investment in 7 years. Round to the nearest percent.

5. Do an internet search on a local bank or financial institution to determine if the interest rate from Problem 4 is reasonable in the current economy. Using three to five sentences, briefly explain why or why not.

CHAPTER 14

Quadratic Equations

Math @ Work

Quadratic functions can be used to describe a variety of situations, from motion to electrical resistors. Quadratic functions are also used to calculate and track company revenue. For any business, generating revenue is a top priority. Without revenue, company owners have no way to pay their workers or produce and improve their product lines.

A key element in generating revenue requires creating a product that appeals to the consumer population. However, the product must also be priced such that people feel that the benefit they receive from the product is worth the expense. The formula for calculating revenue is $R = px$, where p is the price and x is the number of units being sold. The price to charge to receive maximum revenue can be found by calculating the maximum value of the quadratic function derived using the revenue formula.

Suppose you manage the pro shop at a golf resort and you estimate that by charging x dollars to rent a set of golf clubs for the day, golfers will rent $300 - x$ sets of clubs over the course of a week. How would you determine what price will yield maximum revenue?

14.1 Quadratic Equations: The Square Root Method

Solving Quadratic Equations by Factoring

1. Add or subtract terms as necessary so that 0 is on one side of the equation and the ____________________

____

2. Factor completely. (If there are any fractional coefficients, ______________________________________

____.)

3. Set each nonconstant factor equal to ____

4. Check each solution, one ____

PROCEDURE

Watch and Work

Watch the video for Example 1 in the software and follow along in the space provided.

Example 1 Solving Quadratic Equations by Factoring

Solve the quadratic equation by factoring.

$x^2 - 15x = -50$

Solution

Now You Try It!

Use the space provided to work out the solution to the next example.

Example A Solving Quadratic Equations by Factoring

Solve the quadratic equation by factoring.

$x^2 - 10x = -16$

Square Root Property

If $x^2 = c$, then ____________________

If $(x-a)^2 = c$, then ____________________________________

Note: If c is negative $(c < 0)$, then ___________________________________

PROCEDURE

14.1 Exercises

Concept Check

True/False. Determine whether each statement is true or false. If a statement is false, explain how it can be changed so the statement will be true. (**Note:** There may be more than one acceptable change.)

1. Using the square root method on an equation of the form $x^2 = c$, where c is a nonnegative number, will always result in two distinct solutions.

2. Quadratic equations that are not easily solved using factoring might be solved by the square root method.

3. The square of a real number can be negative.

Practice

Solve the following quadratic equation by factoring.

4. $x^2 = -15x - 36$

Solve the following quadratic equations by using the square root method. Write each radical in simplest form.

5. $5x^2 = 245$

6. $x^2 - 75 = 0$

7. $(x-2)^2 = \frac{1}{16}$

Determine the length of side c.

8. $a = 9, b = 12, c = ?$

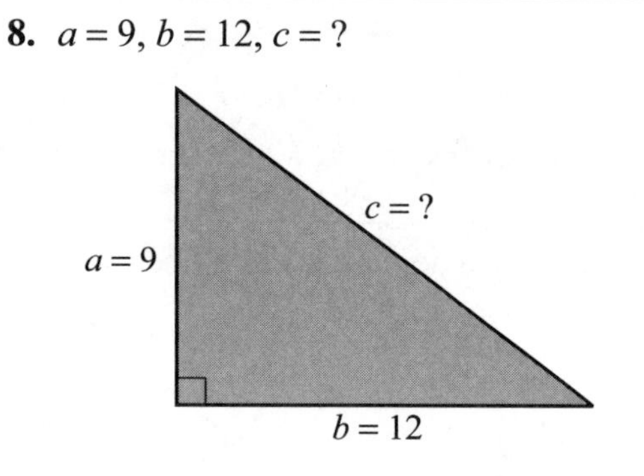

Applications

Solve.

9. ***Right Triangles:*** One leg of a right triangle is three times the length of the other. The length of the hypotenuse is 20 cm. Find the lengths of the legs.

10. ***Falling Objects:*** A ball is dropped from the top of a building that is known to be 144 feet high. The formula for finding the height of the ball at any time is $h = 144 - 16t^2$ where t is measured in seconds. How many seconds will it take for the ball to hit the ground?

14.2 Quadratic Equations: Completing the Square

Solve a Quadratic Equation by Completing the Square

1. If necessary, divide or multiply on both sides of the equation so that the ____________________

   ____

2. If necessary, isolate the ____________________

3. Find the constant that completes ____________________

   Remember that the constant term is the ____________________

   Rewrite the polynomial as ____________________

4. Use the square root property to ____________________

PROCEDURE

Watch and Work

Watch the video for Example 8 in the software and follow along in the space provided.

Example 8 Quadratic Equations with Known Roots

Find a quadratic equation with the given roots.

$x = 5 - \sqrt{2}$ and $x = 5 + \sqrt{2}$

Solution

Now You Try It!

Use the space provided to work out the solution to the next example.

Example A Quadratic Equations with Known Roots

Find a quadratic equation with the given roots.

$x = 2 - 3\sqrt{2}$ and $x = 2+3\sqrt{2}$

14.2 Exercises

Concept Check

True/False. Determine whether each statement is true or false. If a statement is false, explain how it can be changed so the statement will be true. (**Note:** There may be more than one acceptable change.)

1. To get a leading coefficient of 1, multiply both sides of the equation by the reciprocal of the leading coefficient.

2. When solving a quadratic equation, there is either no solution or two solutions.

3. It's possible for the roots of a quadratic equation to be nonreal numbers.

4. The last step of solving a quadratic equation by completing the square is to use the square root property.

Practice

Solve the quadratic equations by completing the square.

5. $x^2 + 4x - 5 = 0$

6. $z^2 + 3z - 5 = 0$

7. $x^2 + x + 2 = 0$

8. $3x^2 + 6x + 18 = 0$

Write a quadratic equation with integer coefficients that has the given roots.

9. $x = \sqrt{7}, x = -\sqrt{7}$

10. $x = 7i, x = -7i$

Applications

Solve.

11. ***Framing:*** A local frame shop determines that the revenue function for their custom framing service is $R(p) = 360p - 4p^2$, where p is the base price in dollars for each custom framing job.

a. Set the function equal to 0 and solve for p using the method of completing the square.

b. What do the solutions from Part **a.** mean?

Writing & Thinking

12. Explain, in your own words, the steps involved in the process of solving a quadratic equation by completing the square.

Name: Date:

14.3 Quadratic Equations: The Quadratic Formula

The Quadratic Formula

For the general quadratic equation ______________________________

____

FORMULA

Watch and Work

Watch the video for Example 2 in the software and follow along in the space provided.

Example 2 The Quadratic Formula

Solve by using the quadratic formula.

$7x^2 - 2x + 1 = 0$

Solution

Now You Try It!

Use the space provided to work out the solution to the next example.

Example A The Quadratic Formula

Solve by using the quadratic formula.

$5x^2 - 3x + 2 = 0$

---

1. The expression $b^2 - 4ac$, the part of the quadratic formula that lies under the radical sign, is called the ___________________.

Discriminant	Nature of Solutions
$b^2 - 4ac > 0$	__________________________
$b^2 - 4ac = 0$	__________________________
$b^2 - 4ac < 0$	__________________________

Table 1

14.3 Exercises

Concept Check

True/False. Determine whether each statement is true or false. If a statement is false, explain how it can be changed so the statement will be true. (**Note:** There may be more than one acceptable change.)

1. The quadratic formula will always work when solving quadratic equations.

2. If the discriminant is a perfect square, the quadratic equation is factorable.

3. When using the quadratic formula, if the discriminant is greater than zero, there are infinite solutions.

4. If the discriminant is less than zero, there is no real solution.

Practice

Find the discriminant and determine the nature of the solutions of each quadratic equation.

5. $x^2 + 6x - 8 = 0$

6. $x^2 - 8x + 16 = 0$

Solve each of the quadratic equations using the quadratic formula.

7. $x^2 + 4x - 4 = 0$

8. $x^2 - 2x + 7 = 0$

9. $3x^2 - 7x + 4 = 0$

Applications

Solve.

10. ***Throwing Objects:*** An orange is thrown down from the top of a building that is 300 feet tall with an initial velocity of 6 feet per second. The distance of the object from the ground can be calculated using the equation $d = 300 - 6t - 16t^2$, where t is the time in seconds after the orange is thrown.

a. On a balcony, a cup is sitting on a table located 100 feet from the ground. If the orange is thrown with the right aim to fall into the cup, how long will the orange fall? Round to the nearest hundredth. (**Hint:** The distance is 100 feet.)

b. If the orange misses the cup and falls to the ground, how long will it take for the orange to splatter on the sidewalk? (**Hint:** What is the height of the orange when it hits the ground?)

c. Approximately how much longer would it take for the orange to fall to the sidewalk than it would for the orange to fall into the cup?

Writing & Thinking

11. Find an equation of the form $Ax^4 + Bx^2 + C = 0$ that has the four roots ± 2 and ± 3. Explain how you arrived at this equation.

12. The surface area of a right circular cylinder can be found using the following formula: $S = 2\pi r^2 + 2\pi rh$, where r is the radius of the cylinder and h is the height. Estimate the radius of a circular cylinder of height 30 cm and surface area 300 cm^2. Explain how you used your knowledge of quadratic equations.

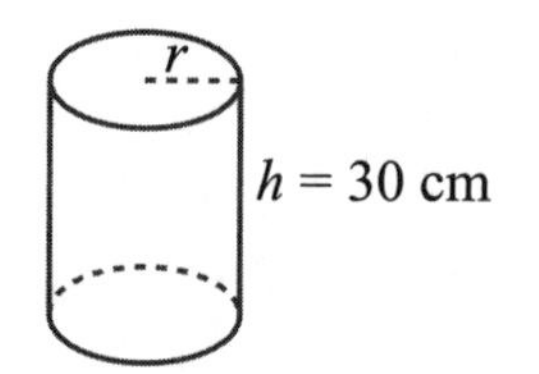

14.4 More Applications of Quadratic Equations

Watch and Work

Vatch the video for Example 5 in the software and follow along in the space provided.

xample 5 Finding the Dimensions of a Box

square piece of cardboard has a small square, 2 in. by 2 in., cut from each corner. The edges are then folded up form a box with a volume of 512 in.^{3} What are the dimensions of the box? (**Hint:** The volume is the product of e length, width, and height: $V = lwh$.)

olution

Now You Try It!

Use the space provided to work out the solution to the next example.

Example A Finding the Dimensions of a Box

A square has sides of length 24 ft. Find the length of the diagonal of the square to the nearest tenth of a foot. (The diagonal of a square is a line connecting opposite corners of the square.)

14.4 Exercises

Concept Check

True/False. Determine whether each statement is true or false. If a statement is false, explain how it can be changed so the statement will be true. (**Note:** There may be more than one acceptable change.)

1. Application problems will always directly tell you which operations to perform.

2. The basic formula for distance-rate-time problems is $d = rt$.

Applications

Solve.

3. ***Swimming Pools:*** The Wilsons have a rectangular swimming pool that is 10 ft longer than it is wide. The pool is completely surrounded by a concrete deck that is 6 ft wide. The total area of the pool and the deck is 1344 ft². Find the dimensions of the pool.

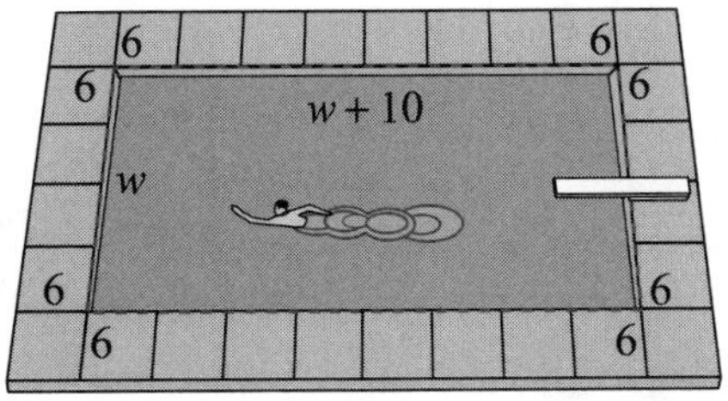

4. ***Group Travel:*** The Ski Club is planning to charter a bus to a ski resort. The cost will be \$900 and each member will share the cost equally. If the club had 15 more members, the cost per person would be \$10 less. How many are in the club now? (**Hint:** If x = number in club now, $\frac{900}{x}$ = cost per person.)

5. ***Boating:*** A small motorboat travels 12 mph in still water. It takes 2 hours longer to travel 45 miles going upstream than it does going downstream. Find the rate of the current. (**Hint:** $12 + c$ = rate going downstream and $12 - c$ = rate going upstream.)

6. ***Product Assembly:*** Two employees together can prepare a large order in 2 hrs. Working alone, one employee takes three hours longer than the other. How long does it take each person working alone?

7. ***Shooting a Projectile:*** The height of a projectile fired upward from the ground with a velocity of 128 ft/sec is given by the formula $h = -16t^2 + 128t$.

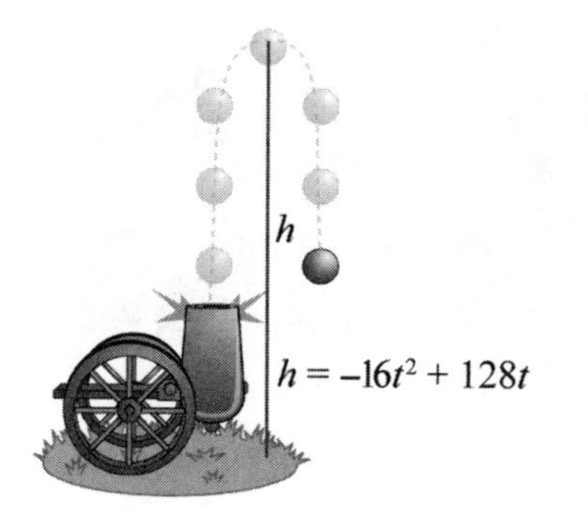

a. When will the projectile be 256 feet above the ground?

b. Will the projectile ever be 300 feet above the ground? Explain.

c. When will the projectile be 240 feet above the ground?

d. In how many seconds will the projectile hit the ground?

8. ***Ladder Height:*** A ladder is 30 ft long and you want to place the base of the ladder 10 ft from the base of a building. About how far up the building (to the nearest tenth of a foot) will the ladder reach?

9. Suppose that you are to solve an applied problem and the solution leads to a quadratic equation. You decide to use the quadratic formula to solve the equation. Explain what restrictions you must be aware of when you use the formula.

14.5 Equations in Quadratic Form

Solving Equations in Quadratic Form by Substitution

1. Look at the ____________________
2. Substitute a ____________________
3. Substitute the ____________________
4. Solve the ____________________
5. Substitute the ____________________

____

PROCEDURE

Watch and Work

Watch the video for Example 6 in the software and follow along in the space provided.

Example 6 Solving Higher-Degree Equations

Solve the equation: $x^5 - 16x = 0$

Solution

Now You Try It!

Use the space provided to work out the solution to the next example.

Example A Solving Higher-Degree Equations

Solve the equation:

$x^5 - 25x = 0$

14.5 Exercises

Concept Check

True/False. Determine whether each statement is true or false. If a statement is false, explain how it can be changed so the statement will be true. (**Note:** There may be more than one acceptable change.)

1. Equations in quadratic form can be solved using the quadratic equation or by factoring.

2. When solving higher-degree equations, you shouldn't factor out any common monomials.

3. The LCM of the denominators is used to clear rational expressions of any fractions.

4. The degree of the first term of an equation in quadratic form must be 2.

Practice

Solve the equations.

5. $x^4 - 13x^2 + 36 = 0$

6. $2x - 9x^{\frac{1}{2}} + 10 = 0$

7. $(2x+3)^2+7(2x+3)+12=0$

8. $\frac{2}{4x-1}+\frac{1}{x+1}=\frac{-x}{x+1}$

Writing & Thinking

9. Consider the following equation: $x-x^{\frac{1}{2}}-6=0$
In your own words, explain why, even though it is in quadratic form, this equation has only one solution.

14.6 Graphing Quadratic Functions

Quadratic Functions

Any function that can be written in the form

____

____

DEFINITION

1. The complete graph of $y = x^2$ a ______________.

2. The point $(0, 0)$ is the "turning point" of the parabola and is called the ____________.

3. The line $x = 0$ (the y-axis) is the ____________________ or ____________________.

4. The graph of a function of the form $y = ax^2 + k$ is a ________________ (or ____________________) of the graph of $y = ax^2$.

5. The shift is __________ k units if $k > 0$ or ____________ $|k|$ units if $k < 0$.

6. The **vertex** is at the point ___.

7. The graph of a function of the form $y = a(x - h)^2$ is a ____________________ (or ____________ ____________) of the graph of $y = ax^2$.

8. The shift is to the ____________ if $h > 0$ and to the __________ if $h < 0$.

Watch and Work

Watch the video for Example 4 in the software and follow along in the space provided.

Example 4 Graphing Quadratic Functions

Find the line of symmetry and vertex for the quadratic function $y = -(x+2)^2 + 1$. Then graph the function, setting up a table of values for x and y as an aid. Choose values of x on each side of the line of symmetry.

Solution

Now You Try It!

Use the space provided to work out the solution to the next example.

Example A Graphing Quadratic Functions

Find the line of symmetry and vertex for the quadratic function $y = 2(x-1)^2 - 3$. Then graph the function.

14.6 Exercises

Concept Check

True/False. Determine whether each statement is true or false. If a statement is false, explain how it can be changed so the statement will be true. (**Note:** There may be more than one acceptable change.)

1. The graph of a quadratic function is a mirror of itself across the line, or axis, of symmetry.

2. The graph of $y = a(x - h)^2$ is a vertical shift (or vertical translation) of the graph of $y = ax^2$.

3. For a quadratic function of the form $y = ax^2$, the bigger $|a|$ is, the wider the opening of the parabola is.

Practice

Solve.

4. Graph the function $y = 2x^2$. Then, without additional computation, graph the following translations.

a. $y = 2x^2 - 3$

b. $y = 2(x - 4)^2$

c. $y = -2(x + 1)^2$

d. $y = -2(x + 2)^2 - 4$

For each of the quadratic functions, determine the line of symmetry and the vertex Then graph the function.

5. $y = 3x^2 - 4$

6. $y = -5(x+2)^2$

7. $y = 4(x-5)^2 + 1$

14.7 More on Graphing Quadratic Functions and Applications

Watch and Work

Watch the video for Example 3 in the software and follow along in the space provided.

Example 3 Graphing Quadratic Functions of the Form $y = ax^2 + bx + c$

For $y = -x^2 - 4x + 2$, find the line of symmetry, vertex, x-intercepts, and y-intercept.

Solution

Now You Try It!

Use the space provided to work out the solution to the next example.

Example A Graphing Quadratic Functions of the Form $y = ax^2 + bx + c$

For $y = -x^2 - 6x + 3$, find the line of symmetry, vertex, x-intercepts, and y-intercept.

Minimum and Maximum Values

For $y = a(x - h)^2 + k$,

1. If $a > 0$, then the parabola opens ______________________________

____

2. If $a < 0$, then the parabola opens ______________________________

____

DEFINITION

14.7 Exercises

Concept Check

True/False. Determine whether each statement is true or false. If a statement is false, explain how it can be changed so the statement will be true. (**Note:** There may be more than one acceptable change.)

1. Quadratic functions of the form $y = a(x - h)^2 + k$ have a line of symmetry at $x = \frac{b}{2a}$.

2. The vertex of a vertical parabola is the lowest point on the parabola.

3. The maximum or minimum value of a quadratic function written in general form can be found by letting $x = -\dfrac{b}{2a}$ and solving for y.

4. When the solutions to a quadratic function are nonreal, the entire graph lies either completely above or below the x-axis.

Practice

Rewrite each quadratic function in the form $y = a(x-h)^2 + k$. Find the line of symmetry, the vertex, the x-intercepts, and the y-intercept. Graph the function.

5. $y = 2x^2 - 4x + 2$

6. $y = 2x^2 - 12x + 16$

For each quadratic function use the formula $x = -\dfrac{b}{2a}$ to find the line of symmetry and the vertex. Then find the x-intercepts and the y-intercept. Graph the function.

7. $y = -3x^2 + 6x - 3$

8. $y = 5x^2 + 10x + 7$

Applications

Use the function $h = -16t^2 + v_0 t + h_0$, where h is the height of the object after time t, v_0 is the initial velocity, and h_0 is the initial height.

9. ***Throwing a Ball:*** A ball is thrown vertically upward from the ground with an initial velocity of 112 ft/s.

   a. When will the ball reach its maximum height?

   b. What will be the maximum height?

Solve.

10. ***Selling Lamps:*** A store owner estimates that by charging x dollars each for a certain lamp, he can sell $40 - x$ lamps each week. What price will give him maximum sales revenue?

Writing & Thinking

11. Discuss the following features of the general quadratic function $y = ax^2 + bx + c$.

   a. What type of curve is its graph?

   b. What is the value of x at its vertex?

   c. What is the equation of the line of symmetry?

   d. Does the graph always cross the x-axis? Explain.

12. Discuss the discriminant of the general quadratic equation $ax^2 + bx + c = 0$ and how the value of the discriminant is related to the graph of the corresponding quadratic function $y = ax^2 + bx + c$.

14.8 Solving Polynomial and Rational Inequalities

To Solve a Polynomial Inequality Algebraically

1. Arrange the terms so that ______________________
2. Factor the polynomial expression, if possible, and ______________________
   (Use the ______________________.)
3. Mark each of these ______________________
   These are the ______________________
4. Test one point from each interval to ______________________
   ____
5. The solution consists of ______________________
6. Mark a bracket for an endpoint that is ______________________
   ____

PROCEDURE

Watch and Work

Watch the video for Example 4 in the software and follow along in the space provided.

Example 4 Solving Polynomial Inequalities Using the Quadratic Formula

Solve the inequality by using the quadratic formula and a number line. Then graph the solution set on a number line.

$$x^2 - 2x - 1 > 0$$

Solution

Now You Try It!

Use the space provided to work out the solution to the next example.

Example A Solving Polynomial Inequalities Using the Quadratic Formula

Solve the inequality.

$$x^2 - x - 5 < 0$$

To Solve a Rational Inequality

1. Simplify the inequality so that one side is ______________________________

   ____

2. Find the points that cause the factors in the ______________________________
3. Mark each of these ______________________________

   These are the ______________________________

4. Test one point from each interval to ______________________________

   ____

5. The solution consists of ______________________________
6. Mark a bracket for an endpoint that is ______________________________

   ______________ Remember that ______________________________

PROCEDURE

14.8 Exercises

Concept Check

True/False. Determine whether each statement is true or false. If a statement is false, explain how it can be changed so the statement will be true. (**Note:** There may be more than one acceptable change.)

1. When solving a polynomial inequality algebraically, the goal is to get the constants on one side of the inequality and to factor the polynomial on the other side.

2. Test points are used to determine which intervals on the number line satisfy the original inequality.

3. The solution of a polynomial inequality is a single interval.

4. If an endpoint causes the denominator of a rational inequality to be 0, it should be marked with a parenthesis.

Practice

Solve the quadratic (and higher degree) inequalities algebraically. Write the answers in interval notation, and then graph each solution set on a number line. (**Note:** You may need to use the quadratic formula to find endpoints of intervals.)

5. $(x-6)(x+2)<0$

6. $(x+7)(2x-5)\geq 0$

Solve the rational inequalities algebraically. Write the answers in interval notation, and then graph each solution set on a number line.

7. $\dfrac{x+4}{2x}\geq 0$

8. $\dfrac{x+6}{x^2}<0$

Applications

Solve.

9. ***Sales:*** A high school student is selling T-shirts to raise money for the band. She realizes that the number of shirts she sells each week can be modeled by $f(x)=-x^2+12x-17$, where x is the amount she charges per shirt. Solve the inequality $-x^2+12x-17\geq 10$ to find the range she can charge per shirt and sell at least 10 shirts in a week.

10. ***Weather:*** Maria tracked the nighttime temperatures for a week and noticed that the temperature, in Celsius, could be modeled by $f(x)=\frac{1}{2}x^2-4x+6$, where x is the number of hours after midnight. Maria has plants that might die if they are left out when the temperature drops below freezing (0 degrees Celsius). Solve the inequality $\frac{1}{2}x^2-4x+6<0$ to find timeframe in which her plants will be in danger.

Writing & Thinking

11. Use a graphing calculator to graph the rational function $y=\dfrac{x^2+3x-4}{x}$.

a. Use the graph to find the solution set for $y>0$.

b. Use the graph to find the solution set for $y<0$.

c. Explain the effect of $x=0$ on the graph and why $x=0$ is not included in either Parts **a.** or **b.**

12. In your own words, explain why (as in Example 5), when the quadratic formula gives nonreal values, the quadratic polynomial is either always positive or always negative.

Chapter 14 Project

Gateway to the West

An activity to demonstrate the use of quadratic equations in real life.

The Gateway Arch on the St. Louis riverfront in Missouri serves as an iconic monument symbolizing the westward expansion of American pioneers, such as Lewis and Clark. A nationwide competition was held to choose an architect to design the monument and the winner was Eero Saarinen, a Finnish American who immigrated to the United States with his parents when he was 13 years old. Construction began in 1962 and the monument was completed in 1965. The Gateway Arch is the tallest monument in the United States. It is constructed of stainless steel and weighs more than 43,000 tons. Although the arch is heavy, it was built to sway with the wind to prevent it from being damaged. In a 20 mph wind, the arch can move up to 1 inch. In a 150 mph wind, the arch can move up to 18 inches.

1. If you were to place the Gateway Arch on a coordinate plane centered around the y-axis, then the equation $y = -0.00635x^2 + 630$ could be used to model the height of the arch in feet.

a. The general form for a quadratic function is $y = ax^2 + bx + c$. Identify the values for a, b, and c from the Gateway Arch equation.

b. Find the vertex of the Gateway Arch equation.

c. Does the vertex represent a maximum or a minimum? Explain your answer based on the coefficients of the Gateway Arch equation.

d. What is the height of the Gateway Arch at its peak?

e. Write the equation for the axis of symmetry of the Gateway Arch equation.

f. Find the x-intercepts of the Gateway Arch equation. Round to the nearest integer.

2. Using the coordinate plane below and the information from Problem 1, graph the Gateway Arch equation.

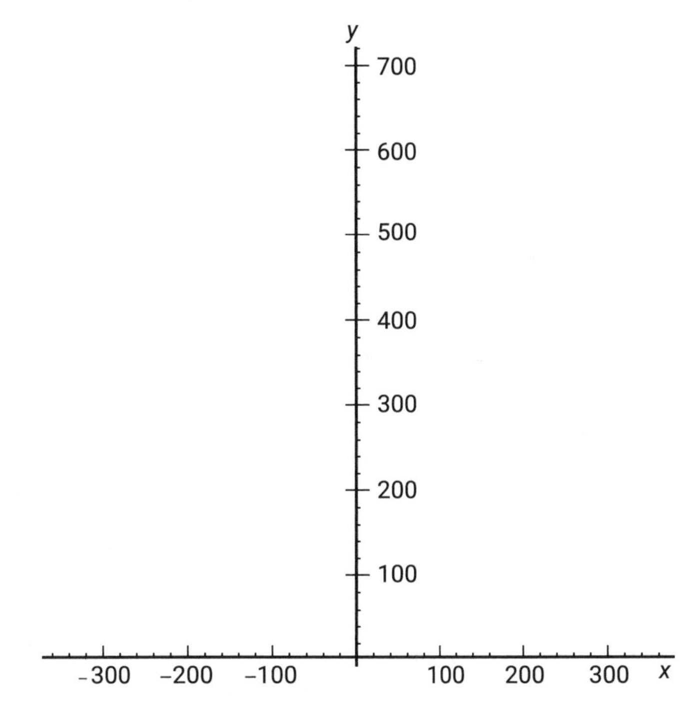

3. How far apart are the legs of the Gateway Arch at its base?

4. The Gateway Arch equation is a mathematical model. Look up the actual values for the height of the Gateway Arch and the distance between the legs of the arch at its base on the internet and describe how they compare to the values calculated using the equation.

CHAPTER 15

Exponential and Logarithmic Functions

Math @ Work

Exponential and logarithmic functions are used in a variety of situations, from describing the intensity of an earthquake to describing the relationship between notes in a piece of music. A common situation where you might find yourself working with exponential functions is calculating the amount of interest owed on your student loans. When making bigger purchases, such as a car or a house, you'll often have to choose between different financing options. Being comfortable with exponential functions can help you make the decision that is best for you and your circumstances.

Suppose you're buying a car that costs $15,000 and the dealership has two financing options available. If you put $5000 down, you can finance the rest at 2.5% interest for 6 years. Alternatively, you could put $0 down and finance the entire purchase at 5.5% for 4 years. What are some factors that might cause you to go with the first option? Under what circumstances might you go with the second option? Which option would result in you paying less overall for the car?

15.1 Algebra of Functions

Algebraic Operations with Functions

If $f(x)$ and $g(x)$ represent two functions and x is a value in the **domain of both functions**, then we define the following operations.

1. Sum of two functions: ______________________

2. Difference of two functions: ______________________

3. Product of two functions: ______________________

4. Quotient of two functions: ______________________

DEFINITION

Watch and Work

Watch the video for Example 3 in the software and follow along in the space provided.

Example 3 Algebraic Operations with Functions with Limited Domains

Let $f(x)=x+5$ and $g(x)=\sqrt{x-2}$. Find the following functions and state the domain of each function.

a. $(f+g)(x)$

b. $\left(\frac{f}{g}\right)(x)$

Solution

Now You Try It!

Use the space provided to work out the solution to the next example.

Example A Algebraic Operations with Functions with Limited Domains

Let $f(x)=x-1$ and $g(x)=\sqrt{x+3}$. Find the following functions.

a. $(f+g)(x)$

b. $\left(\frac{g}{f}\right)(x)$

Remember that when operating with functions, the operations are performed with the y-values for each value of x in the common domain.

15.1 Exercises

oncept Check

rue/False. Determine whether each statement is true or false. If a statement is false, explain how it can be changed so he statement will be true. (**Note:** There may be more than one acceptable change.)

1. One way to find the sum of two functions is to find the algebraic sum of the two expressions.

2. The function $(f+g)(x)$ means the same as $f(x)+g(x)$.

3. If functions do not have the same domain, any algebraic sums, differences, products, and quotients are restricted to portions of the range that are in common.

4. If two functions have graphs that consists of line segments, the sum of the two functions will produce a graph that is a continuous line.

ractice

or the following pairs of functions find, **a.** $(f+g)(x)$, **b.** $(f-g)(x)$, **c.** $(f \cdot g)(x)$, and **d.** $\left(\frac{f}{g}\right)(x)$.

5. $f(x)=x+2,\ g(x)=x-5$

6. $f(x)=x^3+6x,\ g(x)=x^2+6$

Find the indicated functions and state their domains in interval notation.

7. If $f(x) = \sqrt{2x - 6}$ and $g(x) = x + 4$, find $(f + g)(x)$.

8. Find $f(x) \cdot g(x)$ given that $f(x) = 3x + 2$ and $g(x) = x - 7$.

Graph each pair of functions and the sum of these functions on the same set of axes.

9. $f(x) = x^2$ and $g(x) = -1$

9. $f(x) = x + 1$ and $g(x) = x^2 - 1$

Writing & Thinking

10. Explain why, in general, $(f - g)(x) \neq (g - f)(x)$ if $f(x) \neq g(x)$.

15.2 Composition of Functions and Inverse Functions

Composite Function

For two functions *f* and *g*, the composite function $f \circ g$ is defined as follows.

____

The domain of $f \circ g$ consists of those values of ____

____

DEFINITION

Watch and Work

Watch the video for Example 2 in the software and follow along in the space provided.

Example 2 Compositions

Form the compositions **a.** $(f \circ g)(x)$ and **b.** $(g \circ f)(x)$ if $f(x) = 5x + 2$ and $g(x) = 3x - 7$.

Solution

Now You Try It!

Use the space provided to work out the solution to the next example.

Example A Compositions

If $f(x) = x^2$ and $g(x) = x^2 + 2$ form the compositions $(f \circ g)(x)$ and $(g \circ f)(x)$.

One-to-One Functions

A function is a one-to-one function (or 1-1 function) if ____________________

____

DEFINITION

Horizontal Line Test

A function is one-to-one if ____________________

____

DEFINITION

Inverse Functions

If f is a one-to-one function with ordered pairs of the form (x, y), then ____________________

____

DEFINITION

To Determine whether Two Functions are Inverses

If f and g are one-to-one functions and

____

____

____

That is, ____________________

PROCEDURE

To Find the Inverse of a One-to-One Function

1. ____________________
2. ____________________
3. In the new equation, ____________________
4. ____________________

PROCEDURE

15.2 Exercises

Concept Check

True/False. Determine whether each statement is true or false. If a statement is false, explain how it can be changed so the statement will be true. (**Note:** There may be more than one acceptable change.)

1. The vertical line test is used to determine whether a graph represents a vertical line.

2. In a one-to-one function, each x-value corresponds to exactly one y-value.

3. The horizontal line test is used to determine whether a graph of a function is one-to-one.

4. The notation $f^{-1}(x)$ means $\frac{1}{f(x)}$.

Practice

Find the indicated function values for each function given.

5. $g(y) = 5y^2 + 4$

a. $g(x-2)$

b. $g(3n^2)$

Form the compositions $f(g(x))$ and $g(f(x))$ for pair of functions..

6. $f(x) = \frac{1}{\sqrt{x}}, \quad g(x) = x^2 - 4$

Show that the given one-to-one functions are inverses of each other. Then graph both functions on the same set of axes and show the line $y = x$ as a dotted line on each graph. (You may use a calculator as an aid in finding the graphs.)

7. $f(x) = 3x + 1$ and $g(x) = \dfrac{x-1}{3}$

Find the inverse of the given function. Then graph both functions on the same set of axes and show the line $y = x$ as a dotted line on the graph.

8. $f(x) = 2x - 5$

Using the horizontal line test, determine which of the graphs are graphs of one-to-one functions. If the graph represents a one-to-one function, graph its inverse by reflecting the graph of the function across the line $y = x$. (**Hint:** If a function is one-to-one, label a few points on the graph and use the fact that the x- and y-coordinates are interchanged on the graph of the inverse.)

9.

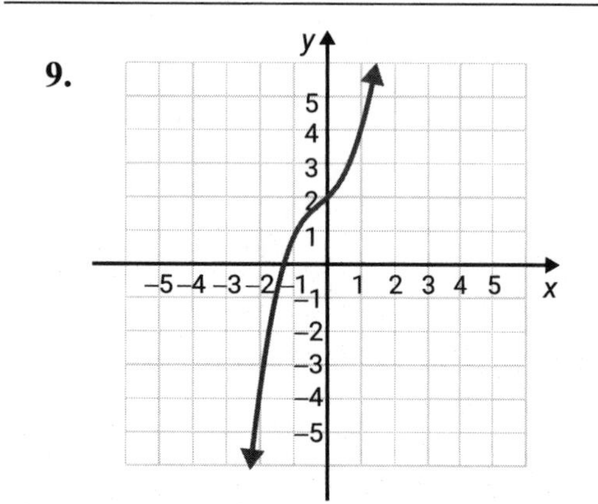

Writing & Thinking

10. Explain briefly why a function must be one-to-one to have an inverse.

15.3 Exponential Functions

Exponential Functions

An exponential function is a function of the form ______________

____

DEFINITION

General Concepts of Exponential Functions

For $b > 1$:

1. __________

2. ____

3. ______________________________

4. ____

For $0 < b < 1$:

1. __________

2. ____

3. ______________________________

4. ____

DEFINITION

Compound Interest

Compound interest on a principal P invested at an annual interest rate r (in decimal form) for t years that is compounded ____

____

____

FORMULA

Watch and Work

Watch the video for Example 4 in the software and follow along in the space provided.

Example 4 Application: Calculating Compound Interest

What will be the value of a principal investment of $1000 invested at 6% for 3 years if interest is compounded monthly (12 times per year)?

Solution

Now You Try It!

Use the space provided to work out the solution to the next example.

Example A Application: Calculating Compound Interest

Find the value of $2000 invested at $r = 5\%$ for 2 years if the interest is compounded monthly.

The Number *e*

The number *e* is defined to be

____

DEFINITION

Continuously Compounded Interest

Continuously compounded interest on a principal *P* invested at an annual interest rate *r* for *t* years can be calculated using the following formula where ____

____

FORMULA

15.3 Exercises

Concept Check

True/False. Determine whether each statement is true or false. If a statement is false, explain how it can be changed so that the statement will be true. (**Note:** There may be more than one acceptable change.)

1. For all exponential functions $f(x) = x^b$, $b < 0$.

2. The function $f(x) = 5^x$ is an example of an exponential growth model.

3. In an exponential decay function, b^x approaches the x-axis for positive values of x.

4. The number e is defined to be approximately 3.14159.

Practice

Sketch the graph of each exponential function and label three points on each graph.

5. $y = 4^x$

6. $f(x) = 2^{0.5x}$

Find the following function values.

7. If $f(t) = 3 \cdot 4^t$, what is the value of $f(2)$?

Applications

Solve.

8. ***Bacteria:*** A biologist knows that in the laboratory, bacteria in a culture grow according to the function $y = y_0 \cdot 5^{0.2t}$, where y_0 is the initial number of bacteria present and t is time measured in hours. How many bacteria will be present in a culture at the end of 5 hours if there were 5000 present initially?

9. ***Investing:*** Find the value of \$1800 invested at 6% for 3 years if the interest is compounded continuously.

10. ***Prescriptions:*** A cancer patient is given a dose of 50 mg of a particular drug. In five days the amount of the drug in her system is reduced to 1.5625 mg. If the drug decays (or is absorbed) at an exponential rate, find the function that represents the amount of the drug at a given time. (**Hint:** Use the formula $y = y_0 b^{-t}$ and solve for b.)

Writing & Thinking

11. Discuss, in your own words, the symmetrical relationship of the graphs of the two exponential functions $y = 10^x$ and $y = 10^{-x}$.

15.4 Logarithmic Functions

1. The inverse of an exponential function is a __________________________.

Definition of Logarithm (base *b*)

For $b > 0$ and $b \neq 1$,

_______________ is read ____

DEFINITION

2. A logarithm is ____________________.

Basic Properties of Logarithms

For $b > 0$ and $b \neq 1$,

1. $\log_b 1 = 0$ Regardless of the base, the ______________________________
2. $\log_b b = 1$ The logarithm of the base ________________________
3. $x = b^{\log_b x}$ For __________________
4. $\log_b b^x = x$

PROPERTIES

Watch and Work

Watch the video for Example 3 in the software and follow along in the space provided.

Example 3 Solving Logarithmic Equations

Solve by first changing the equation to exponential form: $\log_{16} x = \frac{3}{4}$

Solution

Now You Try It!

Use the space provided to work out the solution to the next example.

Example A Solving Logarithmic Equations

Solve by first changing the equation to exponential form:

$$\log_8 x = \frac{2}{3}$$

15.4 Exercises

Concept Check

True/False. Determine whether each statement is true or false. If a statement is false, explain how it can be changed so that the statement will be true. (**Note:** There may be more than one acceptable change.)

1. Exponential functions of the form $y = b^x$ are one-to-one functions and have inverses.

2. The exponent of an exponential function is the base of its inverse logarithmic function.

3. Exponents are logarithms.

4. The logarithm of the base is always 1.

Practice

5. Express $7^2 = 49$ in logarithmic form.

6. Express $\log_5 125 = 3$ in exponential form.

Solve by first changing each equation to exponential form.

7. $\log_5 \frac{1}{125} = x$

8. $\log_x 121 = 2$

9. $\log_8 8^{3.7} = x$

Graph function and its inverse on the same set of axes.

10. $f(x) = 2^x$

Writing & Thinking

11. Discuss, in your own words, the symmetrical relationship of the graphs of the two logarithmic functions $y = \log_{10} x$ and $y = -\log_{10} x$.

Name: Date:

15.5 Properties of Logarithms

Product Rule for Logarithms

For $b > 0$, $b \neq 1$, and $x, y > 0$,

____

In words, the logarithm of a ____

PROPERTIES

Quotient Rule for Logarithms

For $b > 0$, $b \neq 1$, and $x, y > 0$,

____

In words, the logarithm of a ____

____

PROPERTIES

Power Rule for Logarithms

For $b > 0$, $b \neq 1$, $x > 0$, and any real number r,

____

In words, the logarithm of a ____

____

PROPERTIES

Watch and Work

Watch the video for Example 4 in the software and follow along in the space provided.

Example 4: Using the Properties of Logarithms to Expand Expressions

Use the properties of logarithms to expand each expression as much as possible.

a. $\log_b 2x^3$

b. $\log_b \frac{xy^2}{z}$

c. $\log_b (xy)^{-3}$

d. $\log_b \sqrt{3x}$

Solution

Now You Try It!

Use the space provided to work out the solution to the next example.

Example A Using the Properties of Logarithms to Expand Expressions

Use the properties of logarithms to expand each expression as much as possible.

a. $\log_b 3x^2$

b. $\log_b \dfrac{x^3 y}{z}$

c. $\log_b (mn)^{-2}$

d. $\log_b \sqrt{2a}$

15.5 Exercises

Concept Check

True/False. Determine whether each statement is true or false. If a statement is false, explain how it can be changed so the statement will be true. (**Note:** There may be more than one acceptable change.)

1. The properties of exponents are used to prove the properties of logarithms.

2. The power rule for logarithms states that the exponent r must be a positive integer.

3. The log of a product does not equal the product of the logs.

4. The expression $\log_5 \frac{4}{3}$ is equivalent to $\frac{\log_5 4}{\log_5 3}$.

Practice

Use the properties of logarithms to expand each expression as much as possible.

5. $\log_b 5x^4$

6. $\log_6 \sqrt[3]{\frac{x^2}{y}}$

Use the properties of logarithms to write each expression as a single logarithm of a single expression.

7. $\frac{1}{2}\log_b 25 + \log_b 3 - \log_b x$

8. $\frac{1}{2}(\log_5 x - \log_5 y)$

9. $-\frac{2}{3}\log_2 x - \frac{1}{3}\log_2 y + \frac{2}{3}\log_2 z$

Writing & Thinking

10. Prove the quotient rule for logarithms: For $b > 0$, $b \neq 1$, and $x, y > 0$, $\log_b \frac{x}{y} = \log_b x - \log_b y$.

Name: Date:

15.6 Common Logarithms and Natural Logarithms

1. Whenever the base notation is omitted, the base is ________________________.

$$\log_{10} x = \log x$$

2. The logarithm (with base greater than 1) of any number between ______________ will always be negative.

3. There are no logarithms of _______________ numbers or _____.

   If $\log x = N$, then $x =$ _______ and the number x is called the _______________________.

4. The notation for natural logarithms is shortened to ____________________

5. If $\ln x = N$, then $x =$ ______ and the number x is called the inverse ln of N.

Watch and Work

Watch the video for Example 6 in the software and follow along in the space provided.

Example 6 Using a Calculator to find the Inverse ln of *N*

Use a TI-84 Plus graphing calculator to find the inverse ln of N for each expression. (That is, find the value of x.)

a. $\ln x = 3$

b. $\ln x = -1$

c. $\ln x = -0.1$

d. $\ln x = 50$

Solution

Now You Try It!

Use the space provided to work out the solution to the next example.

Example A Application: Finding the Mean

Use a TI-84 Plus graphing calculator to find the inverse ln of N for each expression, accurate to the nearest ten-thousandth.

a. $\ln x = 40$

b. $\ln x = -\frac{1}{100}$

15.6 Exercises

Concept Check

True/False. Determine whether each statement is true or false. If a statement is false, explain how it can be changed so the statement will be true. (**Note:** There may be more than one acceptable change.)

1. Whenever the base of a logarithm is omitted, it is understood to be 1.

2. Logarithms of negative numbers or 0 do not exist.

3. Common logarithms have an inverse while natural logarithms do not.

4. Given $\log x = 4$ the inverse log of 4 is $x = 10^4 = 10{,}000$.

Practice

Use a calculator to evaluate the logarithms accurate to the nearest ten-thousandths place.

5. $\log 173$

6. $\log 0.0061$

Use a calculator to find the value of x in each equation accurate to the nearest ten-thousandths place.

7. $\log x = 2.31$

8. $\ln x = -8.3$

Writing & Thinking

9. Explain the difference in the meaning of the expressions $\log x$ and $\ln x$.

10. The function $y = \log x$ is defined only for $x > 0$. Discuss the function $y = \log(-x)$. That is, does this function even exist? If it does exist, what is its domain? Sketch its graph and the graph of the function $y = \log x$.

15.7 Logarithmic and Exponential Equations and Change-of-Base

Properties of Real Exponents

If a and b are positive real numbers and x and y are any real numbers, then

1. ________
2. ________
3. ________
4. ________
5. ________
6. ________
7. ________

PROPERTIES

Properties of Equations with Exponents and Logarithms

For $b > 0$ and $b \neq 1$,

1. If $b^x = b^y$, then ________
2. If $x = y$, then ________
3. If ________ then ________
4. If $x = y$, then ________ ________

PROPERTIES

Change-of-Base Formula

For $a, b, x > 0$ and $a,\ b \neq 1$,

____

FORMULA

Watch and Work

Watch the video for Example 4 in the software and follow along in the space provided.

Example 4 Change-of-Base

Use the change-of-base formula to evaluate the following expressions.

a. $\log_2 3.42$

b. $\log_3 0.3333$

Solution

Now You Try It!

Use the space provided to work out the solution to the next example.

Example A Change-of-Base

Use the change-of-base formula to evaluate $\log_2 4$

15.7 Exercises

Concept Check

True/False. Determine whether each statement is true or false. If a statement is false, explain how it can be changed so the statement will be true. (**Note:** There may be more than one acceptable change.)

1. The change of base formula is $\log_b x = \dfrac{\log_a b}{\log_a x}$, for $a, b, x > 0$ and $b \neq 1$.

2. If the terms in an exponential equation all have the same base, there is no need to use logarithms to solve the equation.

3. Exponential equations with different bases can be solved by taking either the log of both sides or the ln of both sides.

4. If $5^x = 5^y$, then x is equal to y.

Practice

Use the properties of exponents and logarithms to solve each of the equations. If necessary, use a calculator and round answers to the nearest ten-thousandth.

5. $3^7 \cdot 3^{-2} = 3^x$

6. $7^{3x} = 49^4$

7. $5^{1-x} = 1$

8. $\log(x-15) = 2 - \log x$

Use the change-of-base formula to evaluate each of the expressions or solve the equations. Round answers to the nearest ten-thousandth.

9. $\log_3 12$

10. $\log_{15} 739$

Writing & Thinking

11. Explain, in your own words, why $7 \cdot 7x \neq 49x$ when $x \neq 1$. Show each of the expressions $7 \cdot 7x$ and $49x$ as a single exponential expression with base 7.

15.8 Applications: Exponential and Logarithmic Functions

Watch and Work

Watch the video for Example 5 in the software and follow along in the space provided.

Example 5 Application: Half-life of Radium

The half-life of a substance is the time needed for the substance to decay to one-half of its original amount. The half-life of radium-226, a common isotope of radium, is 1600 years. If 10 grams are present today, how many grams will remain in 500 years?

Solution

Now You Try It!

Use the space provided to work out the solution to the next example.

Example A Application: Half-life of Radium

The half-life of plutonium is 24,100 years. If 12 grams are present today, how many grams will remain in 10,000 years?

15.8 Exercises

Concept Check

True/False. Determine whether each statement is true or false. If a statement is false, explain how it can be changed so the statement will be true. (**Note:** There may be more than one acceptable change.)

1. In Newton's law of cooling, the variable C is the constant temperature of the medium surrounding the cooling object.

2. The formula $A = A_0 e^{-0.1t}$ is used for skin healing, where t is measured in hours.

Applications

Solve. If necessary, round answers to the nearest hundredth (unless otherwise specified).

3. ***Investing:*** If Kim invests $2000 at a rate of 7% compounded continuously, what will be her balance after 10 years?

4. ***Battery Reliability:*** The reliability of a certain type of flashlight battery is given by $f = e^{-0.03x}$, where f is the fractional part of the batteries produced that last x hours. What fraction of the batteries produced are good after 40 hours of use?

5. ***Dosages:*** The concentration of a drug in the blood stream is given by $C = C_0 e^{-0.8t}$, where C_0 is the initial dosage and t is the time in hours elapsed after administering the dose. If 20 mg of the drug is given, how much time elapses until 5 mg of the drug remains?

6. ***Baking:*** The temperature of a carrot cake is 350° when it is removed from the oven. The temperature in the room is 72°. In 10 minutes, the cake cools to 280°. How long will it take for the cake to cool to 160°?

7. ***Half-life:*** Radioactive iodine has a half-life of 60 days. If an accident occurs at a nuclear plant and 30 grams of radioactive iodine are present, in how many days will 1 gram be present? (Round k to at least 7 decimal places.)

Chapter 15 Project

The Ups and Downs of Population Change

An activity to demonstrate the use of exponential functions in real life.

A mid-sized city with a population of 125,000 people experienced a population boom after several tech startups produced very successful products. The population of the city grew by 7% per year for several years. The population of the city can be modeled by the equation $P = P_0e^{rt}$, where P is the future population, P_0 is the initial population of the city, r is the rate of growth, and t is the time passed, in years.

1. Write the equation of the model for the city's population growth in terms of t.

2. Find the population of the city and determine the number of new citizens after 5 years of growth. Round your answer to the nearest person.

3. Find the population of the city and determine the number of new citizens after 10 years of growth. Round your answer to the nearest person.

4. Compare the growth in population after 5 years and after 10 years. Is the growth after 10 years twice the growth after 5 years? Explain your answer.

After 10 years, the population became too large for the city infrastructure (such as roads and freeways) to handle. The city raised taxes to improve the infrastructure, but the updates took a few years. Due to the traffic congestion in the city and the increased taxes, several of the large companies moved to neighboring cities. This loss of jobs in the city lead to a decline in population at a rate of 5% per year.

5. Write the equation of the model for the city's population decay in terms of t.

6. It took 4 years to finish the roadway project in the city and people continued to move away at a steady rate. Find the population of the city when the project was complete.

7. What percent of the population moved away during those 4 years? Round your answer to the nearest percent.

8. Write the equation of the model for the city's population growth in terms of t.

After the city infrastructure was updated, a few companies moved back to the city. As a result, the population started to grow by 2% per year.

9. At this rate of growth, how long would it take for the city population to reach 250,000 citizens? Round your answer to the nearest year.

CHAPTER 16

Conic Sections

Math @ Work

Halley's Comet is visible from Earth every 74 to 79 years. The comet has an elliptical orbit around the Sun, just like the orbits of the planets in our solar system. Ellipses, along with parabolas, hyperbolas, and circles, are conic sections. Conic sections occur naturally all around us. If you've seen a shadow on a wall cast by a circular lamp shade, you've seen a hyperbolic curve. If you've seen a fountain with a jet of water going upward and then falling back down, you've seen a parabola.

Not only do conic sections occur naturally, they are also useful in a variety of applications. Early camera makers discovered that circular lenses did the most efficient job of focusing light onto film. Car headlights use a parabolic shape to concentrate the light beam to improve safety. Medical specialists have used the ellipse to create a device that treats kidney stones with shock waves instead of surgery. The hyperboloid is the design standard for all nuclear cooling towers and some coal-fired power plants because its structure can withstand high winds, it requires less material than other shapes, and it maximizes cooling efficiency.

Suppose the town you live in is holding an architectural competition to design a public monument that will be part of the downtown renovation project. You submit a proposal to build a circular fountain at the center of town. You've mapped the town on a coordinate grid and found the origin, where you plan to place the center of the fountain. If your proposal states that one point on the edge of the fountain will be at the point $(0,5)$, is this enough information to find the equation of the circle? What is another point on the edge of the fountain?

16.1 Translations and Reflections

1. The formula

$$\frac{\qquad}{\qquad}$$

is called the **difference quotient**. A geometric interpretation of the difference quotient is the ____________ of a line through two points on the ____________________________.

Watch and Work

Watch the video for Example 2 in the software and follow along in the space provided.

Example 2 Finding the Difference Quotient $\frac{f(x+h)-f(x)}{h}$

Find the difference quotient for the function $f(x)=2-6x$.

Solution

Now You Try It!

Use the space provided to work out the solution to the next example.

Example 3 Finding the Difference Quotient $\frac{f(x+h)-f(x)}{h}$

Find the difference quotient for the function $f(x)=8x-12$.

Horizontal and Vertical Translations

Given the graph of $y=f(x)$, the graph of ______________________

1. ______________________

2. ______________________

of the graph of ______________

Draw the graph of $y=f(x)$ in relation to (h, k) as if (h, k) ______________________

This new graph will be the ______________________

PROCEDURE

2. The graph of $y=-f(x)$ is ______________________ of the graph of $y=f(x)$.

16.1 Exercises

Concept Check

True/False. Determine whether each statement is true or false. If a statement is false, explain how it can be changed so the statement will be true. (**Note:** There may be more than one acceptable change.)

1. Algebraically, the two functions $f(x+h)$ and $f(x)+h$ represent the same thing.

2. Translation changes the shape of the graph of a function.

3. The figure generated by choosing a point on a line and rotating that line around the point in a circular fashion is called a circular cone.

4. More than one translation cannot be applied to a function at the same time.

Practice

Evaluate the function at the given expressions.

5. For $g(x)=2-x^2$, find:

a. $g(\sqrt{2})$

b. $g(a-1)$

c. $g(x+h)$

d. $\dfrac{g(x+h)-g(x)}{h}$

Find and simplify the difference quotient, $\dfrac{f(x+h)-f(x)}{h}$, for the given function.

6. $f(x)=2x-3$

The graph of $y = \sqrt{x}$ is given along with a few points as aids. Graph the functions using your understanding of reflections and translations with no additional computations.

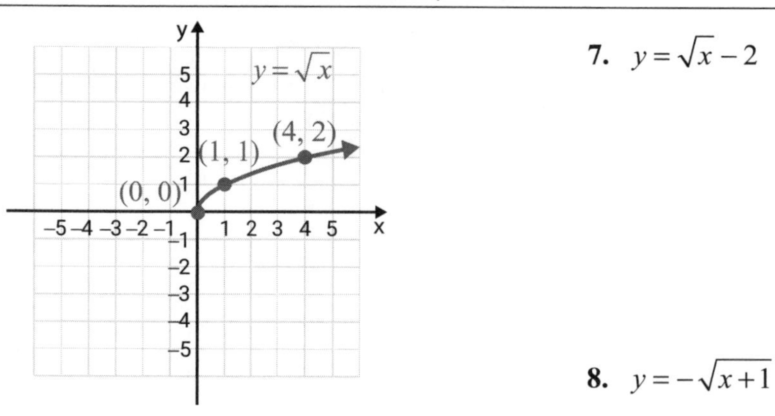

7. $y = \sqrt{x} - 2$

8. $y = -\sqrt{x+1}$

9. $y = 5 + \sqrt{x+2}$

Applications

Solve.

10. ***Landscaping:*** A landscaper is planning a scenic walkway to go in the large space between the front entrance of a building and the main road that runs in front of the building. He sketches a graph of the area. On it, the edge of the walkway is modeled by $y = x^4 - 5x^2 + 2x$. In order to get a permit, however, the city's building department tells the landscaper that he must move the walkway 10 meters up, farther away from the road. Write the function that represents the new placement of the curve.

Writing and Thinking

11. Explain, in your own words, how the graph of the function $y = f(x-h)+k$ represents a horizontal and a vertical shift of the graph of the function $y = f(x)$.

Name: Date:

16.2 Parabolas as Conic Sections

Equations of Horizontal Parabolas

Equations of horizontal parabolas (parabolas that open to the left or right) can be written in the form

____

The parabola opens to the ______________________

The vertex is ____________

The line ____________________

DEFINITION

1. By setting $x =$ ___ and solving the following quadratic equation

$$0 = ay^2 + by + c,$$

we can determine the ____________ (the points, if any, where the graph intersects the y-axis).

Watch and Work

Watch the video for Example 1 in the software and follow along in the space provided.

Example 1 Sketching Horizontal Parabolas

For $x = y^2 - 6y + 4$, find the vertex, the y-intercepts, and the line of symmetry. Then sketch the graph.

Solution

Now You Try It!

Use the space provided to work out the solution to the next example.

Example A Sketching Horizontal Parabolas

For $x = y^2 - 4y + 1$, find the vertex, the y-intercepts, and the line of symmetry. Then sketch the graph.

16.2 Exercises

Concept Check

True/False. Determine whether each statement is true or false. If a statement is false, explain how it can be changed so the statement will be true. (**Note:** There may be more than one acceptable change.)

1. Not all parabolas are functions.

2. Parabolas open down if $a > 0$ and open up if $a < 0$.

3. The line $x = h$ is the line of symmetry for a horizontal parabola.

Practice

For the given equations, **a.** find the vertex, **b.** find the *y*-intercept, **c.** find the line of symmetry, and **d.** sketch the graph.

4. $x = y^2 + 4$

5. $x = (y-2)^2$

6. $x = y^2 - 8y + 16$

7. $x = -y^2 - 6y - 9$

Writing and Thinking

8. For $x = ay^2 + by + c$ we know that the graph of the parabola opens to the right if $a > 0$ and to the left if $a < 0$. Discuss which values of a will cause the parabola to be wider and which will cause it to be narrower than the graph of $x = y^2$.

Name: Date:

16.3 Distance Formula, Midpoint Formula, and Circles

The Pythagorean Theorem

In a right triangle, if c is the length of the ____________________

____

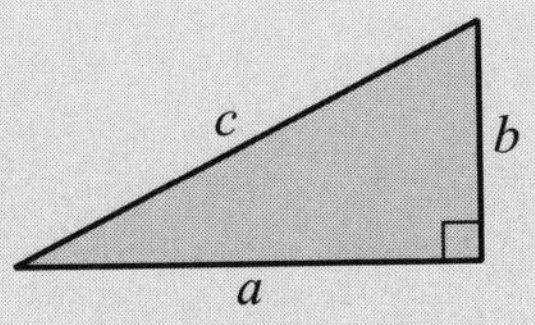

THEOREM

The Distance Formula

For two points $P(x_1, y_1)$ and $Q(x_2, y_2)$ in a plane, the ____________________

____

FORMULA

Midpoint Formula

The formula for the midpoint between two ____________________

____

FORMULA

Circle, Center, Radius, and Diameter

A **circle** is the set of all points in a plane that ____________________

____________ is called the **center** of the circle.

____________ is called the **radius** of the circle.

The distance from ____________________

____________ is called the **diameter** of the circle.

DEFINITION

Equation of a Circle

The equation of a circle with radius r and center at (h, k) is

____

If the center is at the origin, $(0, 0)$,

____

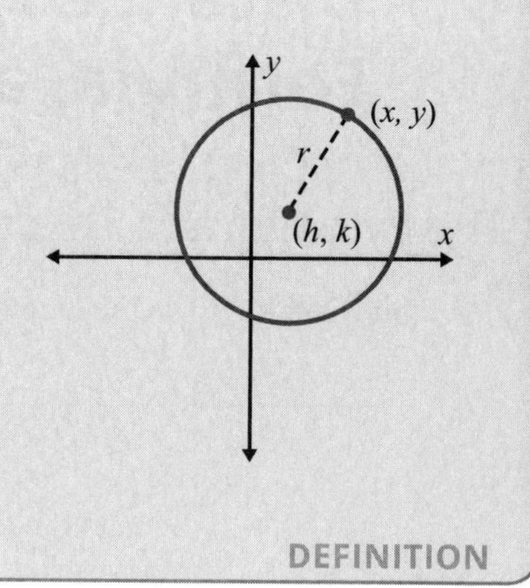

DEFINITION

Watch and Work

Watch the video for Example 5 in the software and follow along in the space provided.

Example 5 Finding the Equation of a Circle

Find the equation of the circle with center at $(5, 2)$ and radius 3. Is the point $(5, 5)$ on the circle?

Solution

Now You Try It!

Use the space provided to work out the solution to the next example.

Example A Finding the Equation of a Circle

Find the equation of the circle with center at $(3, -2)$ and radius 4. Is the point $(7, -2)$ on the circle?

16.3 Exercises

Concept Check

True/False. Determine whether each statement is true or false. If a statement is false, explain how it can be changed so the statement will be true. (**Note:** There may be more than one acceptable change.)

1. The Pythagorean Theorem is used to derive the distance formula.

2. The distance from the center of a circle to any point on the circle is called the diameter of the circle.

3. To find the equation of a circle with center at $(2, 5)$ and radius 6, the distance formula can be used.

4. The hypotenuse of a right triangle is the longest side of the triangle.

Practice

Find the distance between the two given points and the coordinates of the midpoint of the line segment joining the two points.

5. $(2, 4), (6, 7)$

6. $(5, -2), (7, -5)$

Find equations for each of the circles.

7. Center $(0, 0)$; $r = 4$

8. Center $(5, 0)$; $r = \sqrt{2}$

Write $x^2 + y^2 = 9$ in standard form. Find the center and radius of the circle and then sketch the graph.

9. $x^2 + y^2 = 9$

Applications

Solve.

10. ***Distance:*** The roadways in Descartesville are laid out such that streets run east to west and avenues run north to south. The north to south avenues and the east to west streets are numbered sequentially, beginning with 1. For example, a person standing on the corner of 1st Street and 1st Avenue may be considered to be at standing at $(1, 1)$. A person begins at the corner of 1st Street and 2nd Avenue and walks to the corner of 9th Street and 8th Avenue. If the person were able to walk directly from the beginning corner to the ending corner, the distance traveled would be the same as the distance of how many blocks?

16.4 Ellipses and Hyperbolas

Ellipse

An **ellipse** is the set of all points in a plane for which ______________________________ ______________.

______________________ is called a **focus** (plural foci).

The **center** of an ellipse is ______________________________.

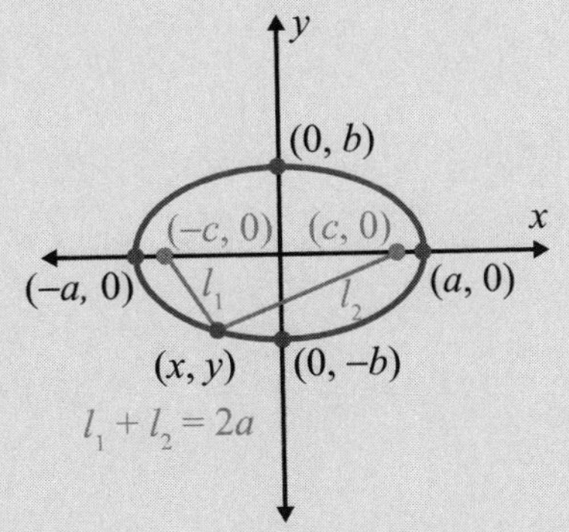

y, x, $(0, b)$, $(-c, 0)$, $(c, 0)$, $(-a, 0)$, $(a, 0)$, l_1, l_2, (x, y), $(0, -b)$

$l_1 + l_2 = 2a$

The graph above is of an ellipse with its center at the origin $(0, 0)$, foci at _____ and _____ x-intercepts at _____ and _____ and y-intercepts at _____ and _____ (where $a^2 > b^2$).

DEFINITION

Equation of an Ellipse

The standard form for the equation of an ellipse with its center at the origin is

____

The points $(a, 0)$ and $(-a, 0)$ are the ______________________________.

The points $(0, b)$ and $(0, -b)$ are the ______________________________.

When $a^2 > b^2$:

1. The segment of length ______________________________.
2. The segment of ______________________________.

When $b^2 > a^2$:

1. The segment of length ______________________________.
2. The segment of length ______________________________.

DEFINITION

Watch and Work

Watch the video for Example 2 in the software and follow along in the space provided.

Example 2 Equation of an Ellipse—Major Axis Vertical

Graph the ellipse $b^2 = 9$.

Solution

Now You Try It!

Use the space provided to work out the solution to the next example.

Example A Equation of an Ellipse—Major Axis Vertical

Graph the ellipse $\frac{x^2}{4} + \frac{y^2}{16} = 1$.

Hyperbola

A **hyperbola** is the set of all points in a plane such that the ______________________________

______________________________.

______________________________ is called a **focus** (plural foci).

The **center** of a hyperbola is ______________________________.

The graph of a hyperbola with its center at the origin $(0, 0)$, foci along the x-axis at ____ and ____ and x-intercepts at ____ and ____. There are no y-intercepts.

DEFINITION

Equations of Hyperbolas

In general, there are two standard forms for equations of hyperbolas with their centers at the origin.

1. ______________

x-intercepts (vertices) at ______________

No ______________

Asymptotes: ______________

The curves "open" ______________

2. ______________

y-intercepts (vertices) at ______________

No ______________

Asymptotes: ______________

The curves "open" ______________

DEFINITION

Ellipse with Center at (h, k)

The equation of an ellipse with its center at __________ is

____

where a and b are ______________________________________

DEFINITION

Hyperbola with Center at (h, k)

The equation of a hyperbola with its center at __________ is

____

where a and b are ______________________________________

DEFINITION

16.4 Exercises

Concept Check

True/False. Determine whether each statement is true or false. If a statement is false, explain how it can be changed so the statement will be true. (Note: There may be more than one acceptable change.)

1. An ellipse's center is the point midway between the two foci.

2. The foci of an ellipse lie on the ellipse.

3. The midway point between the foci of a hyperbola is called the origin.

Practice

Write each of the equations in standard form. Then sketch the graph. For hyperbolas, graph the asymptotes as well.

4. $x^2 + 9y^2 = 36$

5. $x^2 - y^2 = 1$

6. $4x^2 - 9y^2 = 36$

7. $4x^2 + 3y^2 = 12$

Use your knowledge of translations to graph each of the following equations. These graphs are ellipses and hyperbolas with centers at points other than the origin.

8. $\frac{(x-2)^2}{25}+\frac{(y-1)^2}{9}=1$

9. $\frac{(x+5)^2}{1}-\frac{(y+2)^2}{16}=1$

16.5 Nonlinear Systems of Equations

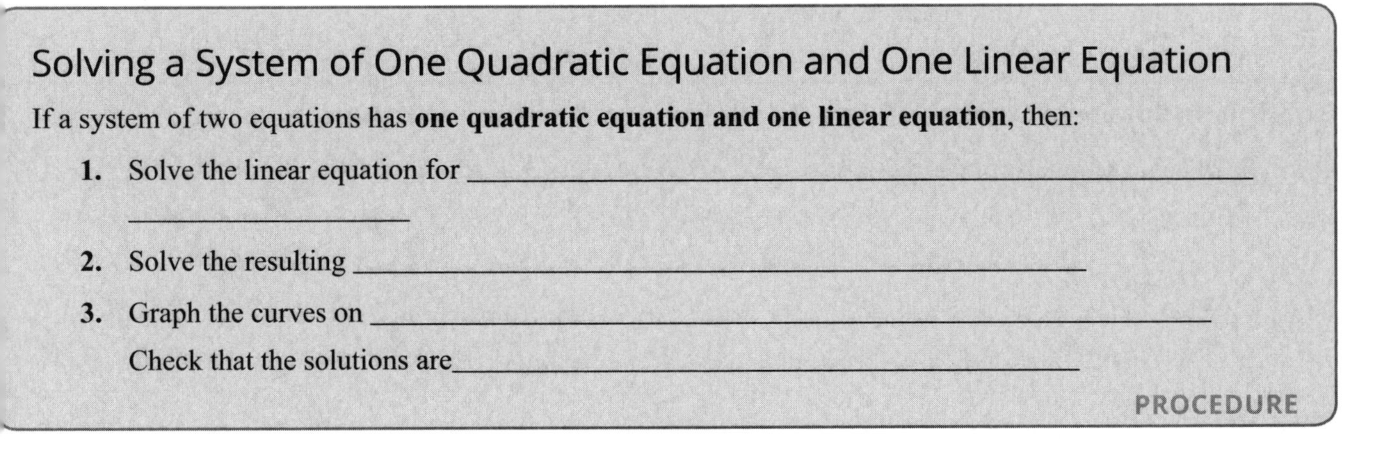

Solving a System of One Quadratic Equation and One Linear Equation

If a system of two equations has **one quadratic equation and one linear equation**, then:

1. Solve the linear equation for ______________________________

   ____
2. Solve the resulting ______________________________
3. Graph the curves on ______________________________

   Check that the solutions are______________________________

PROCEDURE

Watch and Work

Watch the video for Example 1 in the software and follow along in the space provided.

Example 1 Solving a System of One Quadratic and One Linear Equation

Solve the following system of equations and graph both curves on the same set of axes.

A circle and a line: $\begin{cases} x^2 + y^2 = 25 & \text{Circle} \\ x + y = 5 & \text{Line} \end{cases}$

Solution

Now You Try It!

Use the space provided to work out the solution to the next example.

Example A Solving a System of One Quadratic and One Linear Equation

Solve the system and graph both curves.

$$\begin{cases} x - y = 6 \\ x^2 + y^2 = 36 \end{cases}$$

Solving a System of Two Quadratic Equations

If a system of two equations has **two quadratic equations**, then:

1. The method used depends on ______________________________
2. Substitution may work or ______________________
3. Graph the curves on ____

   Check that the solutions are reasonable and ____________________________

PROCEDURE

16.5 Exercises

Concept Check

True/False. Determine whether each statement is true or false. If a statement is false, explain how it can be changed so the statement will be true. (**Note:** There may be more than one acceptable change.)

1. Equations that have at least one second-degree term are called linear equations.

2. Graphing a system of equations is useful for approximating solutions.

3. Graphing a system of equations will determine the exact number of solutions.

4. If a system of equations has two quadratic equations, the curves must intersect.

Practice

Solve each system of equations. Sketch the graphs.

5. $\begin{cases} y = x^2 + 1 \\ 2x + y = 4 \end{cases}$

6. $\begin{cases} x^2 + 3y^2 = 12 \\ x = 3y \end{cases}$

7. $\begin{cases} x^2 + y^2 = 9 \\ x^2 - y + 3 = 0 \end{cases}$

8. $\begin{cases} 2x^2 - 3y^2 = 6 \\ 2x^2 + y^2 = 22 \end{cases}$

Chapter 16 Project

What's in a Logo?

An activity to demonstrate the use of conic sections in real life.

You constantly see company logos in your day to day life. Companies use logos for brand recognition and to advertise their products. Take a look around you as you go throughout your day and you'll see company logos everywhere. Your kitchen appliances have logos on them, your car has a logo on it, your smartphone has a logo, and logos appear in all forms of advertisements. To make a logo clean and easy to remember, a company may use conic sections in the design.

Use the internet to look up or research the logos mentioned in the problems that follow. A graphing calculator or graphing application will be needed to recreate the logos indicated in the problems. If you don't have a graphing calculator, you can use the free graphing application located at www.desmos.com.

1. Some logos designs, such as the Target logo, are based on concentric circles. Circles are concentric when they share the same center. Perform an internet search for other logos that use concentric circles. Draw three of these logos.

2. The Target logo consists of three concentric circles. Use your graphing tool to recreate the Target logo. Sketch the recreation on the coordinate plane and write the equations you used.

3. The logo for the Olympic Games consists of five circles that all have the same radius. Use your graphing tool to recreate the Olympic Games logo. Sketch the recreation on the coordinate plane and write the equations you used. Use the equation $x^2 + y^2 = 4$ to represent the center ring on the top row.

4. The Toyota logo consists of three ellipses. Use your graphing tool to recreate the Toyota logo. Sketch the recreation on the coordinate plane and write the equations you used.

5. Find an existing logo or create your own logo that consists of more than one type of conic section. Describe the conic sections used in the logo and provide a sketch of the logo.

6. Recreate the logo by graphing conic sections. Create a simpler version of the logo, if necessary. Sketch your version of the logo on the coordinate plane and write the equations you used.

Math@Work

Math@Work

Basic Inventory Management

As an business manager you will need to evaluate the company's inventory several times per year. While evaluating the inventory, you will need to ensure that enough of each product will be in stock for future sales based on current inventory count, predicted sales, and product cost. Let's say that you check the inventory four times a year, or quarterly. You will be working with several people to get all of the information you need to make the proper decisions. You need the sales team to give you accurate predictions of how much product they expect to sell. You need the warehouse manager to keep an accurate count of how much of each product is currently in stock and how much of that stock has already been sold. You will also have to work with the product manufacturer to determine the cost to produce and ship the product to your company's warehouse. It's your job to look at this information, compare it, and decide what steps to take to make sure you have enough of each product in stock for sales needs. A wrong decision can potentially cost your company a lot of money.

Suppose you get the following reports: an inventory report of unsold products from the warehouse manager and the report on predicted sales for the next quarter (three months) from the sales team.

Unsold Products	
Item	**Number in Stock**
A	5025
B	150
C	975
D	2150

Predicted Sales	
Item	**Expected Sales**
A	4500
B	1625
C	1775
D	2000

Suppose the manufacturer gives you the following cost list for the production and shipment of different amounts of each inventory item.

Item	Amount	Cost	Amount	Cost	Amount	Cost
A	500	$875	1000	$1500	1500	$1875
B	500	$1500	1000	$2500	1500	$3375
C	500	$250	1000	$400	1500	$525
D	500	$2500	1000	$4250	1500	$5575

1. Which items and how much of each item do you need to purchase to make sure the inventory will cover the predicted sales?

2. If you purchase the amounts from Problem 1, how much will this cost the company?

3. By ordering the quantities you just calculated, you are ordering the minimum of each item to cover the expected sales. If the actual sales during the quarter are higher than expected, what might happen? How would you handle this situation?

4. Which skills covered in this chapter were necessary to help you make your decisions?

Math@Work

Hospitality Management: Preparing for a Dinner Service

As the manager of a restaurant you will need to make sure everything is in place for each meal service. This means that you need to predict and prepare for busy times, such as a Friday night dinner rush. To do this, you will need to obtain and analyze information to determine how much of each meal is typically ordered. After you estimate the number of meals that will be sold, you need to communicate to the chefs how much of each item they need to expect to prepare. An additional aspect of the job is to work with the kitchen staff to make sure you have enough ingredients in stock to last throughout the meal service.

You are given the following data, which is the sales records for the signature dishes during the previous four Friday night dinner services.

Week	Meal A	Meal B	Meal C	Meal D
1	30	42	28	20
2	35	38	30	26
3	32	34	26	26
4	30	32	28	22

Meal C is served with a risotto, a type of creamy rice. The chefs use the following recipe, which makes 6 servings of risotto, when they prepare Meal C. (**Note:** The abbreviation for tablespoon is T and the abbreviation for cup is c.)

$5\frac{1}{2}$ c chicken stock	$2\frac{1}{3}$ T chopped shallots	$\frac{1}{2}$ c red wine
$1\frac{1}{2}$ c rice	2 T chopped parsley	$4\frac{3}{4}$ c thinly sliced mushrooms
2 T butter	2 T olive oil	$\frac{1}{2}$ c Parmesan cheese

1. For the past four Friday night dinner services, what was the average number of each signature meal served? If the average isn't a whole number, explain why you would round this number either up or down.

2. Based on the average you obtained for Meal C, calculate how much of each ingredient your chefs will need to make the predicted amount of risotto.

3. The head chef reports the following partial inventory: $10\frac{3}{4}$ c rice, $15\frac{3}{4}$ c mushrooms, and 10 T shallots. Do you have enough of these three items in stock to prepare the predicted number of servings of risotto?

4. Which skills covered in this and the previous chapter helped you make your decisions?

ath@Work

ookkeeper

s a bookkeeper, you will often receive bills and receipts for various purchases or expenses from employees of the company u work for. You will need to split the bill by expense code, assign costs according to customer, and reimburse an employee r their out-of-pocket spending. To do this you will need to know the company's reimbursement policies, the expense codes r different spending categories, and which costs fall into a particular expense category.

uppose two employees from the sales department recently completed sales trips. Employee 1 flew out of state and visited two stomers, Customer A and Customer B. This employee had a preapproved business meal with Customer B and was traveling r three days. Employee 2 drove out of state to visit Customer C. This employee stayed at a hotel for the night and then drove ack the next day. The expenses for the two employees are as follows.

Employee 1	
Flight and Rental Car	$470.50
Hotel	$278.88
Meals	$110.56
Business Meal	$102.73
Presentation Materials	$54.86

Employee 2	
Miles Driven	578.5 miles
Fuel	$61.35
Hotel	$79.60
Meals	$53.23
Presentation Materials	$67.84

he expense categories used by your company to track spending are: Travel (includes hotel, flights, mileage, etc.), Meals usiness), Meals (travel), and Supplies. Traveling employees are reimbursed up to $35 per day for meals while traveling and r all preapproved business meals. They also receive $0.565 per mile driven with their own car.

. How much will you reimburse each employee for travel meals? Did either employee go over their allowed meal reimbursement amount?

. What were the total expenses for each employee?

. The company you work for keeps track of how much is spent on each customer. When a sales person visits multiple customers during one trip, the tracked costs are split between the customers. Fill in this table according to how much was spent on each customer for the different expense categories. (**Note:** For meals, only include the amount the employee was reimbursed.)

Expense	Customer A	Customer B	Customer C
Travel			
Meals (business)			
Meals (travel)			
Supplies			
Total			

Math@Work

Pediatric Nurse

s a pediatric nurse working in a hospital setting, you will be responsible for taking care of several patients during your work day. ou will need to administer medications, set IVs, and check each patient's vital signs (such as temperature and blood pressure). /hile doctors prescribe the medications that nurses need to administer, it is important for nurses to double check the dosage nounts. Administering the incorrect amount of medication can be detrimental to the patient's health.

uring your morning nursing round, you check in on three new male patients and obtain the following information.

	Patient A	Patient B	Patient C
Age	10	9	12
Weight (pounds)	81	68.5	112
Blood Pressure	97/58	100/59	116/73
Temperature (°F)	99.7	97.3	101.4
Medication	A	B	A

he following table shows the bottom of the range for abnormal blood pressure (BP) for boys. If either the numerator or e denominator of the blood pressure ratio is greater than or equal to the values in the chart, this can indicate a stage of ypertension.

Abnormal Blood Pressure for Boys by Age	
	Systolic BP / Diastolic BP
Age 9	109/72
Age 10	111/73
Age 11	113/74
Age 12	115/74
Source: http://www.nhlbi.nih.gov/health/public/heart/hbp/bp_child_pocket/bp_child_pocket.pdf	

Medication Directions	
Medication	**Dosage Rate**
A	40 mg per 10 pounds
B	55 mg per 10 pounds

. Do any of the patients have a blood pressure which may indicate they have hypertension? If yes, which patient(s)?

. Use proportions to determine the amount of medication that should be administered to each patient based on weight. Round to the nearest 10 pounds before calculating.

. The average body temperature is 98.6 degrees Fahrenheit. You are supposed to alert the doctor on duty if any of the patients have a temperature 2.5 degrees higher than average. For which patients would you alert a doctor?

. Which skills covered in this chapter and the previous chapters were necessary to help you make your decisions?

Math@Work

Architecture

As a project architect, you will be part of a team that creates detailed drawings of the project that will be used during the construction phase. It will be your job to ensure that the project will meet guidelines given to you by your company, such as square footage requirements and budget constraints. You will also need to meet the design requirements requested by the client.

Suppose you are part of a team that is designing an apartment building. You are given the task to create the floor plan for an apartment unit with two bedrooms and one bathroom. The apartment management company that has contracted your company to do the project has several requirements for this specific apartment unit.

1. One bedroom is the "master bedroom" and must have at least 60 square feet more than the other bedroom.
2. All walls must intersect or touch at 90 degree angles.
3. The kitchen must have an area of no more than 110 square feet.
4. The apartment must be between 1000 square feet and 1050 square feet.

A preliminary sketch of the apartment is shown here.

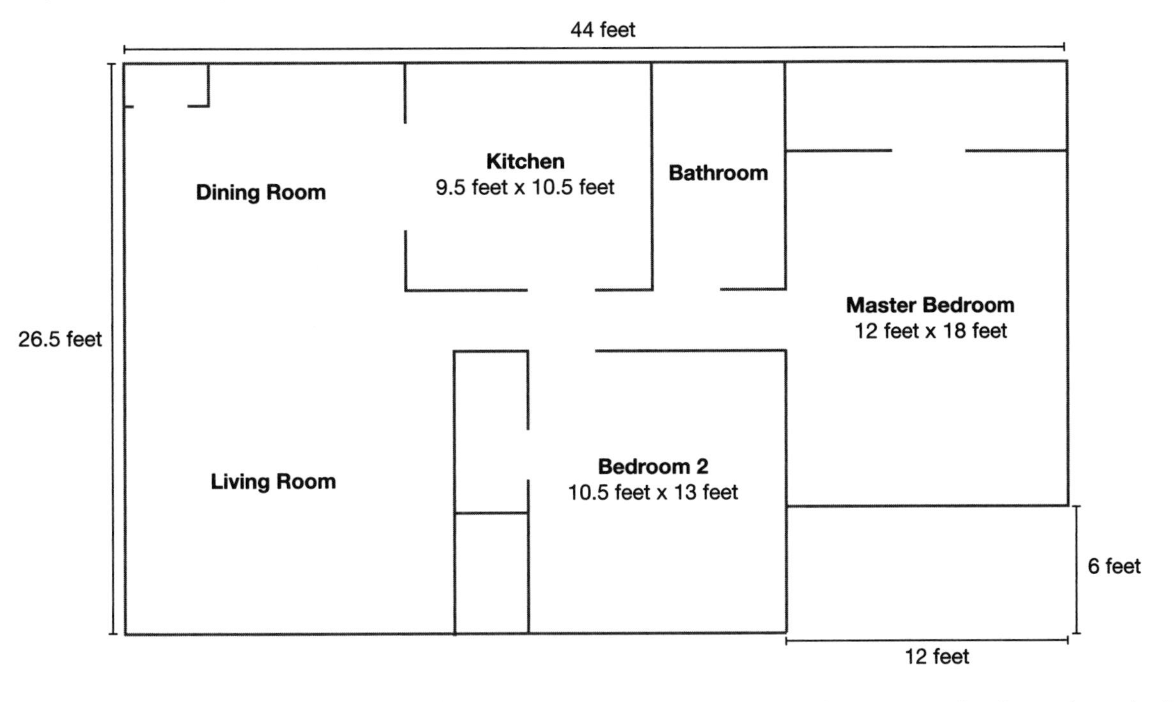

1. Does the apartment have the required total square footage that was requested? Is it over or under the total required?

2. Does the apartment blueprint meet the other requirements given by the client? If not, what does not meet the requirements?

3. For this specific apartment unit, the total construction cost per square foot is estimated to be $75.75. Approximately how much will it cost to construct each two-bedroom apartment based on the floor plan?

Math@Work

Statistician: Quality Control

uppose you are a statistician working in the quality control department of a company that manufactures the hardware sold in its to assemble book shelves, TV stands, and other ready-to-assemble furniture pieces. There are three machines that produce a articular screw and each machine is sampled every hour. A measurement of the screw length is determined with a micrometer, hich is a device used to make highly precise measurements. The screw is supposed to be 3 inches in length and can vary from is measurement by no more than 0.1 inches or it will not fit properly into the furniture. The following table shows the screw ength measurements (in inches) taken each hour from each machine throughout the day. The screw length data from each achine has also been plotted

Screw Length Measurements (in inches)

Sample Time	Machine A	Machine B	Machine C
8 a.m.	2.98	2.92	2.99
9 a.m.	3.00	2.94	3.00
10 a.m.	3.02	2.97	3.01
11 a.m.	2.99	2.96	3.03
12 p.m.	3.01	2.94	3.05
1 p.m.	3.00	2.95	3.04
2 p.m.	2.97	2.93	3.06
3 p.m.	2.99	2.92	3.08
Mean			
Range			

Screw Length Measurements

Screw Length (in inches)
3.10
3.05
3.00
2.95
2.90
2.85
2.80
8 a.m. 9 a.m. 10 a.m. 11 a.m. 12 p.m. 1 p.m. 2 p.m. 3 p.m.
Sample Time
Machine A Machine B Machine C

. Calculate the mean and range of the data for each machine and place them in the bottom two rows of the table.

. If the screw length can vary from 3 inches by no more than 0.1 inch (plus or minus), what are the lowest and highest values for length that will be acceptable? Place a horizontal line on the graph at each of these values on the vertical axis. These are the tolerance or specification limits for screw length.

. Have any of the three machines produced an unacceptable part today? Are any of the machines close to making a bad part? If so, which one(s)?

. Look at the graph and the means from the table that show the average screw length produced by each machine. Draw a bold horizontal line on the graph at 3 to emphasize the target length. Do all the machines appear to be making parts that vary randomly around the target of 3 inches?

. Look at the range values from the table. Do any of the machines appear to have more variability in the length measurements than the others?

. In your opinion, which machine is performing best? Would you recommend that any adjustments be made to any of the machines? If so, which one(s) and why?

Math@Work

Dental Assistant

As a dental assistant, your job duties will vary depending on where you work. Suppose you work in a dental office where you assist with dental procedures and managing patients' accounts. When a patient arrives for their appointment, you will need to review their chart and make sure they are up to date on preventive care, such as X-rays and cleanings. When the patient leaves, you will need to fill out an invoice to determine how much to charge the patient for their visit.

Dental patients generally have a new X-ray taken yearly. Cleanings are performed every 6 months, although some patients have their teeth cleaned more often. The following table shows the date of the last X-ray and cleaning for three patients that are visiting the office today. (**Note:** All dates are within the past year.)

Patient Histories		
Patient	**Last X–ray**	**Last Cleaning**
A	April 15	October 20
B	June 6	January 12
C	October 27	October 27

During Patient A's visit, she received a fluoride treatment and a cleaning. Patient A has no dental insurance. During Patient B's visit, he received a filling on one surface of a tooth. Patient B has dental insurance which pays for 60% of the cost of fillings. During Patient C's visit, he had a cleaning, a filling on one surface of a tooth, and a filling on two surfaces of another tooth. Patient C has dental insurance which covers the full cost of cleanings and 50% of the cost of fillings.

Fee Schedule	
Procedure	**Cost**
Cleaning	$95
Fluoride treatment	$35
Filling, One surface	$175
Filling, Two surfaces	$235
X–ray, Panoramic	$110

1. Using today's date, determine which of the three patients are due for a dental cleaning in the next two months?

2. Using today's date, determine which of the patients will require a new set of X-rays during this visit.

3. Determine the amount each patient will be charged for their visit (without insurance). Don't forget to include the cost of any X-rays that are due during the visit.

4. Use the insurance information to determine the amount that each patient will pay out-of-pocket at the end of their visit.

Math@Work

Financial Advisor

As a financial advisor working with a new client, you must first determine how much money your client has to invest. The client may have a lump sum that they have saved or inherited, or they may wish to contribute an amount monthly from their current salary. In the latter case, you must then have the client do a detailed budget, so that you can determine a reasonable amount that the client can afford to set aside on a monthly basis for investment.

The second piece of information necessary when dealing with a new client is determining how much risk-tolerance they have. If the client is young or has a lot of money to invest, they may be willing to take more risk and invest in more aggressive, higher interest-earning funds. If the client is older and close to retirement, or has little money to invest, they may prefer less-aggressive investments where they are essentially guaranteed a certain rate of return. The range of possible investments that would suit each client's needs and goals are determined using a survey of risk-tolerance.

Suppose you have a client who has a total of $25,000 to invest. You determine that there are two investment funds that meet the client's investment preferences. One option is an aggressive fund that earns an average of 12% interest and the other is a more moderate fund that earns an average of 5% interest. The client desires to earn $2300 this year from these investments.

Investment Type	Principal Invested	·	Interest Rate	=	Interest Earned
Aggressive Fund	x				
Moderate Fund					

To determine the amount of interest earned you know to use the table above and the formula $I = Prt$, where I is the interest earned, P is the principal or amount invested, r is the average rate of return, and t is the length of time invested. Since the initial investment will last one year, $t = 1$.

1. Fill in the table with the known information. If x is the amount invested in the aggressive fund and the total amount to be invested is $25,000, create an expression involving x for the amount that will be left to invest in the moderate fund. Place this expression in the appropriate cell of the table.

2. Determine an expression in x for the interest earned on each investment type by multiplying the principal by the interest rate.

3. Determine the amount invested in each fund by setting up an equation using the expressions in column four and the fact that the client desires to earn $2300 from the interest earned on both investments.

4. Verify that the investment amounts calculated for each fund in the previous step are correct by calculating the actual interest earned in a year for each and making sure they sum to $2300.

5. Why would you not advise your client to invest all their money in the fund earning 12% interest, after all, it has the highest average interest rate?

Math@Work

Market Research Analyst

As a market research analyst, you may work alone at a computer, collecting and analyzing data, and preparing reports. You may also work as part of a team or work directly with the public to collect information and data. Either way, a market research analyst must have strong math and analytical skills and be very detail-oriented. They must have strong critical-thinking skills to assess large amounts of information and be able to develop a marketing strategy for the company. They must also possess good communication skills in order to interpret their research findings and be able to present their results to clients.

Suppose you work for a shoe manufacturer who wants to produce a new type of lightweight basketball sneaker similar to a product a competitor recently released into the market. You have gathered some sales data on the competitor in order to determine if this venture would be worthwhile, which is shown in the table below. To begin your analysis, you create a scatter plot of the data to see the sales trend. (A scatter plot is a graph made by plotting ordered pairs in a coordinate plane in order to show the relationship between two variables.) You determine that the x-axis will represent the number of weeks after the competitors new sneaker went on the market and the y-axis will represent the amount of sales in thousands of dollars.

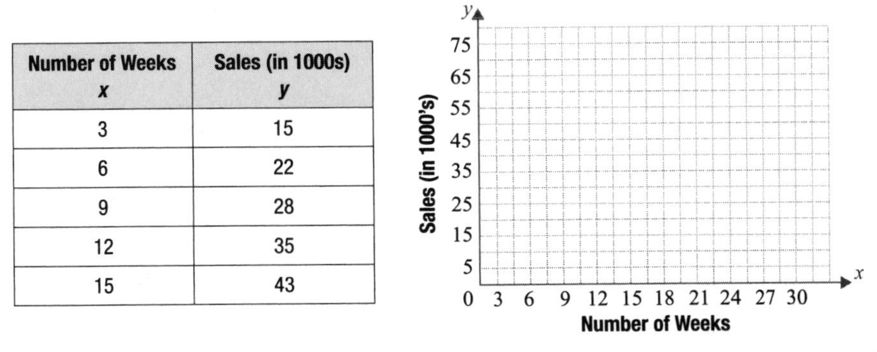

Number of Weeks x	Sales (in 1000s) y
3	15
6	22
9	28
12	35
15	43

1. Create a scatter plot of the sales data by plotting the ordered pairs in the table on the coordinate plane. Does the data on the graph appear to follow a linear pattern? If so, sketch a line that you feel would "best" fit this set of data. (A market research analyst would typically use computer software to perform a technique called regression analysis to fit a "best" line to this data.)

2. Using the ordered pairs corresponding to weeks 9 and 15, find the equation of a line running through these two data points.

3. Interpret the value calculated for the slope of the equation in Problem 2 as a rate of change in the context of the problem. Write a complete sentence.

4. If you assume that the sales trend in sneaker sales follows the model determined by the linear equation in Problem 2, predict the sneaker sales in 6 months. Use the approximation that 1 month is equal to 4 weeks.

5. Give at least two reasons why the assumption made in Problem 4 may be invalid?

Math@Work

Chemistry

As a pharmaceutical chemist, you will need an advanced degree in pharmaceutical chemistry, which combines biology, biochemistry, and pharmaceuticals. In this career, you will most likely spend your day in a lab setting creating new medications or researching their effectiveness. You will often work as part of a team working towards a joint goal. As a result, in addition to strong math skills and an understanding of chemistry, you will need to have good communication and leadership skills. Since you will be working directly with chemicals, you will also need to have a strong understanding of lab safety rules to ensure the safety of not only yourself, but your coworkers as well.

Suppose you work at a pharmaceutical company which creates and produces medications for various skin conditions. You are currently on a team which is developing an acne-controlling facial cleanser. Your team is working on determining the gentlest formula possible that is still effective so that the cleanser can be used on sensitive skin. Half of your team is working with salicylic acid and the other half is working with benzoyl peroxide.

As a part of your work, you will need to keep up on current research. Learning about new chemicals, new methods, and new research will be a continuous part of your life.

1. Perform an Internet search for benzoyl peroxide. How does it work to clean skin and prevent acne?

2. Perform an Internet search for salicylic acid. How does it work to clean skin and prevent acne?

3. Based on your research, which chemical seems better suited to treat acne on sensitive skin?

Another aspect of your career will involve the mixing of chemicals to create new compounds. Having the correct concentrations of chemicals is also important so the resulting solution works as you expect it to. When you don't have the correct concentration of a chemical in stock, it is possible to mix two concentrations together to obtain the desired concentration.

4. Your team wants to create a cleanser with 4% benzoyl peroxide. The lab currently has 2.5% and 10% concentrations of benzoyl peroxide in stock. To create 500 mL of 4% benzoyl peroxide, how much of each concentration should be combined?

Math@Work

Astronomy

Astronomy is the study of celestial bodies, such as planets, asteroids, and stars. While you work in the field of astronomy, you will use knowledge and skills from several other fields, such as mathematics, physics, and chemistry. An important tool of astronomers is the telescope. Several powerful telescopes are housed in observatories around the world. One of the many things astronomers use observatories for is discovering new celestial objects such as a near-Earth object (NEO). NEOs are comets, asteroids, and meteoroids that orbit the sun and cross the orbital path of Earth. The danger presented by NEOs is that they may strike the Earth and result in global catastrophic damage. (**Note:** The National Aeronautics and Space Administration (NASA) keeps track of all NEOs which are a potential threat at the website http://neo.jpl.nasa.gov/risk/)

For an asteroid to be classified as an NEO, the asteroid must have an orbit that partially lies within 0.983 and 1.3 astronomical units (AU) from the sun, where 1 AU is the furthest distance from the Earth to the sun, approximately 9.3×10^7 miles.

Near-Earth Object Distance			
	Minimum		**Maximum**
Distance in AU	0.983 AU	1 AU	1.3 AU
Distance in Miles		9.3×10^7 miles	

Suppose you discover three asteroids that you suspect may be NEOs. You perform some calculations and come up with the following facts. The furthest that Asteroid A is ever from the sun is 81,958,000 miles. The closest Asteroid B is ever to the sun is 12,529,000 miles. The closest Asteroid C is ever to the sun is 92,595,000 miles.

1. To determine if any of the asteroids pass within the range to be classified as an NEO, fill in the missing values from the table.

2. Based on the measurements from Problem 1, do any of the three asteroids qualify as an NEO?

There are two scales that astronomers use to explain the potential danger of NEOs. The Torino Scale is a scale from 0 to 10 that indicates the chance that an object will collide with the Earth. A rating of 0 means there is an extremely small chance of a collision and a 10 indicates that a collision is certain to happen. The Palermo Technical Impact Hazard Scale is used to rate the potential impact hazard of an NEO. If the rating is less than –2, the object poses a very minor threat with no drastic consequences if the object hits the Earth. If the rating is between –2 and 0, then the object should be closely monitored as it could cause serious damage.

Go to the NASA website http://neo.jpl.nasa.gov/risk/ to answer the following questions.

3. Does any NEO have a Torino Scale rating higher than 0? If so, what is the object's designation (or name) and during which year range could a potential impact occur?

4. Which NEO has the highest Palermo Scale rating? During which year range could a potential impact occur?

Math@Work

Math Education

As a math instructor at a public high school, your day will be spent preparing class lectures, grading assignments and tests, and teaching students with a wide variety of backgrounds. While teaching math, it is your job to explain the concepts and skills of math in a variety of ways to help students learn and understand the material. As a result, a solid understanding of math and strong communication skills are very important. Teaching math is a challenge and being able to understand the reasons that students struggle with math and empathize with these students is a critical aspect of the job.

Suppose that the next topics you plan to teach to your algebra students involve finding the greatest common factor and factoring by grouping. To teach these skills, you will need to plan how much material to cover each day, choose examples to walk through during the lecture, and assign in-class work and homework. You decide to spend the first day on this topic explaining how to find the greatest common factor of a list of integers.

1. It is usually easier to teach a group of students a new topic by initially showing them a single method. If a student has difficulty with that method, then showing the student an alternative method can be helpful. Which method for finding the greatest common factor would you teach to the class during the class lecture?

2. On a separate piece of paper, sketch out a short lecture on finding the greatest common factor of a list of integers. Be sure to include examples that range from easy to difficult.

3. While the class is working on an in-class assignment, you find that a student is having trouble following the method that you taught to the entire class. Describe an alternative method that you could show the student.

4. From your experience with learning how to find the greatest common factor of a list of integers, what do you think are some areas that might confuse students and cause them to struggle while learning this topic? Explain how understanding the areas that might cause confusion can help you become a better teacher.

Math@Work

Physics

As an employee of a company that creates circuit boards, your job may vary from designing new circuit boards, setting up machines to mass produce the circuit boards, to testing the finished circuit boards as part of quality control. Depending on your position, you may work alone or as part of a team. Regardless of who you work with, you will need strong math skills to be able to create new circuit board designs and strong communication skills to describe the specifications for a new circuit board design, describe how to set up the production line, or explain why a part is faulty.

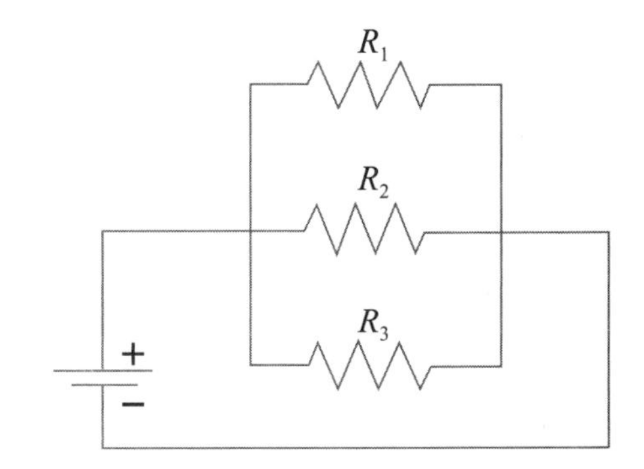

Suppose your job requires you to create new circuit boards for a variety of electronic equipment. The latest circuit board that you are designing is a small part of a complicated device. The circuit board you create has three resistors which run in parallel, as shown in the diagram.

Two of the resistors were properly labeled with their correct resistance, which is measured in ohms. The first resistor has a rating of 2 ohms. The second resistor has a rating of 3 ohms. The third resistor was taken from the supply shelf for resistors of a certain rating, but the resistor was unlabeled. As a result, you are unsure if it has the correct resistance for the current you want to produce. You use an ohmmeter, a device that measures resistance in a circuit, to determine that the total resistance of the circuit you created is $\frac{30}{31}$ ohms.

You know that the equation to determine the total resistance R_t is $\frac{1}{R_t}=\frac{1}{R_1}+\frac{1}{R_2}+\frac{1}{R_3}$, where R_1 is the resistance of the first resistor, R_2 is the resistance of the second resistor, and R_3 is the resistance of the third resistor.

1. Use the formula to determine the resistance of the third resistor given that the total resistance of the circuit is $\frac{30}{31}$ ohms.

2. Was the third resistor on the correct shelf if you took it from the supply shelf that holds resistors with a rating of 7 ohms?

3. What would be the total resistance of the circuit if the third resistor had a rating of 7 ohms?

4. What do you think would happen if the resistance of the unlabeled resistor wasn't determined and the circuit board was sent to the production line to be mass produced?

Math@Work

Forensic Scientist

As a forensic scientist, you will work as part of a team to investigate the evidence from a crime scene. Every case you encounter will be unique and the work may be intense. Communication is especially important because you will need to be clear and honest about your findings and your conclusions. A suspect's freedom may depend on the conclusions your team draws from the evidence.

Suppose the most recent case that you are involved in is a hit-and-run accident. A body was found at the side of the road with skid marks nearby. The police are unsure if the cause of death of the victim was vehicular homicide. Among the case description, the following information is provided to you.

Accident Report	
Date:	June 14
Time:	9:30 pm
Climate:	55 degrees Fahrenheit, partly cloudy, dry
Description of crime scene:	
Victim was found at the side of a road. Body temperature upon arrival is 84.9 °F. Posted speed limit is 30 mph. Road is concrete. Conditions are dry. Skid marks near the body are 88 feet in length.	

Known formulas and data:

A body will cool at a rate of 2.7 °F per hour until the body temperature matches the temperature of the environment. Average human body temperature is 98.6 °F.

Impact Speed and Risk of Death	
Impact Speed	**Risk of Death**
23 mph	10%
32 mph	25%
42 mph	50%
58 mph	90%
Source: 2011 AAA Foundation for Traffic Safety "Impact Speed and Pedestrian's Risk of Severe Injury or Death"	

Braking distance is calculated using the formula $\frac{s}{\sqrt{l}} = k$, where s is the initial speed of the vehicle in mph, l is the length of the skid marks in feet, and k is a constant that depends on driving conditions. Based on the driving conditions on that road for the last 12 hours, $k = \sqrt{20}$.

1. Based on the length of the skid marks, how fast was the car traveling before it attempted to stop? Round to the nearest whole number.

2. Based on the table, what percent of pedestrians die after being hit by a car moving at that speed?

3. Based on the cooling of the body, if the victim died instantly, how long ago did the accident occur? Round to the nearest hour.

4. Can you think of any other factors that should be taken into consideration before determining whether the impact of the car was the cause of death?

Math@Work

Other Careers in Mathematics

Earning a degree in mathematics or minoring in mathematics can open many career pathways. While a degree in mathematics or a field which uses a lot of mathematics may seem like a difficult path, it is something anyone can achieve with practice, patience, and persistence. Three growing fields of study which rely on mathematics are actuarial science, computer science, and operations research. While each of these fields involves mathematics, they require special training or additional education outside of a math degree. A brief description of each career is provided below along with a source to find more information about these careers.

Growing Fields of Study

Actuarial Science: The field of actuarial science uses methods of mathematics and statistics to evaluate risk in industries such as finance and insurance. Visit www.beanactuary.org for more information

Computer Science: From creating web pages and computer programs to designing artificial intelligence, computer science uses a variety of mathematics. Visit computingcareers.acm.org for more information.

Operations Research: The discipline of operations research uses techniques from mathematical modeling, statistical analysis, and mathematical optimization to make better decisions, such as maximizing revenue or minimizing costs for a business. Visit www.informs.org for more information.

There are numerous careers that have not been discussed in this workbook. Exploring career options before choosing a major is a very important step in your academic career. Learning about the career you are interested in before completing your degree can help you choose courses that will align with your career goals. You should also explore the availability of jobs in your chosen career and whether you will have to relocate to another area to be hired. The following web sites will help you find information related to different careers that use mathematics. Another great resource is the mathematics department at your college.

The **Mathematical Association of America** has a website with information about several careers in mathematics. Visit www.maa.org/careers to learn more.

The **Society for Industrial and Applied Mathematics** also has a webpage dedicated to careers in mathematics. Visit www.siam.org/careers to learn more.

The **Occupational Outlook Handbook** is a good source for information on educational requirements, salary ranges, and employability of many careers, not just those that involve mathematics. Visit http://www.bls.gov/ooh/ to learn more.

Answer Key

Chapter 1: Whole Numbers

1.1 Exercises

Concept Check

1. True
3. True

Practice

5. 2: ten thousands, 4: thousands, 6: hundreds, 8 ones
7. 537,082

Applications

9. 8520; 2000

Writing & Thinking

11. Hyphens are used for two-digit numbers larger than 20 that do not end in a zero

1.2 Exercises

Concept Check

1. False; A polygon has three or more sides.
3. False; Borrowing must occur.

Practice

5. 58
7. 144
9. 31 m

Applications

11. $39,100

Writing & Thinking

13. If a sum of digits is greater than 9, the tens digit of the sum should be added to the column to the left; Examples will vary.

1.3 Exercises

Concept Check

1. False; The numbers being multiplied are called factors.
3. True

Practice

5. 2352
7. $y = 7$; Associative property of multiplication
9. 63 square meters

Applications

11. 8928 slices of bread

Writing & Thinking

13. The distributive property distributes multiplication to two (or more) numbers that are being added; Examples will vary.

1.4 Exercises

Concept Check

1. False; If a division problem has a zero remainder...
3. False; $12 \div 0$ is undefined.

Practice

5. Undefined
7. 9

Applications

9. 16 grams

Writing & Thinking

11. To check a division problem, multiply the quotient and divisor, and then add the remainder. The result should equal the original dividend.

1.5 Exercises

Concept Check

1. True
3. True

Practice

5. 220; 223
7. 40,000; 43,680

Applications

9. $40,000; $35,316

Writing & Thinking

11. Estimation uses rounded values to find an approximate sum, difference, product, etc. Answers will vary.

1.6 Exercises

Concept Check

1. True
3. False; Quotient indicates division.

Applications

5. 1103 calories
7. 380 sq in.

Writing & Thinking

9. Answers will vary.

1.7 Exercises

1. False; Evaluating expressions and solving equations are related concepts.
3. False; The solution to the equation $n + 3 = 10$ is 7.

Practice

5. 10 is not a solution
7. $x = 30$
9. $9 = y$

Applications

11. 9 pounds

1.8 Exercises

Concept Check

1. False; Equals 81
3. False; 7^0 is 1.

Practice

5. **a.** 2 **b.** 3 **c.** 8
7. 2

Applications

9. **a.** No. Here it shows that we are only dividing the old trading cards by 6 friends versus both the old and new trading cards by 6 friends.

   b. 522; $\frac{15 \cdot 10 \cdot 20 + 132}{6}$

Writing & Thinking

11. If addition is within parentheses (or other grouping symbols), addition would be performed first.

1.9 Exercises

Concept Check

1. True
3. False; 7605 is divisible by 5.

Practice

5. 3, 5
7. None

Applications

9. 5 people would raise $2480 each; 10 people would raise $1240 each.

Writing & Thinking

11. **a.** 30, 45; Answers will vary.

    b. 9, 12; Answers will vary.

    c. 10, 25; Answers will vary.

1.10 Exercises

Concept Check

1. False; A prime number has exactly 2 factors.
3. False; 231 is a composite number.

Practice

5. Prime
7. 5^3

Applications

9. 1, 2, 3, 4, 6, 8, 12, 24

Writing & Thinking

11. No, some odd numbers are the product of two or more odd prime factors, for example, $3 \cdot 3 = 9$, $3 \cdot 5 = 15$, $3 \cdot 7 = 21$, etc.

Chapter 2: Integers

2.1 Exercises

Concept Check

1. False; If −8 lies to the right of a number on a number line, then −8 is greater than that number.
3. True

Practice

5. 

7. <
9. 4

Applications

11. −4500 meters

Writing & Thinking

13. If y represents a negative number, then $-y$ represents its opposite, a positive number. For example, if $y = -2$, then $-y = -(-2) = 2$.

2.2 Exercises

Concept Check

1. False; When adding numbers with unlike signs, the answer can be negative or positive, depending on numbers used.
3. False; The additive inverse of negative seven is 7.

Practice

5. −15
7. −8
9. 0

Applications

11. −275 ft (275 feet below sea level)

Writing & Thinking

13. Sometimes. Examples will vary.

2.3 Exercises

Concept Check

1. True
3. True

Practice

5. −5
7. −2

Applications

9. 5°

Writing & Thinking

11. Answers will vary. When the absolute value of the number being subtracted is greater than the absolute value of the other number, the difference will be positive.

2.4 Exercises

Concept Check

1. False; The product of zero and an integer is 0.
3. True

Practice

5. 0
7. −4
9. 39

Applications

11. 77

Writing & Thinking

13. $\left(3^2 - 9\right) = 0$ and division by 0 is undefined.

2.5 Exercises

Concept Check

1. True
3. False; In the term "$12a$," 12 is the coefficient.

Practice

5. −5, 3, and 8 are like terms; $7x$ and $9x$ are like terms.
7. $2x^2 + 2x$
9. $-x^2 + 5x - 7; -21$

Applications

11. 0 feet

Writing & Thinking

13. -13^2 is the square of 13 multiplied by −1 while $(-13)^2$ is the square of −13. This means that $-13^2 = -169$ and $(-13)^2 = 169$.

2.6 Exercises

Concept Check

1. True
3. False; Subtraction is indicated by the phrase "five less than a number."

Practice

5. $x + 6$
7. $\frac{x}{2} - 18$
9. **a.** $4n - 6$

   b. $6 - 4n$
11. The product of a number and negative nine

Writing & Thinking

13. The Commutative Property of Addition and Multiplication permits the order of items being added or multiplied to change and still have the same result. This property does not hold true for subtraction or division. Therefore, order is important for subtraction and division problems or the answer will change or be incorrect.

2.7 Exercises

Concept Check

1. True
3. False; When solving an equation of the form $ax + b = c$, if a variable has a constant coefficient other than 1, use the division principle to divide both sides by the coefficient

Practice

5. $x = 5$
7. $y = -4$

Applications

9. $t = -3$ degrees per hour

Writing & Thinking

11. Write the equation.
Simplify.
Divide both sides by 9.
Simplify.

Chapter 3: Fractions, Mixed Numbers, and Proportions

3.1 Exercises

Concept Check

1. False; In $\frac{11}{13}$, the numerator is 11.

3. True

Practice

5.

7.

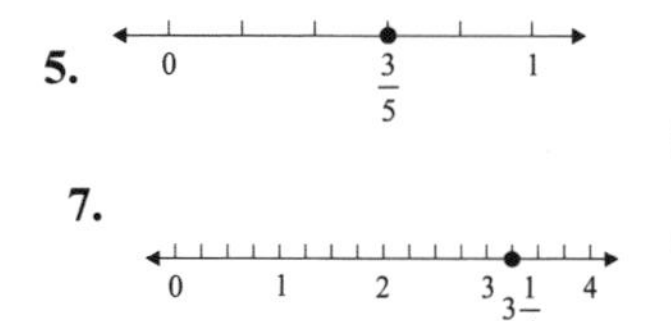

Applications

9. $\frac{6}{35}; \frac{29}{35}$

Writing & Thinking

11. The two parts are the numerator and the denominator. The denominator represents the number of pieces in a whole and the numerator represents the number of these pieces being considered.

3.2 Exercises

Concept Check

1. True

3. False; The statement $\frac{1}{3} \cdot \frac{2}{5} = \frac{2}{5} \cdot \frac{1}{3}$ is an example of the commutative property of multiplication.

Practice

5. $\frac{1}{4}$

7. 1

Applications

9. $\frac{3}{8}$

Writing & Thinking

11. No. If a fraction is less than 1 then its product with another number will be less than that other number. So, if the other number is less than 1, the product will be less than 1. Answers will vary.

3.3 Exercises

Concept Check

1. False; The reciprocal of 1 is 1.

3. False; The reciprocal of -12 is $-\frac{1}{12}$.

Practice

5. $\frac{8}{9}$

7. Undefined

9. -1

Applications

11. **a.** More

b. Less

c. 200 passengers

Writing & Thinking

13. No. For example, $\frac{4}{5} \neq \frac{5}{4}$.

3.4 Exercises

Concept Check

1. True

3. False; The mixed number $4\frac{1}{5}$ is equal to $\frac{21}{5}$.

Practice

5. $2\frac{1}{6}$

7. 4

Applications

9. **a.** $22\frac{1}{2}$ gallons **b.** \$45

11. **a.** The product will always be smaller than the other number if the number is a positive fraction. Examples will vary.

b. The product will always be smaller if it is multiplied by a positive whole number. Examples will vary.

c. If the number is negative, the product will always be greater than the other number. Examples will vary.

3.5 Exercises

Concept Check

1. False; The LCM of 15 and 25 is 75.

3. False; The first five multiples of 4 are 4, 8, 12, 16, and 20.

Practice

5. 30

7. **a.** LCM = 490

b. $490 = 14 \cdot 35 = 35 \cdot 14 = 49 \cdot 10$

9. 15

Applications

11. **a.** 60 minutes

b. 4, 3, and 2 trips, respectively

Writing & Thinking

13. Since the LCM is constructed using the prime factors of each number in the set, by definition, each number will divide the LCM.

3.6 Exercises

Concept Check

1. True

3. False; LCD stands for least common denominator.

5. True

Practice

7. $\frac{17}{21}$

9. $-\frac{3}{20}$

Applications

11. 1 ounce

Writing & Thinking

13. The LCM finds the least common multiple of a set of numbers. The LCD does the same thing for the set of numbers determined by the denominators.

3.7 Exercises

Concept Check

1. True

3. False; LCDs are required when adding or subtracting mixed numbers.

Practice

5. $10\frac{3}{8}$

7. $4\frac{3}{5}$

9. $4\frac{1}{4}$

Applications

11. $4\frac{13}{20}$ parts

3.8 Exercises

Concept Check

1. True
3. True

Practice

5. $\frac{4}{10}$ by $\frac{1}{40}$

7. $\frac{2}{3}$

9. -2

Applications

11. $\frac{85}{9}=9\frac{4}{9}$ feet per hour

Writing & Thinking

13. **a.** Yes; If both fractions are greater than one half, the sum will be greater than one

 b. No; Multiplying a number by a fraction between 0 and 1 results in a product that is less than the original number.

3.9 Exercises

Concept Check

1. True
3. False; The equation $\frac{3}{4}x=\frac{8}{15}$ can be solved by multiplying both sides of the equation by $\frac{4}{3}$.

Practice

5. $x=\frac{1}{3}$
7. $x=-210$

Applications

9. 144 students

Writing & Thinking

11. The LCD is divisible by each of the denominators, so multiplying the equation by it will force the denominators to be cancelled out; Answers will vary.

3.10 Exercises

Concept Check

1. True
3. False; The ratio 8:2 can be reduced to the ratio 4:1.

Practice

5. $\frac{9}{14}$

7. $\frac{\$2\text{ profit}}{\$5\text{ invested}}$

9. Answer: \$6/shirt; \$4/shirt; 5 shirts for \$20

Applications

11. $\frac{9}{41}$

13. Numerator

3.11 Exercises

Concept Check

1. False; A proportion is a statement that two ratios are equal.
3. False; When using proportions to solve a word problem, there are many correct ways to set up the proportion.

Practice

5. True
7. $x=10$

Applications

9. They are the same.

Writing & Thinking

11. A proportion has been set up correctly if the same units are in the same location in both ratios.

3.12 Exercises

Concept Check

1. False; The individual result of an experiment is an outcome.
3. True

Applications

5.

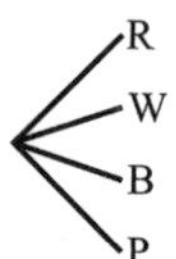

$S=\{R, W, B, P\}$
R = red, W = white,
B = blue, P = purple

7. $\frac{2}{5}$

Writing & Thinking

9. Chance experiments include, but are not limited to, tossing a coin, spinning a bottle, drawing a card from a standard deck of cards, picking numbers in the lottery, choosing straws, and picking colored marbles.

Chapter 4: Decimal Numbers

4.1 Exercises

Concept Check

1. True
3. False; On a number line, any number to the right of another number is larger than that other number.
5. 2.57
7. 6.028
9. **a.** 5

 b. 2

 c. 2, 5, 2

 d. 3.0065

Applications

11. Two and eight-hundred twenty-five ten-thousandths

Writing & Thinking

13. Moving left to right, compare digits with the same place value. When one compared digit is larger, the corresponding number is larger.

4.2 Exercises

Concept Check

1. True
3. False; In subtracting decimal numbers, line up the decimal points and corresponding digits vertically.

Practice

5. 50.085
7. 3.3

Applications

9. **a.** \$94.85

 b. \$5.15

Writing & Thinking

11. Decimal numbers need to be aligned vertically so that numbers with the same place value are being added together. If not, then a 60 may be added to a 7 as if it were a 6 being added to a 7, giving 13,

not the value of 67 that it should be.

4.3 Exercises

Concept Check

1. False; The decimal points do not need to be aligned vertically when multiplying decimal numbers.
3. False; Multiplying by 100 requires that the decimal point be moved 2 places to the right.

Practice

5. −0.112
7. −0.18

Applications

9. \$240.90

Writing & Thinking

11. In multiplication with decimal numbers, placement of the decimal point must be considered. Otherwise, multiplication with whole numbers and decimal numbers are the same.

4.4 Exercises

Concept Check

1. True
3. False; When estimating 16.469÷3.87, the answer would be 5.
5. False; According to the rules for order of operations, multiplication and division should be performed before addition and subtraction.

Practice

7. 20; 26.08
9. 2; 2.05

Applications

11. **a.** 39 pounds
    b. 35.43 pounds

4.5 Exercises

Concept Check

1. True
3. False; The number that appears the greatest number of times in a set of data is the mode.

Practice

5. **a.** 58
   b. 57
   c. 57
   d. 4
7. **a.** \$48,625
   b. \$46,500
   c. \$63,000
   d. \$43,000

Applications

9. 79

Writing & Thinking

11. The mean and median may be the same number as in the set of data: 12, 16, 20. However, more generally, the mean and median are two different numbers as in the set of data: 22, 23, 36. In this set, the median is 23, while the mean is 27.

4.6 Exercises

Concept Check

1. True
3. False; In some cases, fractions can be converted to decimal form without losing accuracy

Practice

5. $-\frac{17}{100}$
7. 6.67
9. 1.64

Applications

11. 17.92 inches

Writing & Thinking

13. For the numerator, write the whole number formed by all the digits of the decimal number, and for the denominator, write the power of 10 that corresponds to the rightmost digit. Reduce the fraction, if possible.

4.7 Exercises

Concept Check

1. False; When an equation is solved, the variable can be on the left side or the right side.

Practice

3. $y = 81.2$
5. $z = -4.16\overline{6}$

Applications

7. \$55.45 (video game); \$17.30 (game guide)

Writing & Thinking

9. Checking your answer at the end will ensure that the decimal point is placed properly in the solution; Answers will vary.

Chapter 5: Percents

5.1 Exercises

Concept Check

1. False; It is possible to have a percent greater than 100%.
3. False; To change from a percent to a decimal, move the decimal point two places to the left and omit the percent sign.

Practice

5. 20%
7. 0.6
9. $1\frac{1}{2}$

Applications

11. 0.0725

Writing & Thinking

13. 100% = 1 so anytime there is a mixed number, which has a value greater than 1, the percentage will be greater than 100%. Proper fractions (numerator is smaller than denominator) have a value less than 1 and therefore the percentage will be less than 100%.

5.2 Exercises

Concept Check

1. True
3. False; In the problem "What is 26% of 720?" the missing number is the amount.

Practice

5. 7.5
7. 15
9. 250

Applications

11. \$97,600

Writing & Thinking

13. Proportions would work for mixed numbers because a mixed number can be rewritten as a fraction. The only additional step required would be to change the mixed number to an improper fraction

and then solve the proportion as normal.

5.3 Exercises

Concept Check

1. False; In order to solve the equation $0.56 \cdot B = 12$ for the base, B, one would divide 12 by 0.56.
3. False; The solution to the problem "50% of what number is 352?" could be found by solving the equation $0.5 \cdot B = 352$.

Practice

5. 7
7. 42
9. 20

Applications

11. $175,000

Writing & Thinking

13. The amount is the number that is often near the word "is." The base is the number that often follows the word "of." The rate is the number written either as a fraction or as a decimal number that has not been identified as the amount or the base, and usually appears before the word "of."

5.4 Exercises

Concept Check

1. True
3. True

Practice

5. **a.** $82.50

   b. $192.50
7. $11,700

Writing & Thinking

9. Sales tax and tips are percentages of some item or service. The percent is the rate, while the cost of the item being purchased is the base. The amount is then the sales tax itself, which is being compared to the base. A sales tax might be 8%, as in "what is 8% of the cost of the item purchased?"

5.5 Exercises

Concept Check

1. True
3. False; Compound interest is earned on the principal and interest earned.

Applications

5. $112.50
7. $189.24

Writing & Thinking

9. The simple interest formula is $I = P \cdot r \cdot t$ where I is interest, P is principal, r is rate, and t is time. Interest is the amount of money paid for the use of money. The principal is the starting amount invested. Rate is the interest rate and should be written as a decimal or fraction. Time is the amount of time, in years, that interest is being earned on the principal. Time can be written as a decimal or fraction. When a decimal is used, it should only be when it is a terminating decimal so that no rounding is required, which could change the value calculated.

5.6 Exercises

Concept Check

1. True
3. True

Applications

5. **a.** Social Science

   b. Chemistry & Physics

   c. About 3300

   d. About 21.2%
7. **a.** February and May

   b. 6 inches

   c. March

   d. 3.58 inches

Writing & Thinking

9. All graphs should be **1.** clearly labeled, **2.** easy to read, and **3.** have appropriate titles.

Chapter 6: Measurement and Geometry

6.1 Exercises

Concept Check

1. True
3. True

Practice

5. 8
7. 13

Applications

9. $93.22

Writing & Thinking

11. Colby would need to know that there are 3 feet in a yard and 5280 feet in a mile.

6.2 Exercises

Concept Check

1. False; To change from smaller units to larger units, division must be used.
3. False; In metric units, a square that is 1 centimeter long on each side is said to have an area of 1 square centimeter.

Practice

5. 0.01977
7. 1300;130 000

Applications

9. 1750 railroad ties

Writing & Thinking

11. Each category of metric units has a base unit. The prefixes determine how many or what fraction of the base unit is being used. For example, the basic unit of length is meter and a millimeter is 1/1000 of a meter, a centimeter is 1/100 of a meter, and a kilometer is 1000 meters.

6.3 Exercises

Concept Check

1. True
3. False; In 1 liter there are 1000 milliliters.
5. True

Practice

7. 6300
9. 2

Applications

11. 20 cups

6.4 Exercises

Concept Check

1. False; Water freezes at 32 degrees Fahrenheit.
3. False; A 5k (km) run is shorter than a 5 mile run.

Practice

5. 77
7. 19.35
9. 72.75

Applications

11. 226.3 km

6.5 Exercises

Concept Check

1. True
3. True
5. True

Practice

7. **a.** Straight
   b. Right
   c. Acute
   d. Obtuse
9. **a.** 150°
   b. Yes; ∠2 and ∠3 are supplementary.
   c. ∠1 and ∠3; ∠2 and ∠4
   d. ∠1 and ∠2; ∠2 and ∠3; ∠3 and ∠4; ∠1 and ∠4
11. Equilateral

Applications

13. **a.** $m\angle Z = 80°$
    b. Acute
    c. $\overline{YZ}$
    d. $\overline{XZ}$ and $\overline{XY}$
    e. No, no angle is 90°

6.6 Exercises

Concept Check

1. **a.** True
   b. False; Not all rectangles have four equal sides.
3. True
5. **a.** F
   b. C
   c. A
   d. E
   e. D
   f. B

Practice

7. 188.4 cm
9. 35 ft

Applications

11. **a.** 4605 ft
    b. 15.7 minutes

Writing & Thinking

13. Some of the polygons are: triangle (3 sides), square (4 sides), rectangle (4 sides), parallelogram (4 sides), and trapezoid (4 sides).

6.7 Exercises

Concept Check

1. False; The $(b + c)$ in the trapezoid area formula represents the sum of the lengths of the two parallel bases.
3. False; The area formula for a triangle is $A = \frac{1}{2}bh$.

Practice

5. 81 ft^2
7. 160 $in.^2$
9. 52.56 km^{20}

Applications

11. 204 square feet

6.8 Exercises

Concept Check

1. True
3. True
5. **a.** C
   b. E
   c. D
   d. B
   e. A

Practice

7. 12.56 mm^3
9. 1017.36 mm^3

Applications

11. 800 ft^3

Writing & Thinking

13. Volume is measured in cubic units. Volume takes up a three-dimensional space and the units can be thought of as small cubes which leads to the concept of cubic units.

6.9 Exercises

Concept Check

1. False; Similar triangles have corresponding sides that are proportional.
3. False; If $\Delta ABC \cong \Delta DEF$, then $AC = DF$.

Practice

5. The triangles are not similar. The corresponding sides are not proportional.
7. $x = 50°; y = 70°$
9. Congruent by SAS

Applications

11. 7.5 feet

Writing & Thinking

13. **a.** $\angle B$ should be equal to $m\angle E$ and $m\angle C$ should be equal to $m\angle F$.
    b. Corresponding sides are proportional, not that they have the same length.

6.10 Exercises

Concept Check

1. True
3. True

Practice

5. 6
7. **a.** 3 and 4
   b. 3.6056
9. $c = 5$

Applications

11. 17.0 inches
13. The radical sign is $\sqrt{\ }$. The radicand is the number under the radical sign. A radical expression includes the radical sign and its radicand. For example, in the expression $\sqrt{36}$ the entire expression is called a radical expression, 36 is the radicand, and the symbol $\sqrt{\ }$ is called the radical sign.

Chapter 7: Solving Linear Equations and Inequalities

7.1 Exercises

Concept Check

1. False; The commutative property of addition allows the order to change.
3. False; The additive identity of all numbers is 0.

Practice

5. $3+7$
7. $4 \cdot 19$
9. $(16+9)+11$

Applications

11. **a.** \$118.25

    b. $\$11 \cdot \left(6\frac{1}{2}\right) + \$11 \cdot \left(4\frac{1}{4}\right)$

    c. Distributive property

7.2 Exercises

Concept Check

1. True
3. True

Practice

5. $x=1$ is not a solution
7. $x=7$

Applications

9. $y=3$
11. 1945 kanji characters

Writing & Thinking

13. **a.** Yes. It is stating that $6+3$ is equal to 9.

    b. No. If we substitute 4 for x, we get the statement $9=10$, which is not true.

7.3 Exercises

Concept Check

1. False; The addition and multiplication principles of equality can be used with decimal or fractional coefficients.
3. True

Practice

5. $x=-3$
7. $x=-\frac{27}{10}$

Applications

9. 14,000 tickets per hour

Writing & Thinking

11. **a.** The 4 should have been multiplied by 3 so that the 3 was distributed over the entire left-hand side of the equation; Correct answer is $x=15$.

    b. 3 should be subtracted from each side, not from each term, and $5x-3$ doesn't simplify to $2x$; Correct answer is $x=\frac{8}{5}$.

7.4 Exercises

Concept Check

1. True
3. False; It is called a contradiction.

Practice

5. $x=-5$
7. $x=-1$
9. Contradiction

Applications

11. 20 guests

Writing & Thinking

13. **a.** $5x+1$

    b. $x=6$

    c. Answers will vary.

7.5 Exercises

Concept Check

1. False; Case matters in formulas.
3. True

Applications

5. \$1030
7. $b=P-a-c$
9. $t=\frac{I}{Pr}$

7.6 Exercises

Concept Check

1. True
3. False; Odd integers are integers that are not even.

Practice

5. $x-5=13-x$; 9
7. $n+(n+1)+(n+2)=93$; 30, 31, 32

Applications

9. **a.** The unknown value is the length of the call in minutes.

    b. $m=20$

    c. The collect call lasted 20 minutes.

Writing & Thinking

11. **a.** $n, n+2, n+4, n+6$

    b. $n, n+2, n+4, n+6$

    c. Yes. Answers will vary.

7.7 Exercises

Concept Check

1. False; The value of r should be written as a decimal number.
3. True

Applications

5. 8.75 hours
7. \$112.50

Writing & Thinking

9. If each equal side is 9 cm long, that would make the perimeter more than 18 cm; Correct answer: 6 cm

7.8 Exercises

Concept Check

1. True
3. False; Only one value in the solution set needs to be checked.

Practice

5. 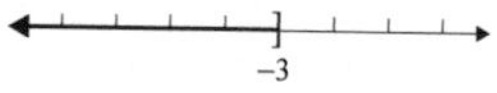

   Half-open interval

7. $(4,\infty)$

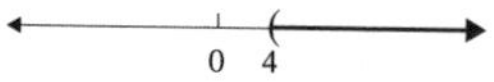

9. $[4,\infty)$

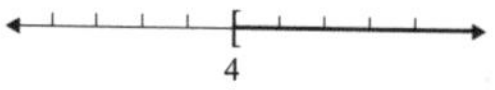

Applications

11. **a.** The student would need a score higher than 102 points, which is not possible. Thus he cannot earn an A in the course.

    b. The student must score at least 192 points to earn an A in the course.

Writing & Thinking

13. **a.** Answers will vary.

    b. Answers will vary.

7.9 Exercises

Concept Check

1. False; The union of two sets contains elements that belong to either one set, the other set, or both sets.
3. True

Practice

5. Union: $\{1, 2, 3, 4, 6, 8\}$; Intersection: $\{2, 4\}$

7. $(-1,\infty)\cap(-\infty,6)$

9. $(-9,1)$

7.10 Exercises

Concept Check

1. False; Equations involving absolute value can have more than one solution.

3. True

Practice

5. No solution

7. No solution

9. $x=-10, \dfrac{18}{7}$

7.11 Exercises

Concept Check

1. False; Only one statement/inequality must be true.

3. False; Must be greater than 2

Practice

5. $(-\infty,\infty)$

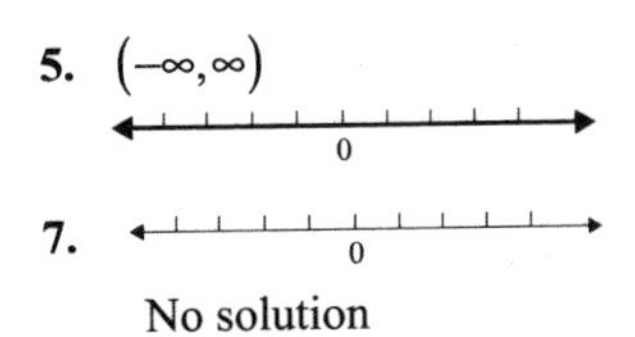

7. No solution

9. $\left(-\infty,-\dfrac{11}{3}\right]\cup[3,\infty)$

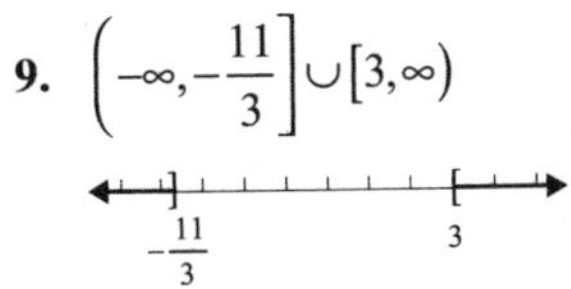

Writing & Thinking

11. a.

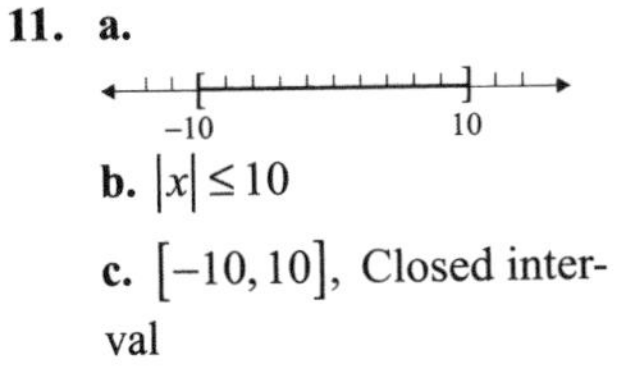

b. $|x|\le 10$

c. $[-10,10]$, Closed interval

Chapter 8: Graphing Linear Equations and Inequalities

8.1 Exercises

Concept Check

1. True

3. True

Practice

5. $\left\{\begin{matrix} A(-5,1), B(-3,3), \\ C(-1,1), D(1,2), \\ E(2,-2) \end{matrix}\right\}$

7. a. $(0,-1)$ b. $(4,1)$ c. $(2,0)$ d. $(8,3)$

9. b, c

Applications

11. a.

D	E
100	85
200	170
300	255
400	340
500	425

b.

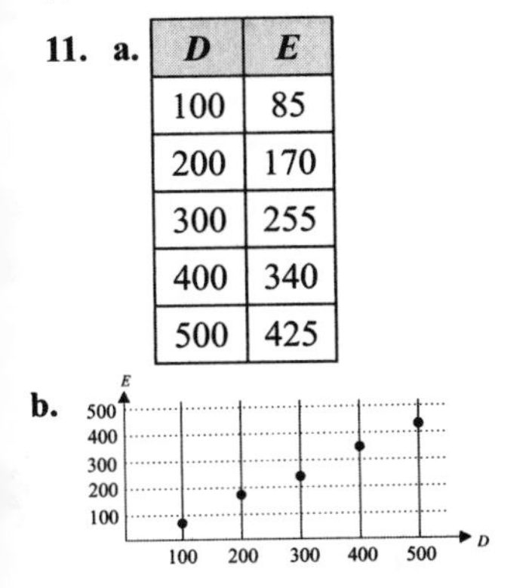

8.2 Exercises

Concept Check

1. True

3. False; Horizontal lines have y-intercepts

Practice

5., 7., 9.

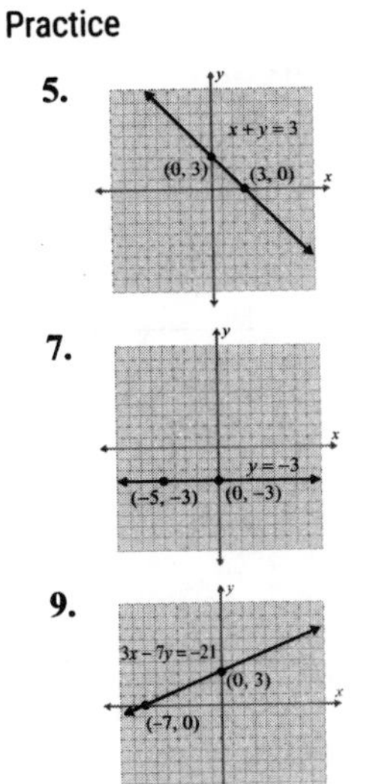

Applications

11. The y-intercept is $(0,30)$, meaning that if a student does no homework at all, the student will get a score of 30 points on the exam.

Writing & Thinking

13. Substitute the x and y values into the equation. Then evaluate both sides to see if the equation is true.

8.3 Exercises

Practice

1. m is undefined

3.

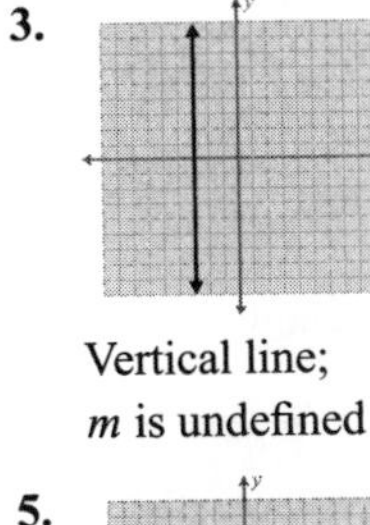

Vertical line; m is undefined

5.

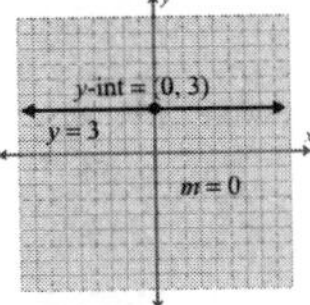

Applications

7. \$4000/year

Writing & Thinking

9. a. For any horizontal line, all of the y values will be the same. Thus the formula for slope will always have 0 in the numerator making the slope of every horizontal line 0.

b. For any vertical line, all of the x values will be the same. Thus the formula for slope will always have 0 in the

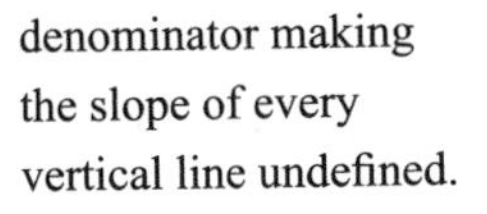

denominator making the slope of every vertical line undefined.

8.4 Exercises

Practice

1. negative reciprocals

3. $y=-\dfrac{1}{7}x+\dfrac{2}{7}$

5. $y=6$

Applications

7. a. $P=100t-5000$
b. 50 tickets

8.5 Exercises

Concept Check

1. True

3. True

Practice

5. $\left\{\begin{matrix}(-5,-4),(-4,-2), \\ (-2,-2),(1,-2),(2,1)\end{matrix}\right\}$; $D=\{-5,-4,-2,1,2\}$; $R=\{-4,-2,1\}$; Function

7. Not a function; $D=(-\infty,\infty)$; $R=(-\infty,\infty)$

9. a. -10 b. 86 c. 86

8.6 Exercises

Concept Check

1. True

3. False; The boundary line is solid if the inequality uses $\leq$ or $\geq$.

Practice

5.

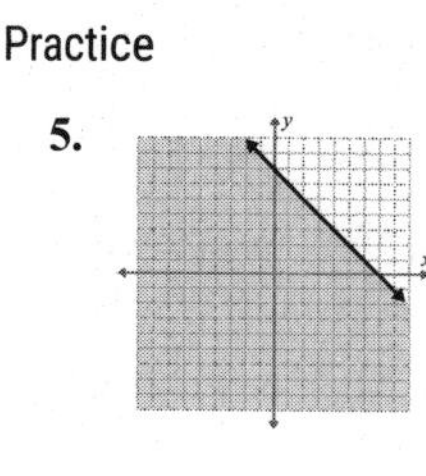

7.

9.

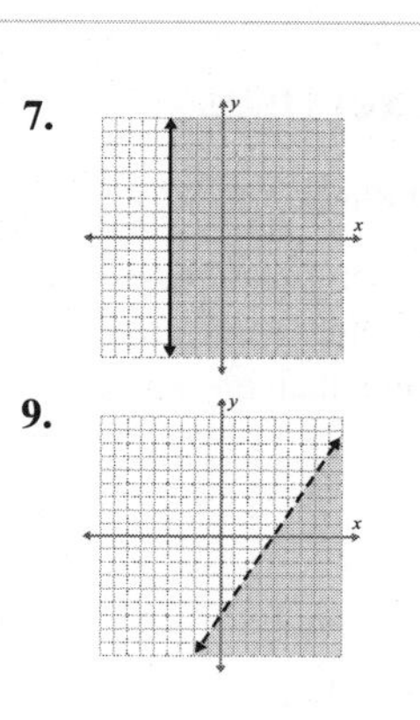

Writing & Thinking

11. Test any point not on the line. If the test point satisfies the inequality, shade the half-plane on that side of the line. Otherwise, shade the other half-plane.

Chapter 9: Systems of Linear Equations

9.1 Exercises

Concept Check

1. False; The solution must be checked in all equations.

3. True

Practice

5. a, c

7. (4, 0)

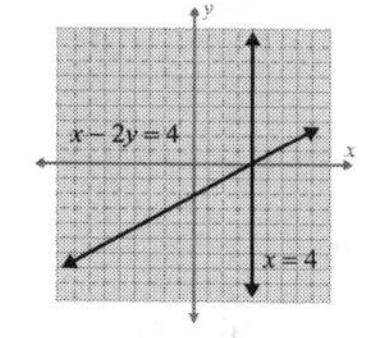

Applications

9.

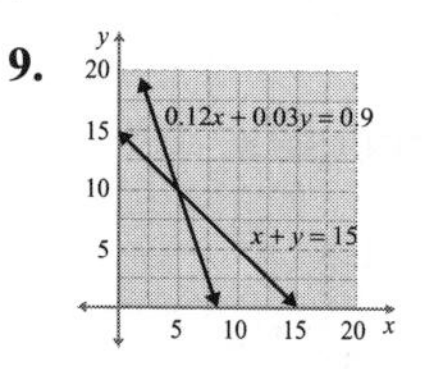

5 gallons of 12%; 10 gallons of 3%

Writing & Thinking

11. The solution to a consistent system of linear equations is a single point, which is easily written as an ordered pair.

9.2 Exercises

Concept Check

1. True

3. True

Practice

5. (2, 4)

7. No solution

Applications

9. 15 m × 10 m

Writing & Thinking

11. Answers will vary.

9.3 Exercises

Concept Check

1. False; The solution always needs to be checked in both original equations.

3. False; Both methods give exact solutions.

Practice

5. No solution

7. $(x, 2x - 4)$

Applications

9. 1200 adults; 3300 children

Writing & Thinking

11. Answers will vary.

9.4 Exercises

Practice

1. 14, 26

Applications

3. Boat, 38 mph; Current, 2 mph

5. 800 adults and 2700 students attended.

9.5 Exercises

Concept Check

1. False; When interest is calculated on an annual basis, $t = 1$

3. True

Practice

5. \$5500 at 6%; \$3500 at 10%

7. 40 liters of 12%; 50 liters of 30%

Writing & Thinking

9. **a.** Answers will vary.

b. Answers will vary.

9.6 Exercises

Concept Check

1. False; Choose 2 equations, eliminate 1 variable.

3. True

Practice

5. (1, 0,1)

7. No solution

Applications

9. 56, 84, 49

Writing & Thinking

11. No. Graphically, the three planes intersect at one point (one solution), or in a line (infinitely many solutions) or, they do not have a common intersection (no solution).

9.7 Exercises

Concept Check

1. False; A matrix that has 3 rows and 5 columns is a 3 x 5 matrix.

3. False; Interchanging two equations in a system of linear equations will not change the solution.

Practice

5. $\begin{bmatrix} 7 & -2 & 7 \\ -5 & 3 & 0 \\ 0 & 4 & 11 \end{bmatrix}$, $\left[\begin{array}{ccc:c} 7 & -2 & 7 & 2 \\ -5 & 3 & 0 & 2 \\ 0 & 4 & 11 & 8 \end{array}\right]$

7. (−1, 2)

Applications

9. 52, 40, 77

Writing & Thinking

11. Solving the second equation for z, we can back substitute into the first equation, eliminating z. The result is the equation $x + 5y = 6$, which means the system has an infinite number of solutions.

.8 Exercises

Concept Check

1. False; When boundary lines are parallel, the solution is either the strip between the boundary lines, a half-plane, or there is no solution.
3. True

Practice

5. 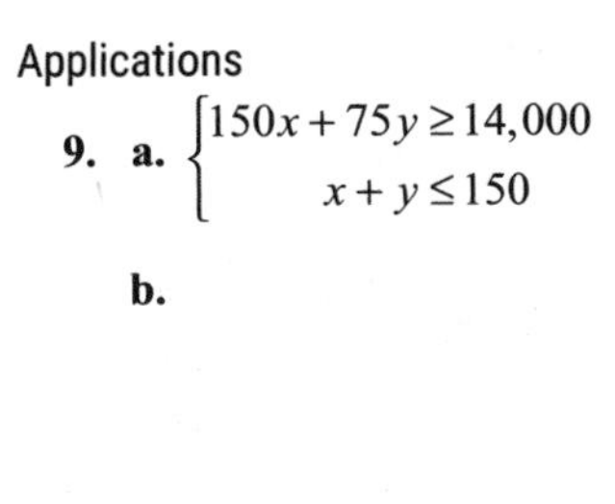

7. 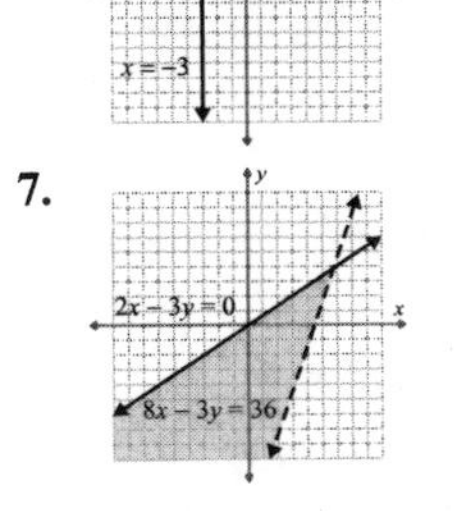

Applications

9. **a.** $\begin{cases} 150x + 75y \geq 14{,}000 \\ x + y \leq 150 \end{cases}$

   b. 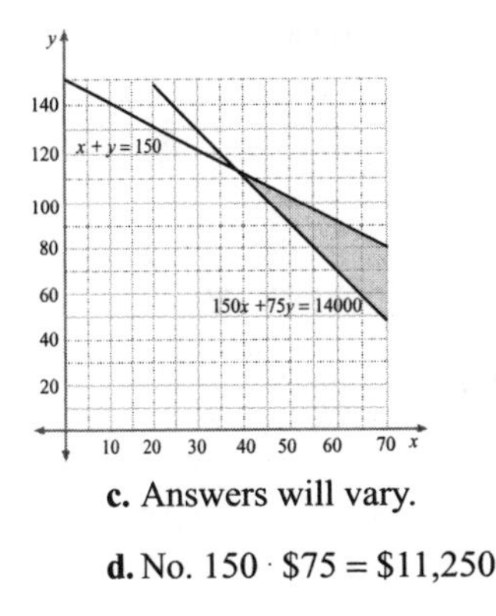

   c. Answers will vary.

   d. No. $150 \cdot \$75 = \$11{,}250$

Chapter 10: Exponents and Polynomials

10.1 Exercises

Concept Check

1. False; If there is no exponent written, the exponent is assumed to be 1.
3. True

Practice

5. y^{11}
7. $\frac{1}{x}$

Applications

9. 2^8 GBs

10.2 Exercises

Concept Check

1. True
3. False; The rules for exponents can be applied in any order, resulting in the same answer.

Practice

5. 64
7. $-\frac{2y^6}{27x^{15}}$
9. $\frac{y^8}{16x^8}$

10.3 Exercises

Concept Check

1. True
3. False; 3.53×10^5 is greater than 8.72×10^{-4}.

Practice

5. 8.6×10^4
7. 0.042
9. $(3 \times 10^{-4})(2.5 \times 10^{-6})$; 7.5×10^{-10}

Applications

11. 4.0678×10^{16} m

10.4 Exercises

Concept Check

1. False; Degree of 0
3. True

Practice

5. $5x^2 - x$; Second-degree binomial; Leading coefficient 5
7. **a.** −4
   b. −16
   c. −10

Applications

9. **a.** Second degree polynomial
   b. 44 feet
   c. 152 feet
   d. 324 feet

Writing & Thinking

11. He was correct. The expression is a sum/difference of monomials.

10.5 Exercises

Concept Check

1. False; All terms are subtracted.
3. False; They are like terms because they have the same variable raised to the same exponent.

Practice

5. $3x^2 + 7x + 2$
7. $x^2 + x + 6$
9. $-15x + 22$

Applications

11. **a.** $3.50x^3 + 4x^2 - 30$ dollars
    b. $6.50x^3 + 4x^2 - 20$ dollars
    c. \$460

Writing & Thinking

13. $(2x + 4) + (x^2 + 2x)$
    $= x^2 + 4x + 4$;
    $(x^9 + x^3) + (x^7 + x^5)$
    $= x^9 + x^7 + x^5 + x^3$
    Answers will vary.

10.6 Exercises

Concept Check

1. False; The distributive property can be used when multiplying any types of polynomials.
3. True

Practice

5. $-4x^8 + 8x^7 - 12x^4$
7. $y^3 + 2y^2 + y + 12$

Applications

9. $V = 10x^2 + 220x + 1200$ cubic inches

10.7 Exercises

Concept Check

1. False; The product will be a binomial.
3. True

Practice

5. $x^2 - 14x + 49$; Perfect square trinomial
7. $9x^2 - 12x + 4$; Perfect square trinomial

Applications

9. **a.** $A(x) = 400 - 4x^2$
   b. $P(x) = 4(20 - 2x)$

Writing & Thinking

11. $(x+5)^2 = x^2 + 2(5x) + 5^2$. Answers will vary.

Practice

5. $y^2 - 2y + 3$
7. $x - 6 + \dfrac{4}{x+4}$
9. $x^3 + 2x + 5 + \dfrac{17}{x-3}$

Applications

11. **a.** $x^2 - 4x - 5$ square inches
b. $x^2 - 3x - 10$ square inches

10.8 Exercises

Concept Check

1. True
3. False; Missing powers should be filled in with zeros.

10.9 Exercises

Concept Check

1. False; The divisor must have a leading coefficient of 1.
3. True

Practice

5. **a.** $x - 9$
b. $c = 3$; $P(3) = 0$
7. **a.** $4x^2 + 3x + 3 + \dfrac{16}{x-1}$
b. $c = 1$; $P(1) = 16$

Applications

9. **a.** $x^2 + 3x - 18$ square inches
b. $x^2 + 10x + 24$ square inches

Chapter 11: Factoring Polynomials

11.1 Exercises

Concept Check

1. False; Variables need to be considered as well.
3. True

Practice

5. 2
7. $7(2x+3)$
9. $(3+a)(x+y)$

Applications

11. **a.** 32 feet
b. $16x(3 - x)$
c. 32 feet
d. Yes. They are equivalent expressions.

11.2 Exercises

Concept Check

1. True
3. True

Practice

5. $(x - 9)(x + 3)$
7. $(x - 12)(x - 2)$

Applications

9. Base $= x + 48$; Height $= x$

Writing & Thinking

11. If the sign of the constant term is positive, the signs in the factors will both be positive or both be negative. If the sign of the constant term is negative, the sign in one factor will be positive and the sign in the ther factor will be negative.

11.3 Exercises

Concept Check

1. False; The middle term should be the sum of the inner and outer products.
3. False; The first step is to multiply *a* and *c*.

Practice

5. $(6x + 5)(x + 1)$
7. Not factorable
9. $2(2x - 5)(3x - 2)$

Writing & Thinking

11. This is not an error, but the trinomial is not completely factored. The completely factored form of this trinomial is $2(x + 2)(x + 3)$.

11.4 Exercises

Concept Check

1. True
3. False; The sum of two squares is not factorable.

Practice

5. $(5 - z)(5 + z)$
7. $(y + 6)(y^2 - 6y + 36)$
9. $4(x - 2)(x^2 + 2x + 4)$

Writing & Thinking

11. **a.** $xy + xy + x^2 + y^2$
$= x^2 + 2xy + y^2$
$= (x + y)^2$
b. $(x + y)(x + y)$
$= (x + y)^2$

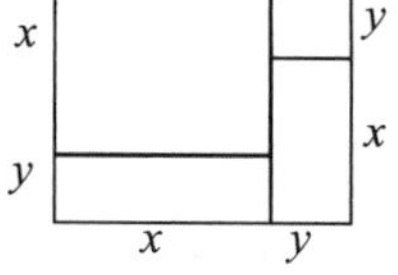

11.5 Exercises

Concept Check

1. False; The first step is to check for a common monomial factor.
3. False; It might be factorable by the grouping method.

Practice

5. $(x + 5)^2$
7. $(5x + 6)(4x - 9)$
9. $(x + 5)(x^2 - 5x + 25)$

11.6 Exercises

Concept Check

1. True
3. True

Practice

5. $x = 2, 9$
7. $x = -4, 6$

Applications

9. **a.** 640 ft; 384 ft
b. 144 ft; 400 ft
c. 7 seconds; $0 = -16(t + 7)(t - 7)$

Writing & Thinking

11. This allows for use of the zero factor property which says that for the product to equal zero one of the factors must equal zero. Answers will vary.

11.7 Exercises

Concept Check

1. False; The sum of the squares of the lengths of the legs is equal to the hypotenuse squared.
3. False: Only with right triangles

Applications

5. $x(x+10)=-25$; $x=-5$, so the numbers are -5 and 5.

7. $w(w+3)=54$; $w=6$, so width is 6 m and length is 9 m.

9. $x^2+(x-49)^2=(x+1)^2$; $x=60$, so height is 60 ft.

Chapter 12: Rational Expressions

12.1 Exercises

Concept Check

1. True
3. False; Rational numbers have zero denominators.

Practice

5. $\frac{3x}{4y}$; $x \neq 0, y \neq 0$
7. $\frac{x-3}{y-2}$; $y \neq -2, 2$

Applications

9. a. $p(x)=\frac{15x+200}{x}$
 b. $35
 c. $x \neq 0$
 d. The variable cannot be negative because you cannot have a negative quantity of people. There would also be a maximum number depending on the size of the room.

Writing & Thinking

11. a. A rational expression is an algebraic expression that can be written in the form $\frac{P}{Q}$ where P and Q are polynomials and $Q \neq 0$.
 b. $\frac{x-1}{(x+2)(x-3)}$ Answers will vary.
 c. $\frac{1}{x+5}$ Answers will vary.

12.2 Exercises

Concept Check

1. False; The reciprocal is $\frac{x+3}{x}$.
3. False; the restriction is 0.

Practice

5. $\frac{x+3}{x}$
7. $\frac{x}{12}$

Applications

9. a. $\frac{x^2-3x-10}{x+3}$
 b. $\frac{x^2+5x+6}{x-5}$
 c. $(x+2)^2=x^2+4x+4$

12.3 Exercises

Concept Check

1. False; The least common denominator needs to be found.
3. True

Practice

5. $\frac{23}{30}$
7. $7(x+5)(x-5)$
9. $33(x-3)$

12.4 Exercises

Concept Check

1. True
3. True

Practice

5. 2
7. $\frac{x^2-x+1}{(x+4)(x-3)}$

Applications

9. a. $\frac{7x^2+3}{x-2}$
 b. $\frac{4x^2+5}{x+2}$
 c. $\frac{22x^3+12x^2+16x-8}{(x+2)(x-2)}$

12.5 Exercises

Concept Check

1. True
3. True

Practice

5. $\frac{4}{5xy}$
7. $\frac{11}{2(1+4x)}$
9. $\frac{-5}{x+1}$

Applications

11. a. $\frac{4r+r^2}{4}$
 b. 0.0609
 c. 0.0609
 d. Yes; They are calculated from different forms of the same expression.
 e. 0.009
 f. Answers will vary. It is larger because the interest is compounded along with the principal.

12.6 Exercises

Concept Check

1. False; It is not a proportion.
3. True

Practice

5. $x \neq \frac{3}{2}, 3$; $x=-3$
7. $x \neq -9, -\frac{1}{4}, 0$; $x=-2, 1$
9. $LK=15$, $JB=5$

Applications

11. 6.8 cups

12.7 Exercises

Concept Check

1. False; The first step is to read carefully.
3. False; Use the formula $t=\frac{d}{r}$

Applications

5. 37.5 miles
7. 50 mph

12.8 Exercises

Concept Check

1. False; Varies directly
3. True

Practice

5. $\frac{7}{3}$
7. 120

Applications

9. 6 in.

Writing & Thinking

11. a. When two variables vary directly, an increase in the value of one variable indicates an increase in the other,

and the ratio of the two quantities is constant.

b. Joint variation is when a variable varies directly with more than one other variable.

c. When two variables vary inversely, an increase in the value of one variable indicates a decrease in the other, and the product of the two quantities is constant.

d. Combined variation is when a variable varies directly or inversely with more than one variable.

Chapter 13: Roots, Radicals, and Complex Numbers

13.1 Exercises

Concept Check

1. False; If the original number is negative, the principal square root will not be the same as the original number.

3. False; The radicand is underneath the radical symbol.

Practice

5. 7

7. 10

9. 0.2

Applications

11. **a.** 4 cm

b. 20 cm

Writing & Thinking

13. Cubing a negative real number is equivalent to multiplying a negative number by itself 3 times. The product of three negative numbers is negative.

13.2 Exercises

Concept Check

1. True

3. True

Practice

5. $9\sqrt{2}$

7. $2x^5y\sqrt{6x}$

9. $-2x^2\sqrt[3]{x^2}$

Applications

11. $\sqrt{3} \approx 1.73$ amperes

Writing & Thinking

13. A cube root has no restrictions as the cube root of a negative number is negative.

13.3 Exercises

Concept Check

1. True

3. False; If x is a real number, then $\sqrt{x^2} = |x|$

Practice

5. $\sqrt[3]{8}$

7. $\dfrac{1}{10}$

9. $a^{\frac{1}{4}}$

Applications

11. 576 ft²

Writing & Thinking

13. No:

$\sqrt[5]{a} \cdot \sqrt{a} = a^{\frac{7}{10}}, \sqrt[5]{a^2} = a^{\frac{2}{5}},$

$a^{\frac{7}{10}} \neq a^{\frac{2}{5}}$

13.4 Exercises

Concept Check

1. False; $\sqrt{a}$ and $\sqrt[3]{a}$ do not have the same index, so they are not like radicals.

3. True

Practice

5. $9\sqrt[3]{x} - \sqrt[3]{y}$

7. 18

9. $13 + 4\sqrt{10}$

Applications

11. $2\sqrt{10} + 4\sqrt{5} + \sqrt{30}$
≈ 20.75 in.

13.5 Exercises

Concept Check

1. True

3. False; You would need to multiply by $\sqrt[3]{a^2}$.

Practice

5. $\dfrac{-3\sqrt{7}}{7}$

7. $\sqrt{6} + 2$

Applications

9. $r = \dfrac{\sqrt{V\pi h}}{\pi h}$

Writing & Thinking

11. Multiply both the numerator and the denominator by the conjugate of the denominator. This works because multiplying the denominator by its conjugate results in an expression with no square roots. Answers will vary.

13.6 Exercises

Concept Check

1. False; Radical equations may also have one or no solution.

3. False; The process needs to be repeated until all radicals have been eliminated.

Practice

5. $x = 3$

7. $x = 4$

Applications

9. 12 ft

13.7 Exercises

Concept Check

1. True

3. True

Practice

5. **a.** $\sqrt{5} \approx 2.2361$

b. 3

c. $5\sqrt{2} \approx 7.0711$

d. 2

7. $(-\infty, \infty)$

9.

y, x, (7, 1), (6, 0), $(-8, \sqrt[3]{-14})$, (−2, −2), (5, −1)

Writing & Thinking

11. The domain of g is $[0, \infty)$, while the domain of f is $(-\infty, \infty)$.

13.8 Exercises

Concept Check

1. True

3. False; $a = c$, $b = d$

Practice

5. Real part: -11, imaginary part: $\sqrt{2}$

7. $x = 6, y = -3$

9. $6+2i$

Writing & Thinking

11. **a.** Yes

b. No

13.9 Exercises

Concept Check

1. False; $i, -i$, 1 and -1

3. False; The product is -1.

Practice

5. $-28-24i$

7. $0-\frac{5}{4}i$

9. $0+i$

Writing & Thinking

11. Given a complex number $(a+bi)$: $(a+bi)(a-bi)$ $= a^2 - abi + abi - b^2i^2$ $= a^2 + b^2$ which is the sum of squares of real numbers. Thus the product must be a positive real number.

Chapter 14: Quadratic Equations

14.1 Exercises

Concept Check

1. True

3. False; The square of a real number cannot be negative.

Practice

5. $x=\pm 7$

7. $x=\frac{7}{4}, \frac{9}{4}$

Applications

9. $2\sqrt{10}$ cm and $6\sqrt{10}$ cm

14.2 Exercises

Concept Check

1. True

3. True

Practice

5. $x=-5, 1$

7. $x=\frac{-1\pm i\sqrt{7}}{2}$ or $-\frac{1}{2}\pm\frac{\sqrt{7}}{2}i$

9. $x^2-7=0$

Applications

11. **a.** $p=0, 90$

b. No income revenue is made if the price is set to \$0 or \$90 per frame.

14.3 Exercises

Concept Check

1. True

3. False; Two real solutions

Practice

5. 68; Two real solutions

7. $x=-2\pm 2\sqrt{2}$

9. $x=1, \frac{4}{3}$

Writing & Thinking

11. $x^4-13x^2+36=0$; multiplied $(x-2)(x+2)(x-3)(x+3)$

14.4 Exercises

Concept Check

1. False; Application problems do not tell you direction which operations to perform.

Applications

3. $(w+12)(w+22)=1344$; $w=20$, the pool is 20 ft by 30 ft

5. $\frac{45}{12-c}-\frac{45}{12+c}=2$; $c=3$, the rate of the current is 3 mph

7. **a.** 4 sec

b. No, the projectile's maximum height is 256 ft.

c. 3 seconds, 5 seconds

d. 8 sec

9. a cannot be equal to zero and b^2 must be greater than or equal to $4ac$ to produce a real solution. Also, all the solutions found may not apply to the problem at hand. You must check that each answer makes sense in the context of the problem.

14.5 Exercises

Concept Check

1. True

3. True

Practice

5. $x=\pm 2, \pm 3$

7. $x=-\frac{7}{2}, -3$

Writing & Thinking

9. Using u-substitution where $u=\sqrt{x}$, we get u values of -2, and 3. However, $\sqrt{x}$ never equals -2, so we are left with only one solution, $x=9$.

14.6 Exercises

Concept Check

1. True

3. False; The bigger $|a|$ is, the narrower the opening of the parabola is.

Practice

5. $x=0$; $(0,-4)$

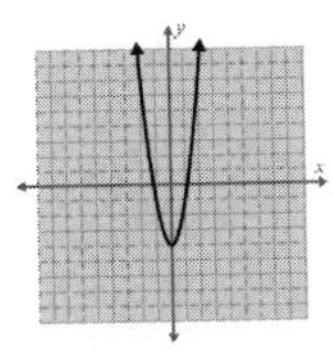

7. $x=5$; $(5,1)$

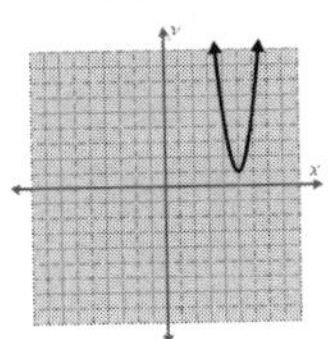

14.7 Exercises

Concept Check

1. False; The line of symmetry is at $x=-\frac{b}{2a}$.

3. True

Practice

5. $y=2(x-1)^2$; $x=1$; Vertex: $(1,0)$; x-int: $(1,0)$; y-int: $(0,2)$

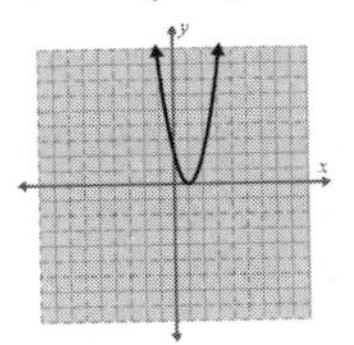

7. $x=1$; Vertex: $(1,0)$; x-int: $(1,0)$; y-int: $(0,-3)$

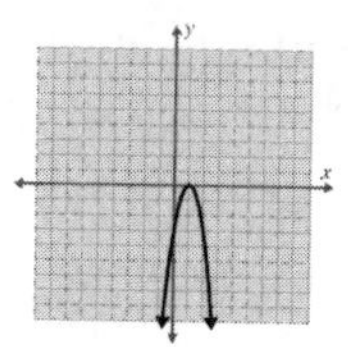

Applications

9. a. 3.5 s

b. 196 ft

Writing & Thinking

11. a. Parabola

b. $x=-\dfrac{b}{2a}$

c. $x=-\dfrac{b}{2a}$

d. No. A graph can be entirely above or below the x-axis.

14.8 Exercises

Concept Check

1. False; The goal is to get 0 on one side of the inequality and factor the other side.

3. False; The solution consists of all intervals where the test points satisfy the original inequality.

Practice

5. $(-2,6)$

7. $(-\infty,-4]\cup[0,\infty)$

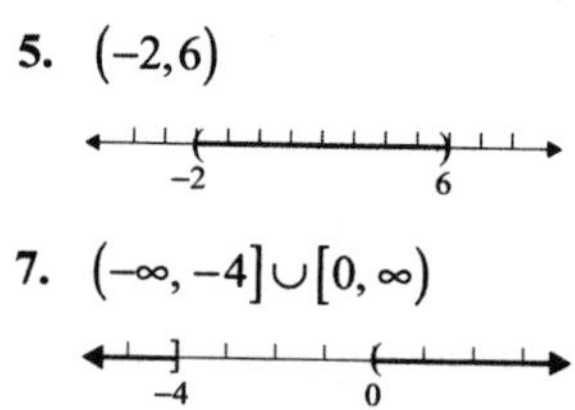

Applications

9. $[3,9]$

Writing & Thinking

11. a. $(-4,0)\cup(1,\infty)$

b. $(-\infty,-4)\cup(0,1)$

c. The function is undefined at $x=0$.

Chapter 15: Exponential and Logarithmic Functions

15.1 Exercises

Concept Check

1. True

3. False; The operations are restricted to the portions of the domain that are in common.

Practice

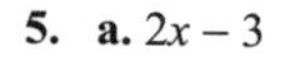

5. a. $2x-3$

b. 7

c. $x^2-3x-10$

d. $\dfrac{x+2}{x-5},\ x\neq 5$

7. $\sqrt{2x-6}+x+4$, $D=[3,\infty)$

9.

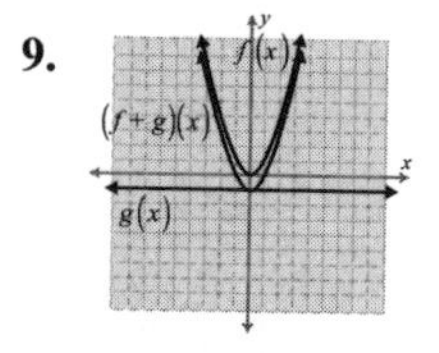

Writing & Thinking

11. In general, subtraction is not commutative. Answers will vary.

15.2 Exercises

Concept Check

1. False; The vertical line test determines whether a graph represents a function.

3. True

Practice

5. a. $5x^2-20x+24$

b. $45n^4+4$

7.

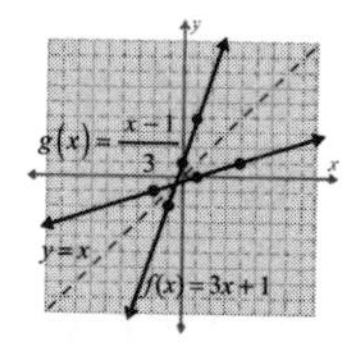

9. One-to-one

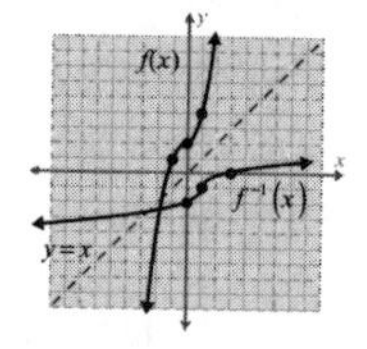

15.3 Exercises

Concept Check

1. False; $b>0$ and $b\neq 1$

3. True

Practice

5.

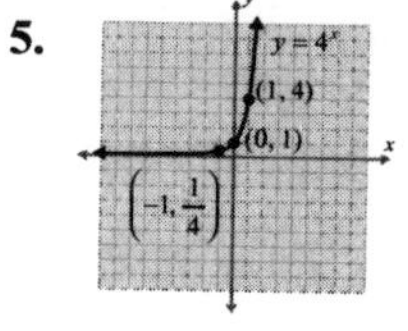

7. 48

Applications

9. $2154.99

Writing & Thinking

11. The graphs are reflections of each other across the y-axis.

15.4 Exercises

Concept Check

1. True

3. True

Practice

5. $\log_7 49=2$

7. $x=-3$

9. $x=3.7$

Writing & Thinking

11. The two functions are symmetric about the x-axis.

15.5 Exercises

Concept Check

1. True

3. True

Practice

5. $\log_b 5+4\log_b x$

7. $\log_b\left(\dfrac{15}{x}\right)$

9. $\log_2\sqrt[3]{\dfrac{z^2}{x^2y}}$

15.6 Exercises

Concept Check

1. False; When the base is omitted, it is understood to be 10.

3. False; Both common logarithms and natural logarithms have inverses.

Practice

5. 2.2380

7. 204.1738

Writing & Thinking

9. $\log x$ is a base 10 logarithm. $\ln x$ is a base e logarithm.

15.7 Exercises

Concept Check

1. False; The change of base formula is $\log_b x = \dfrac{\log_a x}{\log_a b}$.
3. True

Practice

5. $x = 5$
7. $x = 1$
9. 2.2619

Writing & Thinking

11. $7 \cdot 7^x = 7^{1+x}$ and $49^x = 7^{2x}$. Since $1 + x \neq 2x$, in general, $7 \cdot 7^x \neq 49^x$. Answers will vary.

Applications

3. $4027.51
5. 1.73 hours
7. 294.41 days

15.8 Exercises

Concept Check

1. True

Chapter 16: Conic Sections

16.1 Exercises

Concept Check

1. False; The two functions are different.
3. True

Practice

5. **a.** 0

   b. $1 - a^2 + 2a$

   c. $2 - x^2 - 2xh - h^2$

   d. $-2x - h$

7.

9.

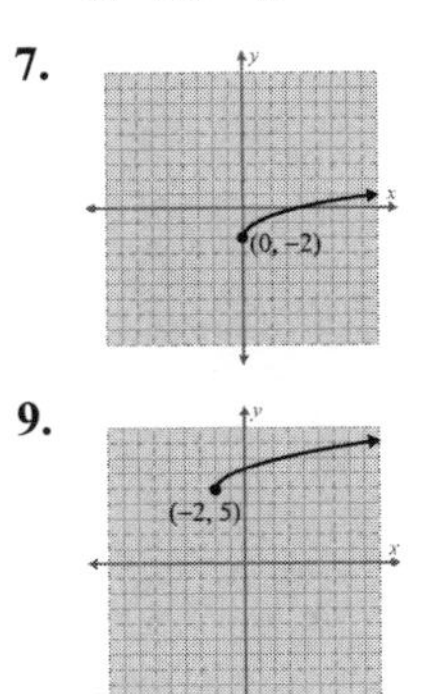

Writing & Thinking

11. The graph of the function $y = f(x-h)+k$ is the graph of the function $y = f(x)$ shifted h units to the right and k units up.

16.2 Exercises

Concept Check

1. True
3. False; $y = k$ is the line of symmetry for a horizontal parabola.

Practice

5. **a.** $(0, 2)$

   b. $(0, 2)$

   c. $y = 2$

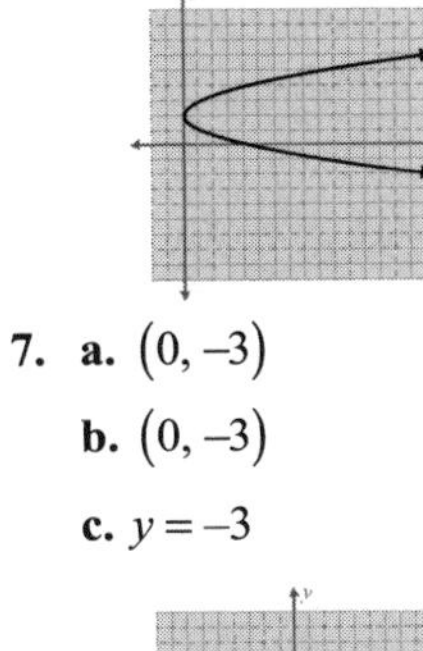

7. **a.** $(0, -3)$

   b. $(0, -3)$

   c. $y = -3$

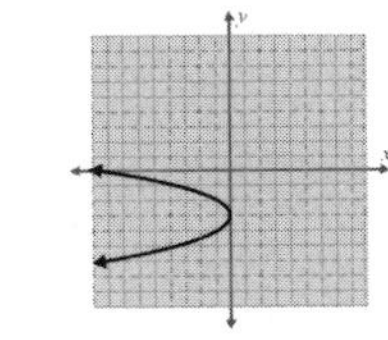

16.3 Exercises

Concept Check

1. True
3. True

Practice

5. 5; $(4, 5.5)$
7. $x^2 + y^2 = 16$
9. $x^2 + y^2 = 9$

   Center: $(0,0)$; $r = 3$

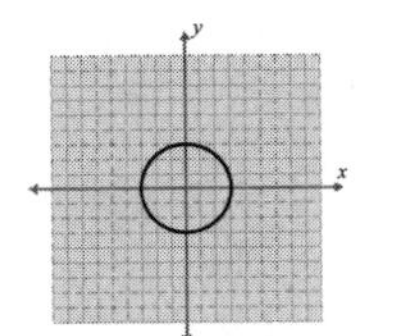

16.4 Exercises

Concept Check

1. True
3. False; The midway point between the foci is the center.

Practice

5. $\dfrac{x^2}{1} - \dfrac{y^2}{1} = 1$

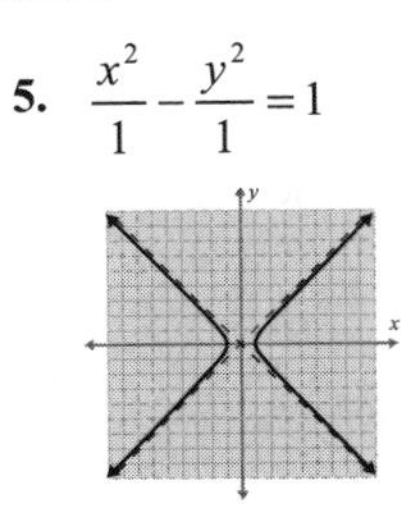

7. $\dfrac{x^2}{3} + \dfrac{y^2}{4} = 1$

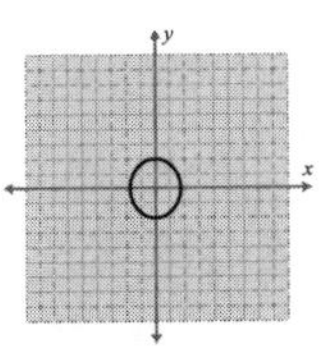

9.

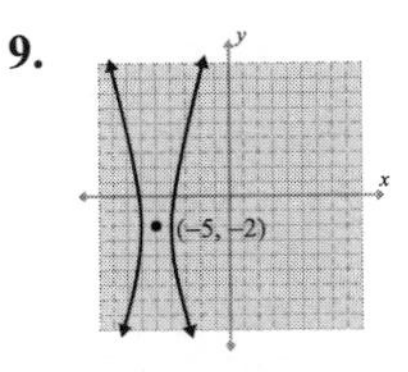

16.5 Exercises

Concept Check

1. False; These equations are called quadratic equations.
3. True

Practice

5. $(-3, 10), (1, 2)$

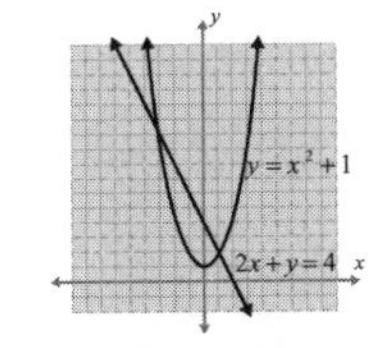

7. $(0, 3)$

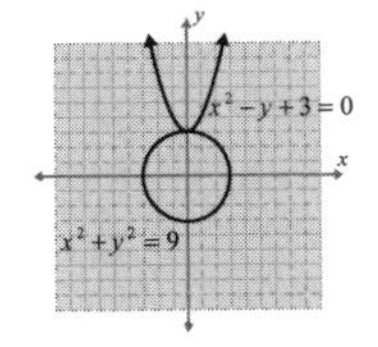

Geometry

P = Perimeter, A = Area, C = Circumference, V = Volume

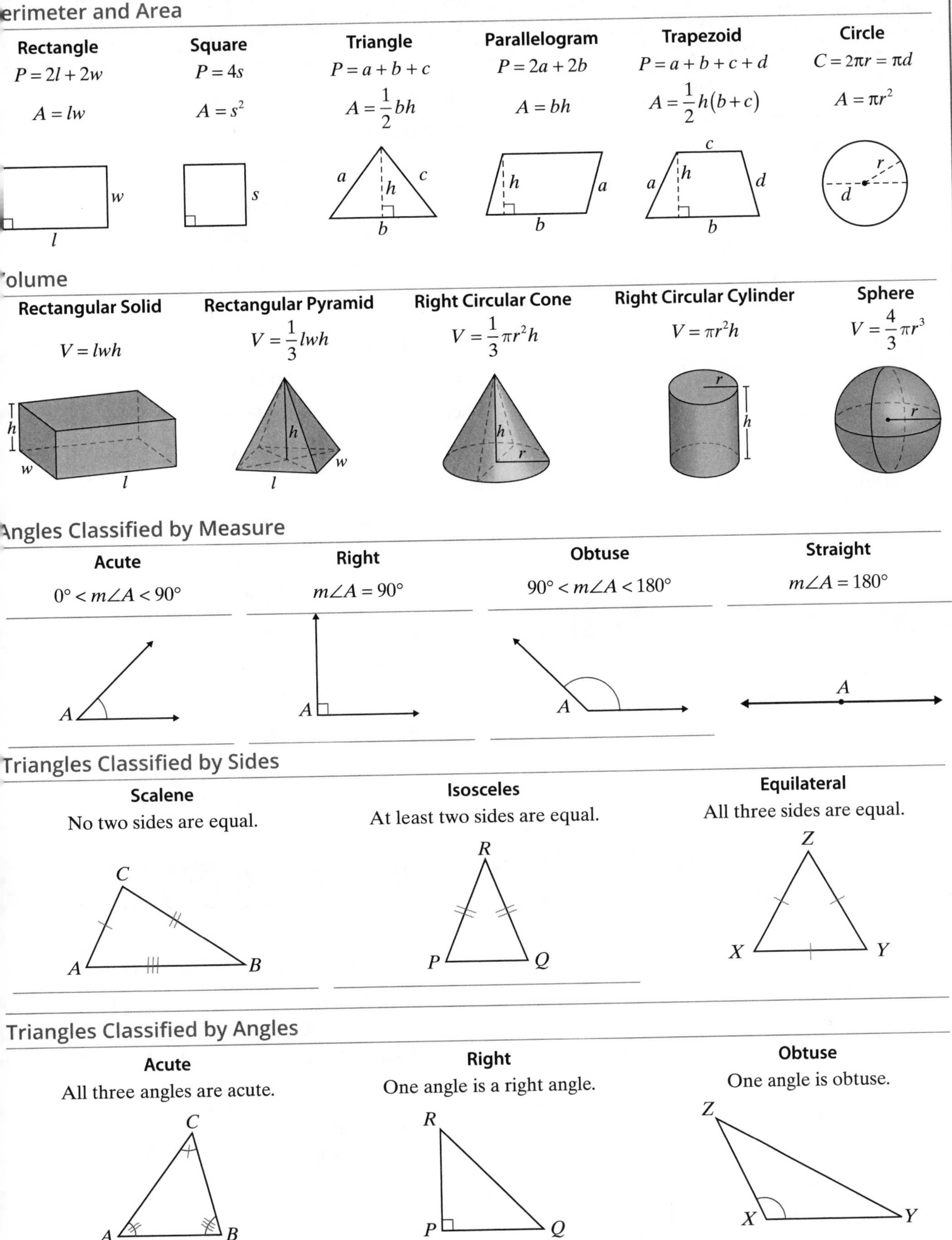

US Customary System of Measurement

Length

12 inches (in.) = 1 foot (ft)
3 feet = 1 yard (yd)
36 inches = 1 yard
5280 feet = 1 mile (mi)

Capacity

1 cup (c) = 8 fluid ounces (fl oz)
2 pints = 1 quart (qt)
2 cups = 1 pint (pt) = 16 fluid ounces
4 quarts = 1 gallon (gal)

Weight

16 ounces (oz) = 1 pound (lb)
2000 pounds = 1 ton (T)

Time

60 seconds (sec) = 1 minute (min)
60 minutes = 1 hour (hr)
24 hours = 1 day
7 days = 1 week

Temperature

Celsius (C) to Fahrenheit (F)

$$F = \frac{9}{5}C + 32$$

Fahrenheit (F) to Celsius (C)

$$C = \frac{5(F - 32)}{9}$$

Metric System of Measurement

Length

1 millimeter (mm)	= 0.001 meter	1 m = 1000 mm
1 centimeter (cm)	= 0.01 meter	1 m = 100 cm
1 decimeter (dm)	= 0.1 meter	1 m = 10 dm
1 meter (m)	= 1.0 meter	
1 dekameter (dam)	= 10 meters	
1 hectometer (hm)	= 100 meters	
1 kilometer (km)	= 1000 meters	

Capacity (Liquid Volume)

1 milliliter (mL)	= 0.001 liter	1 L = 1000 mL
1 liter (L)	= 1.0 liter	
1 hectoliter (hL)	= 100 liters	
1 kiloliter (kL)	= 1000 liters	1 kL = 10 hL

Weight

1 milligram (mg)	= 0.001 gram	1 g = 1000 mg
1 centigram (cg)	= 0.01 gram	
1 decigram (dg)	= 0.1 gram	
1 gram (g)	= 1.0 gram	
1 dekagram (dag)	= 10 grams	
1 hectogram (hg)	= 100 grams	
1 kilogram (kg)	= 1000 grams	1 g = 0.001 kg
1 metric ton (t)	= 1000 kilograms	1 kg = 0.001 t

1t = 1000 kg = 1,000,000 g = 1,000,000,000 mg

US Customary and Metric Equivalents

Length

1 in. = 2.54 cm (exact)	1 cm ≈ 0.394 in.
1 ft ≈ 0.305 m	1 m ≈ 3.28 ft
1 yd ≈ 0.914 m	1 m ≈ 1.09 yd
1 mi ≈ 1.61 km	1 km ≈ 0.62 mi

Area

1 in.^{2} ≈ 6.45 cm^2	1 cm^2 ≈ 0.155 in.^{2}
1 ft^2 ≈ 0.093 m^2	1 m^2 ≈ 10.764 ft^2
1 yd^2 ≈ 0.836 m^2	1 m^2 ≈ 1.196 yd^2
1 acre ≈ 0.405 ha	1 ha ≈ 2.47 acres

Volume

1 in.^{3} ≈ 16.387 cm^3	1 cm^3 ≈ 0.06 in.^{3}
1 ft^3 ≈ 0.028 m^3	1 m^3 ≈ 35.315 ft^3
1 qt ≈ 0.946 L	1 L ≈ 1.06 qt
1 gal ≈ 3.785 L	1 L ≈ 0.264 gal

Mass

1 oz ≈ 28.35 g	1 g ≈ 0.035 oz
1 lb ≈ 0.454 kg	1 kg ≈ 2.205 lb

Notation and Terminology

xponents

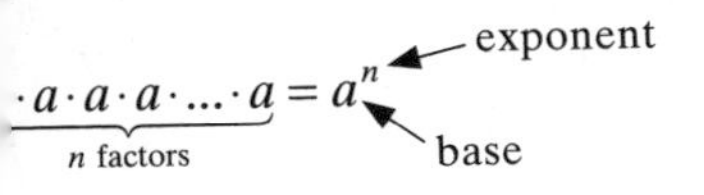

ractions

a ← numerator

b ← denominator

east Common Multiple (LCM)

iven a set of whole numbers, the smallest number that s a multiple of each of these whole numbers.

atios

$\frac{a}{b}$ or $a{:}b$ or a to b A comparison of two quantities by division.

Proportions

$\frac{a}{b} = \frac{c}{d}$ A statement that two ratios are equal.

Greatest Common Factor (GCF)

Given a set of integers, the largest integer that is a factor (or divisor) of all of the integers.

Types of Numbers

Natural Numbers (Counting Numbers):

$\mathbb{N} = \{1, 2, 3, 4, 5, 6, \ldots\}$

Whole Numbers: $\mathbb{W} = \{0, 1, 2, 3, 4, 5, 6, \ldots\}$

Integers: $\mathbb{Z} = \{\ldots, -4, -3, -2, -1, 0, 1, 2, 3, 4, \ldots\}$

Rational Numbers: A number that can be written in the form $\frac{a}{b}$ where a and b are integers and $b \neq 0$.

Irrational Numbers: A number that can be written as an infinite nonrepeating decimal.

Real Numbers: All rational and irrational numbers.

Complex Numbers: All real numbers and the even roots of negative numbers. The standard form of a complex number is $a + bi$, where a and b are real numbers, a is called the real part and b is called the imaginary part.

Absolute Value

$|a|$ The distance a real number a is from 0.

Equality and Inequality Symbols

$=$ "is equal to"

$\neq$ "is not equal to"

$<$ "is less than"

$>$ "is greater than"

$\leq$ "is less than or equal to"

$\geq$ "is greater than or equal to"

Sets

The **empty set** or **null set** (symbolized $\varnothing$ or $\{\ \}$): A set with no elements.

The **union** of two (or more) sets (symbolized $\cup$): The set of all elements that belong to either one set or the other set or to both sets.

The **intersection** of two (or more) sets (symbolized $\cap$): The set of all elements that belong to both sets.

The word **or** is used to indicate union and the word **and** is used to indicate intersection.

Algebraic and Interval Notation for Intervals

Type of Interval	Algebraic Notation	Interval Notation	Graph
Open Interval	$a < x < b$	(a, b)	a b
Closed Interval	$a \leq x \leq b$	$[a, b]$	a b
Half-open Interval	$\begin{cases} a \leq x < b \\ a < x \leq b \end{cases}$	$[a, b)$ $(a, b]$	a b a b
Open Interval	$\begin{cases} x > a \\ x < b \end{cases}$	(a, ∞) $(-\infty, b)$	a b
Half-open Interval	$\begin{cases} x \geq a \\ x \leq b \end{cases}$	$[a, \infty)$ $(-\infty, b]$	a b

Radicals

The symbol $\sqrt{\ }$ is called a **radical sign.**

The number under the radical sign is called the **radicand**.

The complete expression, such as $\sqrt{64}$, is called a **radical** or **radical expression**.

In a cube root expression $\sqrt[3]{a}$, the number 3 is called the index. In a square root expression such as $\sqrt{a}$, the index is understood to be 2 and is not written.

The Imaginary Number *i*

$i = \sqrt{-1}$ and $i^2 = \left(\sqrt{-1}\right)^2 = -1$

Formulas and Theorems

Percent

$\frac{P}{100} = \frac{A}{B}$ (the percent proportion),

where

$P\% =$ percent (written as the ratio $\frac{P}{100}$)
$B =$ base (number we are finding the percent of)
$A =$ amount (a part of the base)

$R \cdot B = A$ (the basic percent equation),
where

$R =$ **rate** or percent (as a decimal or fraction)
$B =$ **base** (number we are finding the percent of)
$A =$ **amount** (a part of the base)

Profit

Profit: The difference between selling price and cost.

profit = selling price − cost

Percent of Profit:

1. Percent of profit **based on cost**: $\frac{\text{profit}}{\text{cost}}$
2. Percent of profit **based on selling price**: $\frac{\text{profit}}{\text{selling price}}$

Interest

Simple Interest: $I = P \cdot r \cdot t$

Compound Interest: $A = P\left(1+\frac{r}{n}\right)^{nt}$

Continuously Compounded Interest: $A = Pe^{rt}$

where

$I =$ interest (earned or paid)

$A =$ amount accumulated

$P =$ principal (the amount invested or borrowed)

$r =$ annual interest rate in decimal or fraction form

$t =$ time (one year or fraction of a year)

$n =$ the number of times per year interest is compounded

$e = 2.718281828459\ldots$

The Pythagorean Theorem

In a right triangle, the square of the length of the hypotenuse is equal to the sum of the squares of the lengths of the two legs: $c^2 = a^2 + b^2$

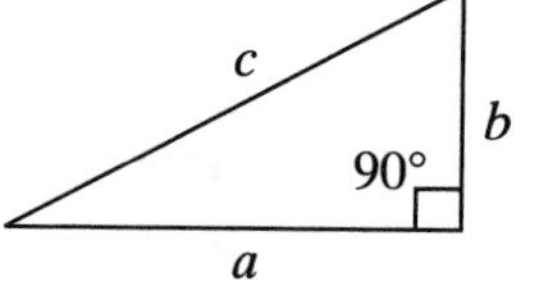

Probability of an Event

probability of an event

$$= \frac{\text{number of outcomes in event}}{\text{number of outcomes in sample space}}$$

Distance-Rate-Time

$d = rt$ The **distance traveled** d equals the product of the rate of speed r and the time t.

Special Products

1. $x^2 - a^2 = (x+a)(x-a)$: Difference of two squares
2. $x^2 + 2ax + a^2 = (x+a)^2$: Square of a binomial sum
3. $x^2 - 2ax + a^2 = (x-a)^2$: Square of a binomial difference
4. $x^3 + a^3 = (x+a)(x^2 - ax + a^2)$: Sum of two cubes
5. $x^3 - a^3 = (x-a)(x^2 + ax + a^2)$: Difference of two cubes

Change-of-Base Formula for Logarithms

For $a, b, x, > 0$ and $a, b \neq 1$, $\log_b x = \frac{\log_a x}{\log_a b}$.

Distance Between Two Points

The distance d between points $P(x_1, y_1)$ and $Q(x_2, y_2)$ is $d = \sqrt{(x_2 - x_1)^2 + (y_2 - y_1)^2}$.

Midpoint Formula

The midpoint between points $P(x_1, y_1)$ and $Q(x_2, y_2)$ is $\left(\frac{x_1 + x_2}{2}, \frac{y_1 + y_2}{2}\right)$.

rinciples and Properties

roperties of Addition and Multiplication

roperty	Addition	Multiplication
ommutative roperty	$a+b=b+a$	$ab=ba$
ssociative roperty	$(a+b)+c=a+(b+c)$	$a(bc)=(ab)c$
dentity	$a+0=0+a=a$	$a\cdot 1=1\cdot a=a$
nverse	$a+(-a)=0$	$a\cdot\frac{1}{a}=1\ (a\neq 0)$

ero Factor Law: $a\cdot 0=0\cdot a=0$

istributive Property: $a(b+c)=a\cdot b+a\cdot c$

ddition (or Subtraction) Principle of Equality

$=B$, $A+C=B+C$, and $A-C=B-C$ have the me solutions (where A, B, and C are algebraic xpressions).

ultiplication (or Division) Principle of Equality

$=B$, $AC=BC$, and $\frac{A}{C}=\frac{B}{C}$ have the same solutions where A and B are algebraic expressions and C is any onzero constant, $C\neq 0$).

roperties of Exponents

or nonzero real numbers a and b and integers m and n:

he exponent 1	$a=a^1$
he exponent 0	$a^0=1$
he product rule	$a^m\cdot a^n=a^{m+n}$
he quotient rule	$\frac{a^m}{a^n}=a^{m-n}$
egative exponents	$a^{-n}=\frac{1}{a^n}$
ower rule	$(a^m)^n=a^{mn}$
ower of a product	$(ab)^n=a^nb^n$
ower of a quotient	$\left(\frac{a}{b}\right)^n=\frac{a^n}{b^n}$

ero Factor Property

a and b are real numbers, and $a\cdot b=0$, then $a=0$ or $=0$ or both.

Properties of Rational Numbers (or Fractions)

If $\frac{P}{Q}$ is a rational expression and P, Q, R, and K are polynomials where $Q, R, S\neq 0$, then

The Fundamental Principle	$\frac{P}{Q}=\frac{P\cdot K}{Q\cdot K}$
Multiplication	$\frac{P}{Q}\cdot\frac{R}{S}=\frac{P\cdot R}{Q\cdot S}$
Division	$\frac{P}{Q}\div\frac{R}{S}=\frac{P}{Q}\cdot\frac{S}{R}$
Addition	$\frac{P}{Q}+\frac{R}{Q}=\frac{P+R}{Q}$
Subtraction	$\frac{P}{Q}-\frac{R}{Q}=\frac{P-R}{Q}$

Properties of Radicals

If a and b are positive real numbers, n is a positive integer, m is any integer, and $\sqrt[n]{a}$ is a real number then

1. $\sqrt[n]{ab}=\sqrt[n]{a}\cdot\sqrt[n]{b}$
2. $\sqrt[n]{\frac{a}{b}}=\frac{\sqrt[n]{a}}{\sqrt[n]{b}}$
3. $\sqrt[n]{a}=a^{\frac{1}{n}}$
4. $a^{\frac{m}{n}}=\left(a^{\frac{1}{n}}\right)^m=(a^m)^{\frac{1}{n}}$ or, in radical notation, $a^{\frac{m}{n}}=(\sqrt[n]{a})^m=\sqrt[n]{a^m}$

Properties of Logarithms

For $b>0, b\neq 1, x, y>0$, and any real number r,

1. $\log_b 1=0$
2. $\log_b b=1$
3. $x=b^{\log_b x}$
4. $\log_b b^x=x$
5. $\log_b xy=\log_b x+\log_b y$ **The product rule**
6. $\log_b\frac{x}{y}=\log_b x-\log_b y$ **The quotient rule**
7. $\log_b x^r=r\cdot\log_b x$

Properties of Equations with Exponents and Logarithms

For $b>0, b\neq 1$,

1. If $b^x=b^y$, then $x=y$.
2. If $x=y$, then $b^x=b^y$.
3. If $\log_b x=\log_b y$, then $x=y$ ($x>0$ and $y>0$).
4. If $x=y$, then $\log_b x=\log_b y$ ($x>0$ and $y>0$).

Equations and Inequalities

Linear Equation in x (First-Degree Equation in x)

$ax + b = c$, where a, b, and c are real numbers and $a \neq 0$.

Types of Equations and their Solutions

Conditional: Finite Number of Solutions

Identity: Infinite Number of Solutions

Contradiction: No Solution

Linear Inequalities

Linear Inequalities have the following forms where a, b, and c are real numbers and $a \neq 0$:

$ax + b < c$ and $ax + b \leq c$

$ax + b > c$ and $ax + b \geq c$

Compound Inequalities

The inequalities $c < ax + b < d$ and $c \leq ax + b \leq d$ are called **compound linear inequalities**.

(This includes $c < ax + b \leq d$ and $c \leq ax + b < d$ as well.)

Absolute Value Equations

For statements 1 and 2, $c > 0$:

1. If $|x| = c$, then $x = c$ or $x = -c$.

2. If $|ax + b| = c$, then $ax + b = c$ or $ax + b = -c$.

3. If $|a| = |b|$, then either $a = b$ or $a = -b$.

4. If $|ax + b| = |cx + d|$, then either $ax + b = cx + d$ or $ax + b = -(cx + d)$.

Absolute Value Inequalities

For $c > 0$:

1. If $|x| < c$, then $-c < x < c$.

2. If $|ax + b| < c$, then $-c < ax + b < c$.

3. If $|x| > c$, then $x < -c$ **or** $x > c$.

4. If $|ax + b| > c$, then $ax + b < -c$ **or** $ax + b > c$.

(These statements hold true for $\leq$ and $\geq$ as well.)

Quadratic Equation

An equation that can be written in the form $ax^2 + bx + c = 0$, where a, b, and c are real numbers and $a \neq 0$.

Quadratic Formula

The solutions of the general quadratic equation $ax^2 + bx + c = 0$, where $a \neq 0$, are $x = \dfrac{-b \pm \sqrt{b^2 - 4ac}}{2a}$.

The Discriminant

The expression $b^2 - 4ac$, the part of the quadratic formula that lies under the radical sign, is called the **discriminant**.

If $b^2 - 4ac > 0$, there are two real solutions.

If $b^2 - 4ac = 0$, there is one real solution, $x = -\dfrac{b}{2a}$.

If $b^2 - 4ac < 0$, there are two nonreal solutions.

Systems of Linear Equations

Systems of Linear Equations (Two Variables)

The system is...

consistent, and the equations are **independent**. (One solution)

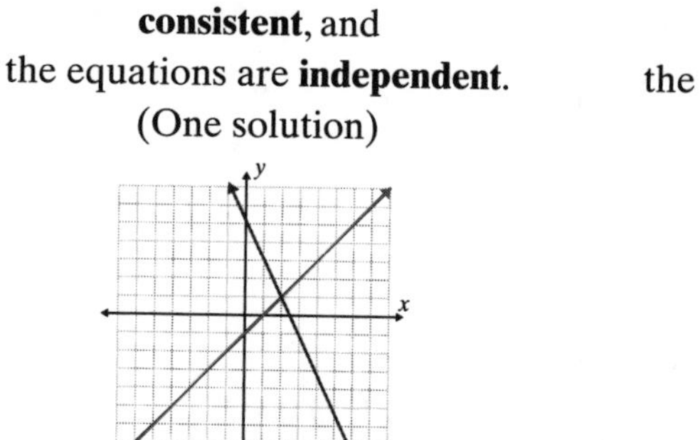

inconsistent, and the equations are **independent**. (No solution)

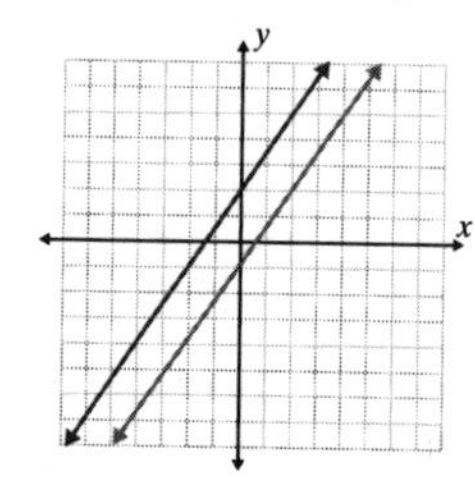

consistent, and the equations are **dependent**. (Infinite number of solutions)

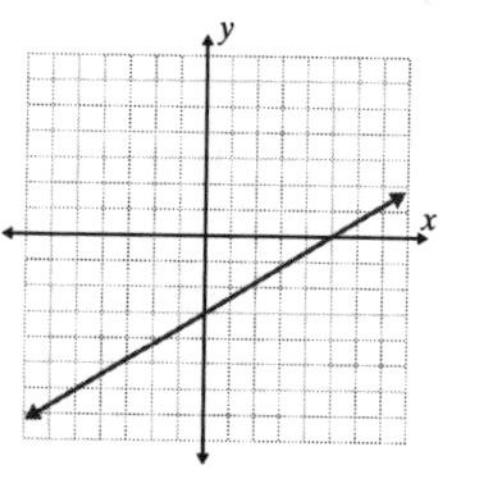

unctions

unction, Relation, Domain, and Range

relation is a set of ordered pairs of real numbers.

he **domain** D of a relation is the set of all first coordi-ates in the relation.

he **range** R of a relation is the set of all second coordi-ates in the relation.

function is a relation in which each domain element as exactly one corresponding range element.

ne-to-One Functions

function is a **one-to-one function** if for each value of in the range there is only one corresponding value of x the domain.

Algebraic Operations with Functions

1. $(f+g)(x)=f(x)+g(x)$
2. $(f-g)(x)=f(x)-g(x)$
3. $(f\cdot g)(x)=f(x)\cdot g(x)$
4. $\left(\frac{f}{g}\right)(x)=\frac{f(x)}{g(x)}$
5. $(f\circ g)(x)=f(g(x))$

Inverse Functions

If f is a one-to-one function with ordered pairs of the form (x, y), then its **inverse function**, denoted as f^{-1}, is also a one-to-one function with ordered pairs of the form (y, x).

If f and g are one-to-one functions and $f(g(x))=x$ for all x in D_g and $g(f(x))=x$ for all x in D_f, then f and g are **inverse functions**.

raphs of Functions

he Cartesian Coordinate System

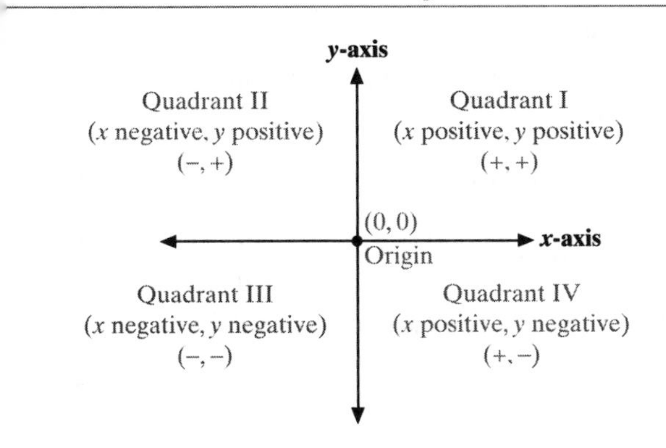

inear Functions (Lines)

tandard form:

$Ax+By=C$ Where A and B do not both equal 0

lope of a line:

$m=\frac{y_2-y_1}{x_2-x_1}$ Where $x_1 \neq x_2$

lope-intercept form:

$y=mx+b$ With slope m and y-intercept $(0, b)$

oint-slope form:

$y-y_1=m(x-x_1)$ With slope m and point (x_1, y_1) on the line

Horizontal line, slope 0: $y=b$

Vertical line, undefined slope: $x=a$

Parallel lines have the same slope.

Perpendicular lines have slopes that are negative reciprocals of each other.

Quadratic Functions (Parabolas)

Parabolas of the form $y=ax^2+bx+c$:

1. Vertex: $\left(-\frac{b}{2a}, f\left(-\frac{b}{2a}\right)\right)$.
2. Line of Symmetry: $x=-\frac{b}{2a}$

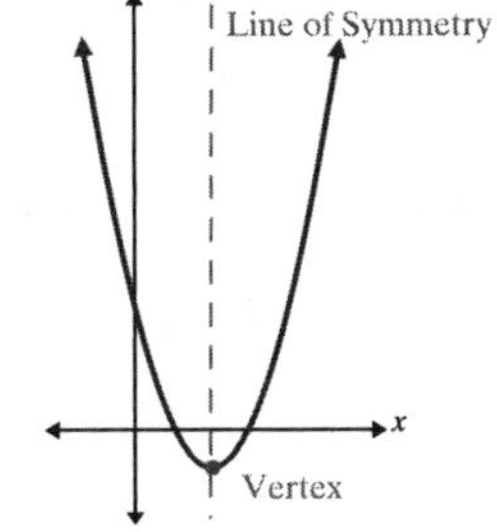

Parabolas of the form $y=a(x-h)^2+k$:

1. Vertex: (h, k)
2. Line of Symmetry: $x=h$
3. The graph is a horizontal shift of h units and a vertical shift of k units of the graph of $y=ax^2$.

In both cases:

1. If $a>0$, the parabola "opens upward."
2. If $a<0$, the parabola "opens downward."

Conic Sections

Equations of a Horizontal Parabola

$x = ay^2 + by + c$ or $x = a(y-k)^2 + h$ where $a \neq 0$.

The parabola opens left if $a < 0$ and right if $a > 0$.

The vertex is at (h, k).

The line $y = k$ is the line of symmetry.

Equation of an Ellipse

The standard form for the equation of an ellipse with its center at the origin is $\frac{x^2}{a^2} + \frac{y^2}{b^2} = 1$.

The points $(a, 0)$ and $(-a, 0)$ are the x-intercepts (called vertices).

The points $(0, b)$ and $(0, -b)$ are the y-intercepts (called vertices).

When $a^2 > b^2$:

- The segment of length $2a$ joining the x-intercepts is called the major axis.
- The segment of length $2b$ joining the y-intercepts is called the minor axis.

When $b^2 > a^2$:

- The segment of length $2b$ joining the y-intercepts is called the major axis.
- The segment of length $2a$ joining the x-intercepts is called the minor axis.

The standard form for the equation of an ellipse with its center at (h, k) is $\frac{(x-h)^2}{a^2} + \frac{(y-k)^2}{b^2} = 1$.

Equation of a Circle

The equation of a circle with radius r and center (h, k) is $(x-h)^2 + (y-k)^2 = r^2$.

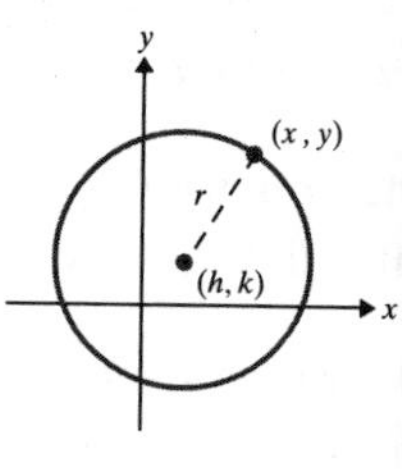

Equation of a Hyperbola

In general, there are two standard forms for equations of hyperbolas with their centers at the origin.

1. $\frac{x^2}{a^2} - \frac{y^2}{b^2} = 1$

x-intercepts (vertices) at $(a, 0)$ and $(-a, 0)$

No y-intercepts

Asymptotes: $y = \frac{b}{a}x$ and $y = -\frac{b}{a}x$

The curves "open" left and right.

2. $\frac{y^2}{b^2} - \frac{x^2}{a^2} = 1$

y-intercepts (vertices) at $(0, b)$ and $(0, -b)$

No x-intercepts

Asymptotes: $y = \frac{b}{a}x$ and $y = -\frac{b}{a}x$

The curves "open" up and down.

The equation of a hyperbola with its center at (h, k) is $\frac{(x-h)^2}{a^2} - \frac{(y-k)^2}{b^2} = 1$ or $\frac{(y-k)^2}{b^2} - \frac{(x-h)^2}{a^2} = 1$

Notes

Notes

Notes

Notes

Notes

Notes

Notes

Notes

Notes

Notes

Notes

Notes